A PEOPLE'S HISTORY
OF SCIENCE

CLIFFORD D. CONNER

A PEOPLE'S HISTORY OF SCIENCE

MINERS, MIDWIVES,
AND "LOW MECHANICKS"

NATION BOOKS · NEW YORK

A People's History of Science:
Miners, Midwives and "Low Mechanicks"
Published by
Nation Books
An Imprint of Avalon Publishing Group Inc.
245 West 17th St., 11th Floor
New York, NY 10011

AVALON
publishing group incorporated

Copyright © 2005 by Clifford D. Conner
First printing November 2005
Nation Books is a co-publishing venture of the
Nation Institute and Avalon Publishing Group Incorporated.

Library of Congress Cataloging-in-Publication Data is available.
ISBN 1-56025-748-2

9 8 7 6 5 4 3 2

Book design by Pauline Neuwirth, Neuwirth & Associates, Inc.

Printed in the United States of America
Distributed by Publishers Group West

Dedicated to the memory of Paul Siegel
(1916–2004)

In the beginning was the word.
—The Gospel according to St. John

In the beginning was the word? . . .
No, in the beginning was the *deed*.
—Goethe, *Faust*

CONTENTS

CONTENTS

· 6 ·
WHO WERE THE WINNERS
IN THE SCIENTIFIC REVOLUTION?

· 7 ·
THE "UNION OF CAPITAL AND SCIENCE"

· 8 ·
THE SCIENTIFIC-INDUSTRIAL COMPLEX

BIBLIOGRAPHY

INDEX

ACKNOWLEDGMENTS

THE ADJECTIVE "PEOPLE'S" in the title of this book may prompt the suspicion that I am one of those authors who "wave some dimension of history in front of their readers, claiming that it is the forgotten grand arcanum, the key to Clio's mysteries."[1] To the contrary, the elements of a people's history of science have already been created by a great number of historians whose research broke the ground and planted the seeds that were essential to the making of this book. My task, as I saw it, was to synthesize their findings and present them as a coherent narrative suitable for consumption by an audience of nonspecialists.

Among those authors to whom I am most indebted are the originators of people's history of science, Boris Hessen and Edgar Zilsel; some key continuators, including J. D. Bernal and Joseph Needham; and some present-day practitioners, including William Eamon, Steven Shapin, Pamela H. Smith, Deborah Harkness, Adrian Desmond, and Stephen Pumphrey. Others whose work I have found very helpful are Eva Germaine Rimington Taylor, Silvio Bedini, Derek J. de Solla Price, Roger Hahn, and Roy Porter. I am especially fortunate to be able to count among my graduate school mentors two pioneers of the social history of science, James Jacob and Margaret Jacob. The Jacobs introduced me to the ideas that led me to write this book but should not be held responsible for what I have made of them.[2]

Science for the People, a periodical that appeared throughout the 1970s and '80s, helped shape my understanding of the relation between science and society. This book also owes an obvious debt to previous people's histories, most notably A. L. Morton's *A People's History of England* and Howard Zinn's *A People's History of the United States.* Peter Linebaugh is another people's historian whose work has served to inspire me. As for Howard Zinn, I am grateful not only for the example provided by his writings but for the warmth of the encouragement he has given me from the time I first submitted the proposal for this book to him.

My deepest thanks are due to a number of scholars (who also happen to be good personal friends) for reading all or parts of draft versions of this book and providing suggestions that enriched the final product. In addition to the aforementioned James Jacob, they include anthropologist Kim Sonderegger, art historian Greta Berman, and my dear, departed friend, the literary scholar Paul Siegel.

I also thank Ken Silverman, biographer of Cotton Mather, for pointing me toward sources of information regarding Onesimus; and Rod Holt, one of the pioneers of the personal computer revolution, for sharing with me his insights into early Apple history.

Without the special talents and efforts of literary agents Sam Stoloff and Frances Goldin, this book would probably still be in manuscript form, languishing in the piles of unread drafts in publishers' offices. I owe my introduction to them and to Howard Zinn to another close friend, Jeff Mackler, whom I met more than three decades ago when we were both active in the trade union and antiwar movements. Jeff is, among other things, currently a national leader of the efforts to win justice for Mumia Abu-Jamal. In that capacity, he works closely with Frances Goldin, Mumia's literary agent, and was kind enough to recommend me to her as a client. I thank Sam Stoloff not only for representing me to publishers but also for a number of valuable suggestions that have improved the text. Warmest thanks also to Carl Bromley and Ruth Baldwin, my editors at Nation Books, for doing what editors do, and doing it so well.

ACKNOWLEDGMENTS

My greatest appreciation is for my spouse, Marush Conner. The Cajuns have a term of endearment that perfectly expresses what she is to me—she is *ma toute-toute:* research collaborator, ballerina, travel coordinator, French-language resource, life companion, confidante, best friend, sublime playmate, and everything else that makes life worth while.

---

[1] Roy Porter and Mikulás Teich, eds., *The Scientific Revolution in National Context*, p. 6.
[2] The works I have drawn on by the authors mentioned in this paragraph and the next are listed in the bibliography.

1

WHAT SCIENCE?
WHAT HISTORY?
WHAT PEOPLE?

WE ALL KNOW the history of science that we learned from grade-school textbooks: how Galileo used his telescope to show that the earth was not the center of the universe; how Newton divined gravity from the falling apple; how Einstein unlocked the mysteries of time and space with a simple equation. This history is made up of long periods of ignorance and confusion, punctuated once an age by the "Eureka!" of a brilliant thinker who puts it all together. In this traditional heroic account, a few Great Men with Great Ideas tower over the rest of humanity, and it is to them that we owe science in its entirety.

The legend of Pythagoras exemplifies the seemingly timeless tendency to attribute all scientific creation to individual hero-savants. With regard to ancient Greek and Roman commentaries on their semimythical predecessor, Walter Burkert observed that "to a later age it seemed natural to retroject their own notion of 'wisdom' upon the great figures of the past and to impute to them that which from a modern point of view is 'science.'"[1] Unfortunately, that practice is still all too much in evidence today.

What I am presenting here, by contrast, is a *people's* history of science that aims to show how ordinary humans participated in creating science in profound ways. It is a history not only *of the people* but *for the people* as well; its intended audience is not confined to professional scientists or historians of science but includes anyone with an interest in the origins of scientific knowledge. And because I have drawn on the collective efforts of many predecessors, it might not be far-fetched to say that in a sense it is also *by the people*.

My central aim is to demonstrate a much, much greater contribution to the production and propagation of scientific knowledge on the part of anonymous masses of humble people—the common people—than is generally recognized or acknowledged. Isaac Newton's ability to "see further" should not be attributed, as he claimed, to his sitting "on the shoulders of giants," but rather to his standing on the backs of untold thousands of illiterate artisans (among others).[2]

It would be absurd, of course, to claim that the formulation of quantum theory or the structure of DNA can be credited directly to artisans or peasants, but if modern science is likened to a skyscraper, then those twentieth-century triumphs are the sophisticated filigrees at its pinnacle that are supported by—and could not exist apart from—the massive foundation created by humble laborers. If science is understood in the fundamental sense of *knowledge of nature*, it should not be surprising to find that it originated with the people closest to nature: hunter-gatherers, peasant farmers, sailors, miners, blacksmiths, folk healers, and others forced by the conditions of their lives to wrest the means of their survival from an encounter with nature on a daily basis.

A few brief examples—all of which will be explained in detail in the chapters below—can illustrate this contention. Virtually every plant and animal species we eat today was domesticated by experimentation and de facto genetic engineering practiced by preliterate ancient peoples. We are infinitely more in debt to pre-Columbian Amerindians than to modern plant geneticists for the scientific knowledge underlying food production. Even in relatively recent times, when American plantation owners wanted to grow rice, they found themselves compelled to buy African slaves with knowledge of the ecology of rice plants.

Likewise, the science of medicine began with and continues to draw on knowledge of plants' therapeutic properties discovered by prehistoric peoples. Amerindians demonstrated to Europeans the efficacy of the bark of the cinchona tree in treating malaria, and an African slave named Onesimus introduced the practice of inoculation against smallpox to North America. Credit for the discovery of vaccination that usually goes to Dr. Edward Jenner belongs instead to a farmer named Benjamin Jesty. Furthermore, until the nineteenth century the advance of medical science owed more to semiliterate barber-surgeons, apothecaries, and "irregular" healers than to university-trained medical scholars, whose influence tended to retard the acquisition of new medical knowledge. It was a Swiss pig-gelder named Jakob Nufer who in the 1580s performed the first recorded cesarean section.

The geography and cartography of the Americas and the Pacific Ocean are founded on the knowledge of the native peoples. Captain John Smith acknowledged that his celebrated map of the Chesapeake Bay area "was had by information of the Savages," and Captain Cook's maps of the Pacific Islands were derived from information given to him by an indigenous navigator named Tupaia. Anonymous sailors and fishermen were the original source of scientific data regarding tides, ocean currents, and prevailing winds; when Benjamin Franklin produced the first chart of the Gulf Stream, he acknowledged that it was entirely based on what he had learned from "simple" whalers.

Chemistry, metallurgy, and the materials sciences in general originated in knowledge produced by ancient miners, smiths, and potters. Mathematics owes its existence and a great deal of its development to surveyors, merchants, clerk-accountants, and mechanics of many millennia. And finally, the empirical method that characterized the Scientific Revolution of the sixteenth and seventeenth centuries, as well as the mass of scientific data on which it built, emerged from the workshops of European artisans.

The "folk" wisdom and lore of early societies was not an inferior kind of knowledge about nature that later was simply canceled out and replaced by more accurate scientific knowledge. Science as it exists today was *created out of* folk and artisanal sources; it became what it is by drawing heavily on those sources. Knowledge, philosopher of science Karl Popper maintained, has for the most part advanced through the modification of earlier knowledge.

It might be argued that the approach I have outlined here cannot produce a balanced account of the origins of modern science. But the historical record has long been severely unbalanced by the gentlemen's historians and by the nature of history's dependence on written documentation, with power relationships determining who did the writing. No history of science could be less balanced than the traditional romantic narratives of Newtons, Darwins, and Einsteins transforming the world by the force of their unique brainpower. I am deliberately "bending the stick in the other direction," searching for the voices of the voiceless and sifting the record for very scarce evidence. The purpose of the sifting, however, is not to salvage some peripheral aspects of the development of science, but to demonstrate how the scarce evidence illuminates the hidden core of that development.

A selective approach does not necessarily lead to an unbalanced conclusion. Although my focus is on the activities of anonymous, ordinary people, I do not believe I have overvalued their importance in the process of generating scientific knowledge. It is not my contention that the familiar Great Men of Science played no role or were unimportant, but that their achievements were predicated on prior contributions of artisans, merchants, midwives, and tillers of the soil—most of whom have never been thought of as great and many of whom were not men.

Attempts to integrate women fully into the traditional heroic narrative are ultimately unlikely to be satisfying, not because women have ever been genetically inferior to men in intellect but because of the social barriers that historically have denied women education and entrée into scientific professions.[3] The contributions of women receive considerably more recognition in a people's history of science because women constitute half of the people, but even here parity is elusive due to the traditional exclusion of women from many craft occupations. But if women did not have much input in the development of oceanography because few were seafarers, they made up for it in the medical sciences in their roles as local healers and midwives.

The contributions to the history of science made by socially subordinate and illiterate people have not left the paper trail that historians customarily depend on for evidence. Lynn White, for example, blames "the neglect which historians have lavished upon the rustic and his works and days" on the fact that the peasant "has seldom been literate":

> Not only histories but documents in general were produced by social groups which took the peasant and his labours largely for granted. Therefore while our libraries groan with data on the ownership of land, there is an astonishing dearth of information about the various, and often changing, methods of cultivation which made the land worth owning.[4]

The heroes of the traditional story of eighteenth-century scientific agriculture are "improving landlords" such as Jethro Tull and Thomas "Turnip" Townshend, whose spirit of experimentation was allegedly the driving force behind the great leap forward. But as T. S. Ashton explained in his classic study of the Industrial Revolution, "Tull was a crank, and his importance in the history of agriculture has been vastly exaggerated." As for Viscount Townshend's role in the introduction of the turnip as a field crop, "recent research has shown that he was the popularizer, rather than the originator, of this." There was no single originator; it was a collective accomplishment. Agricultural experiments "were being made by obscure farmers in many parts of the country"; then, "knowledge of the new methods was spread at tenants' dinners, shearing feasts, and the more frequent meetings of many local farmers' clubs." On the larger manors it was not the wealthy landlords but their humble tenants whose hands were actually in the soil and the manure, experimenting with new crops and procedures. The new agronomic knowledge, "like every major innovation, was the work of many hands and brains."[5]

The development of the peasants' knowledge that "made the land worth owning" is impossible to trace by archival research, and the same is true of much of the scientific knowledge produced by artisans who were unable to read and write. In recent decades, however, historians have begun to tap the methodology of anthropology and other disciplines to show that a great deal can be learned about the past—and not only the "prehistoric" past—without recourse to written sources. Furthermore, some assertions not supportable by documentation may be deemed valid due to the lack of any viable alternative explanation—for example, the proposition that anonymous sailors and fishermen were the original source of scientific data regarding ocean currents and prevailing winds.[6]

The chronological scope of this survey of scientific history is as vast as it could possibly be—from the Paleolithic to the postmodern—but it is markedly weighted toward a particular period: the fourteenth through seventeenth centuries of the Common Era, which encompass the origins of what has come to be known as *modern* science.[7] The geographical scope is likewise unlimited but somewhat skewed toward the western end of the Eurasian landmass. My view of history is not Eurocentric, but because this subject is inextricably entangled with the European imperial conquest of the rest of the globe, it therefore demands a disproportionate amount of attention to activities that occurred in Europe.

WHAT PEOPLE?

WHO ARE *THE people* who are both subject and object of this people's history? Occupational categories—such as artisans, merchants, and so on—give a good first approximation, but how shall they be collectively identified? I prefer to avoid denominating them as *the lower classes* or *the inferior orders*, except in quotation marks indicating irony, because those designations reflect the point of view of the privileged few who have always defined themselves as superior to everyone else. *Common people* and *ordinary people* are slightly pejorative but perhaps not too objectionable; *the masses, the working class*, and *the proletariat* are not intrinsically worthless terms, but they suffer from overuse and an unfortunate association with discredited ideology of the Stalinist period. *The social majority* is not an insulting term, but its class-neutrality renders it a bit bland.

The laboring classes, the poor, or any synonyms thereof are inadequate because they do not include all elements of *the people* of the eighteenth century or earlier. In particular, they leave merchants, master craftsmen, and other members of the incipient capitalist class out of account.[8] In France in the era of the Great Revolution, *le peuple* was handily defined by a legal category: the Third Estate, which included everyone who was not a member of the clergy or the nobility. Perhaps *the subordinate* or *dominated classes*—though somewhat awkward expressions—best convey the social relationships that need to be indicated. *The crowd?* Maybe. *The mob?* Definitely not. *The many?* However defined, it is their interests that form the vantage point from which the origin and development of science are evaluated in this book.

As for who *the people* are not: Those who defined themselves as "upper class," "noble," and "persons of quality" were, by definition, superior in social power to those whom they excluded from those categories. But their self-designations also carry an implication of moral superiority that would strike a dissonant note in a people's history. They can be adequately denominated, and their social position acknowledged, by such terms as *the dominant class, the ruling class, the privileged orders, the elite,* and so forth.

Very few of the traditional Heroes of Science were born into the ruling class per se. Some were indeed aristocrats or royals— Robert Boyle, Tycho Brahe, and Prince Henry "the Navigator" come quickly to mind—but most were co-opted into the circles of privilege as highly placed servants by means of university positions (Newton, Galileo) or other forms of patronage (Galileo, Bacon). "The educated elite," historian of science William Eamon explains, "became a kind of intellectual aristocracy."[9] What defined members of the scientific elite, then, is not the blueness of their blood but their status as professional intellectuals.

A very sharply defined social barrier has always—at least since the origin of social differentiation in the misty dawn of prehistory—been based on the distinction between manual and intellectual work. People who work with their hands have long been looked down upon as inferiors by those who make their livings without getting their hands dirty. The class distinction between educated scribes and illiterate craftsmen in the earliest civilizations is clearly evident in the occupational guidance an ancient Egyptian father of about 1100 B.C.E. gave his son, to "give thy heart to letters" and thereby avoid manual labor. "I have seen the metal worker at his toil before a blazing furnace," the father warned. "His fingers are like the hide of the crocodile, he stinks more than the eggs of fish. And every carpenter who works or chisels, has he any more rest than the ploughman?"[10]

In the ages of Plato and Francis Bacon, the disdain for manual labor was expressed openly and frequently by professional intellectuals (in imitation of their aristocratic patrons) and was given an extensive ideological foundation. "The arts which we call mechanical," says Xenophon, a contemporary of Plato's, are "generally held in bad repute." Furthermore:

> States also have a very low opinion of them—and with justice. For they are injurious to the bodily health of workmen and overseers, in that they compel them to be seated and indoors, and in some cases also all the day before a fire. And when the body grows effeminate, the mind also becomes weaker and weaker. And the mechanical arts, as they are called, will not let men unite with them care for friends and State, so that men engaged in them must ever appear to be both bad friends and poor defenders of their country. And there are States, but more particularly such as are most famous in war, in which not a single citizen is allowed to engage in mechanical arts.[11]

This was an attitude with staying power. "Scholastic depreciation of 'rustics' and 'the crowd' became particularly virulent in the thirteenth century," Eamon reported, "as the educated elite attempted to reinforce its status and to set itself above the herd of ordinary men."[12] The distinction is less commented on today in societies that lay claim to democratic values, but no one, I think, could credibly deny its continued existence.[13]

In addition to taking pride in never soiling their hands through labor, another mark of the scientific elite during the period of the rise of modern science was "literacy," which in early modern Europe meant not merely knowing how to read and write, but being able to do so in Latin. "Knowledge of the Latin language" was "the skill that alone distinguished the learned from the vulgar, the elite from the popular."[14]

Yet another of the primary characteristics of the subjects of this people's history is anonymity. The names of many university-trained scholars who earned a place in scientific history are immortalized by their published writings, but the names of most illiterate and semiliterate artisans are usually recorded, if at all, only in birth, baptismal, marriage, and death records that give no clue as to their role in the creation of natural knowledge.

There are, however, notable exceptions. Some artisans wrote in the vernacular and published manuals and "books of secrets" under their own names.[15] At least two tradesmen have even been treated on occasion as bona fide Great Men of Science: John Harrison, for solving the most pressing scientific puzzle of his age, how to measure longitude at sea;[16] and Antony van Leeuwenhoek, the "father of protozoology and bacteriology."[17] Many artists and architects who contributed to the history of science—the Michaelangelos, the Leonardos, the Brunelleschis—attained great fame and aristocratic patronage, but they were essentially craftsmen nonetheless—"superior manual laborers," as Edgar Zilsel called them.[18] There is also one prominent figure whose battles against the scientific elite arguably qualify him as a "people's scientist": Theophrastus Bombastus von Hohenheim, better known as Paracelsus. But despite the exceptions, it remains the rule that the names of most of those whose contributions form the subject of this book are lost to history. In any event, the main concern of a people's history is the scientific accomplishments not of individuals but of occupational groups.

While acknowledging that some individuals defy categorization, it is useful to recognize working with hands, anonymity, not writing in Latin, and lack of patronage as identifying features that separate the social majority from the elite scholars. In the second half of the seventeenth century, science took its first significant steps toward professionalization, a process that continued over the following three centuries until it reached the point where virtually all scientific activities were being carried out by professional scientists. As the difficulty of attaining new knowledge about nature increased, only well-funded government- and corporate-sponsored research teams could afford to pursue it.

By the twentieth century, science had become the exclusive domain of highly specialized elites. The last two chapters of this book examine the rise of Big Science and consider whether the era when people without Ph.Ds were able to make direct contributions to science has come to a definitive end.

WHAT SCIENCE?

THE MEANING OF *science* is not as simple to pin down as might first be thought. The Latin *scientia* was a generic term encompassing all forms of knowledge, but in recent centuries *science* has come to include only certain kinds of technical knowledge. A few years ago the British journal *Nature* described the systematic efforts of a number of scientists to codify the meaning of the word in some precise way that would make it possible clearly to distinguish science from pseudoscience, but they were unable to produce a satisfying definition.[19] I have followed the example set by J. D. Bernal's masterful *Science in History*, which he began by saying, "Science throughout is taken in a very broad sense and nowhere do I attempt to cramp it into a definition." This nondogmatic approach is necessary because in the last resort it is the people who are the ultimate judges of the meaning and value of science. Where science has been kept a mystery in the hands of a selected few, it is inevitably linked with the interests of the ruling classes and is cut off from the understanding and inspiration that arise from the needs and capacities of the people.[20]

At the very least science must be recognized as both a body of knowledge and a process of obtaining that knowledge. Let us, therefore, take the most uncomplicated approach possible and for the purposes of this book simply consider science to be *knowledge about nature* and the associated *knowledge-producing activities*.

As for the kinds of activities that produce scientific knowledge, the primary focus in this book is on *empirical* as opposed to *theoretical* processes. It is my contention that the foundations of scientific knowledge owe far more to experiment and "hands-on" trial-and-error procedures than to abstract thought. Benjamin Farrington made the point very well:

> In its origin science is not in fact so divorced from practical ends as histories have sometimes made out. Textbooks, right down from Greek times, have tended to obscure the empirical element in the growth of knowledge by their ambition of presenting their subjects in a logical orderly development. This is, perhaps, the best method of exposition; the mistake is to confuse it with a record of the genesis of theory. Behind Euclid's definition of a straight line as "one that lies evenly between the points on it" one divines the mason with his level.[21]

This is a broad, inclusionary concept of science that may not be palatable to readers conditioned to think of science in positivistic ways, with physics as the paradigmatic science against which all other fields are measured.[22] Theoretical physicists have frequently belittled such disciplines as botany and paleontology by likening their intellectual content to that of stamp collecting.[23] The implication of that condescending slur is that physics is "more scientific" than are less theory-driven disciplines—a reflection, once again, of the ancient prejudice that proclaims intellectual labor more honorable than manual labor. But what physicists do is not typical of what most other scientists do. The methodologies of biology, anthropology, ecology, psychology, and sociology have very little in common with the abstractions of theoretical physics, and yet the general ideology of modern science places physics on a pedestal as the model science which all other sciences should strive to emulate.

The "imperialism of physics"[24] was in no small part the creation of American governmental policy. Because of their role in developing the atomic bomb, a few "aristocrats of physics" emerged in the post–World War II period as the primary spokesmen for American science. It was they

> who implanted their values, including disdain for the social and behavioral sciences, on government science policy for decades. The social and behavioral sciences were . . . arrogantly dismissed as the "soft sciences" by the reigning physicists of postwar science (who regarded themselves, along with chemists, mathematicians, and biologists, as practitioners of the "hard sciences").[25]

Giving physics a privileged place among the sciences reinforces the idea that science must be "value-free," especially with respect to social problems.[26] In physics, the ideal of objectivity is equated with neutrality; by that definition, scientists are expected to be neutral and dispassionate with regard to the subject of their inquiry. Neutrality may be a workable stance for physicists to adopt, but in sciences that are closer to social concerns—such as medicine, anthropology, psychology, sociology, and political economy—the appeal to neutrality operates in support of the status quo, which is underpinned by racist, sexist, or bourgeois assumptions of which the scientists themselves are often unaware.[27] In recent decades, for example, the feminist movement has exposed a strong antifemale bias embedded in the traditional wisdom of elite medical science, which has for millennia been highly detrimental to the health of women. That the medical profession long considered being female a pathological condition per se is reflected in the word "hysteria," which derives from the Greek for "uterus."[28]

A people's history of science obviously cannot be constrained by the narrow *physics über alles* concept of science; equal status herein is accorded to all fields of natural knowledge, and no invidious comparisons of "hard" and "soft" or "exact" and "inexact" sciences are entertained. The concept of value-free science and its rise to a dominant position in the ideology of modern science have a history of their own that is further considered in chapter 6.

To challenge, as I do, the notion that science should be restricted to theoretical endeavors is to take sides in the "science wars" that have been roiling intellectual circles for a number of years. The traditionalists who portray science as "pure theory" do so in order to place it beyond criticism. That view of science is frequently an adjunct to reactionary political views because it supposedly offers a source of unchallengeable authority, like religion, and thereby serves as a support for authoritarianism. But many open-minded scholars, radical feminists, and environmental activists reject that notion and refuse to bow down before a deified Science.

SCIENCE AND TECHNOLOGY

ANOTHER IDEOLOGICAL COROLLARY of the intellectuals' contempt for manual labor is the "remarkably widespread wrong idea" that science is rigidly distinct from and supersedes technology in historical importance.[29] Our twenty-first-century perspective prompts us to think of technology as "applied science," a notion based on the facile assumption that scientific theory has always been a precondition of technological advance. Historically the opposite has most often been true: Although technology and science have always been closely associated endeavors, it is technology that has driven the growth of scientific knowledge.[30] The beginning of science, to paraphrase

Goethe, was not the *word* but the *deed*—not the proclamations of brilliant theorists but the creative handiwork of ordinary people. As technologies develop and become more sophisticated, the scientific knowledge generated at earlier stages is continuously incorporated into subsequent practice, and in that sense technology can indeed be said to exhibit the character of "applied science." The relation is one of cumulative mutual reinforcement, with the initial impulses coming from the technology side.

Until the last century or two, the process of gaining knowledge of nature has generally been more a product of hands than brains; that is, of empirical trial-and-error procedures rather than theoretical application. "Science," archaeologist V. Gordon Childe declared, "originated in, and was at first identical with, the practical crafts."[31] Social anthropologist Claude Lévi-Strauss dismissed the notion that any of the "great arts of civilization"—pottery, weaving, metallurgy, agriculture, and the domestication of animals—could have come about as a "fortuitous accumulation of a series of chance discoveries." They provide evidence, he argued, that "Neolithic man" was "the heir of a long scientific tradition":

> Each of these techniques assumes centuries of active and methodical observation, of bold hypotheses tested by means of endlessly repeated experiments. . . . There is no doubt that all these achievements required a genuinely scientific attitude, sustained and watchful interest and a desire for knowledge for its own sake. For only a small proportion of observations and experiments (which must be assumed to have been primarily inspired by a desire for knowledge) could have yielded practical and immediately useful results.[32]

"The crafts first uncovered aspects of nature upon which philosophies were later to be built," and they remained the primary fount of nature-knowledge through the ages.[33] Science progressed in the early modern era—that is, from about 1450 through 1750—through analysis of the inventions and innovations produced by artisans, many of whom were illiterate. In the Scientific Revolution of the sixteenth and seventeenth centuries, practical advances came first and theory followed along behind—usually far behind. This relationship persisted through the Industrial Revolution of the late eighteenth century. Not until well into the nineteenth century, with the mass production of chemical dyes and the advent of the electrical industry, were the long-anticipated hopes of major technologies based on theoretical science finally realized.[34] And even in the twentieth century, artisans were still capable of momentous scientific contributions. In 1903—the era of relativity and quantum theory—it was not theoretical physicists but two bicycle mechanics named Wright who gave the critical impulse to the science of aerodynamics. It was only with World War II and the Manhattan Project that theory began generally to lead the way in scientific discovery.[35]

Rigidly separating the histories of science and technology serves to reinforce the fallacious notion that science arose from the realm of pure thought, floating in the clouds above the world of mundane human pursuits. An undistorted picture of the development of modern science requires recognition and acceptance of its entwinement with technology, often to the point of their not being recognizable as distinct entities.

Consider again, for example, the practice of navigation, which is generally classified as a technology and thought of as more art than science. It may seem odd to call a ship's pilot's activities scientific, but the historical progress of navigation depended entirely on the growth of an underlying body of nature-knowledge: of the oceans' tides, currents, and prevailing winds; of the characteristics of the earth's magnetic field; and of astronomical phenomena. Pioneering navigators in every age and in every part of the globe, as they sailed far from the sight of land, laid the foundations of hydrography and made essential contributions to the scientific disciplines of oceanography, meteorology, physical geography, cartography, and astronomy, among others.

Richard Westfall, a prominent historian of science, downplayed the contributions made by ordinary seamen to the nautical sciences, contending that it was "never the practical tarpaulins who sailed before the mast, but always the astronomers and mathematicians, who taught the navigators."[36] Although it cannot be denied that professional mathematicians eventually contributed to the improvement of navigational practice, in the beginning the land-bound scholars were dependent on data supplied to them by mariners, as is amply demonstrated in chapter 4.

The work of mathematicians depended in an even more fundamental way on the prior activities of seafaring craftsmen. It was a specific need of navigators for simplified computational methods that stimulated John Napier's invention of logarithms, and it was the problem of finding a method of determining longitude at sea that led Isaac Newton to his formulation of the law of universal gravitation.[37] These important scientific advances were not driven primarily by the idle curiosity of isolated thinkers; they illustrate the collective, social nature of knowledge creation. It is not mere coincidence that Napier's and Newton's inspirations occurred in an island nation in an era of rapidly growing oceanic trade. In these cases, it was sailors who constituted the active element by posing the technical problems that elicited the mathematicians' responses. The universality of this relationship is evidenced by the first-century geographer Strabo's report that the Phoenicians, who were "superior to all peoples of all times" in seamanship, were "philosophers in the sciences of astronomy and arithmetic, having begun their studies with practical calculations and with night-sailings."[38]

The counterposition of "mathematicians and astronomers" to "practical tarpaulins" depends on another hidden elitist assumption: that mathematical endeavors were the exclusive province of ivory-tower theoreticians. But that ignores the surveyors, mapmakers, instrument makers, navigators, and mechanics whose activities represented the leading edge, in the fifteenth and sixteenth centuries, of innovation in practical mathematics. The English mathematician John Wallis observed in his autobiography that "Mathematicks" even as late as the

1630s "were scarce looked upon as *Accademical* studies, but rather *Mechanical*; as the business of *Traders, Merchants, Seamen, Carpenters, Surveyors of Lands*, or the like."[39]

The historical priority of technology over theoretical science is most generally exemplified by the central theme of this book, which is that *artisans contributed not only the mass of empirical knowledge that furnished the raw material of the Scientific Revolution, but the empirical method itself*. The science of the preceding era had been based almost entirely on the authority of ancient authors, Aristotle above all. In university and other elite settings, questions of natural knowledge were investigated by looking for answers in books or—in the rare instances when ancient authorities were challenged—by abstract *a priori* reasoning, not by directly interrogating nature. The habit of experimentalism that came to characterize modern science was a product of the craftsmen's shops, as is demonstrated in chapter 5.

WHAT HISTORY?

HISTORIANS IN GENERAL have succeeded in displacing the encomiastic tradition—the Great Man Theory of History—as the predominant viewpoint of the educated reading public, but historians of science—in spite of a great deal of effort and good scholarship—have been less successful. "Science," Derek de Solla Price lamented, "seems tied to its heroes more closely than any other branch of learning."[40] Although few people today would agree with Carlyle's famous dictum that "the history of the world is but the biography of great men," many continue to believe that the Scientific Revolution was the creation of a very few extremely talented geniuses: "from Copernicus to Newton."

Part of the problem is that although the public understanding of history in general has been strongly influenced by professional historians, the way most people conceive of the history of science has been shaped not by historians of science but by scientists themselves, who often hold and propagate distorted conceptions of their predecessors' practices.[41] Scientists have a guild interest in portraying their forerunners as heroes, because it adds to the heroic stature of their profession and enhances their view of their own place in the scheme of things.

More important, most scientists are not professional historians; their primary concerns are not historical. Their interest in their science's path of development is secondary to their interest in the science itself. They therefore often unwittingly adopt a tunnel-vision view of their discipline's past, focusing only on the narrow lineage of successes and ignoring all the false starts and dead ends as uninteresting because they did not "lead anywhere." Tunnel-vision history of science may be of some use as a teaching tool in elementary science courses, but it does not constitute valid history. Its projection of present-day concerns onto the past gives a falsified and misleading picture of the way science has developed in real life.[42]

Some very capable historians of science have labored mightily in recent decades to overcome the idealized picture produced by scientists and by their own predecessors.[43] The scholarship of the best of the new generation of historians has provided the basis for this book, as the notes and bibliography attest. But, unfortunately, the academic field of history of science continues to concentrate most of its attention on a few scientific luminaries—a symptom, perhaps, of the cult of celebrity that afflicts our culture as a whole. The bulk of its books and articles are products of a "Galileo industry," a "Newton industry," a "Darwin industry," an "Einstein industry," and so forth.[44] The Great Men of Science, like the proverbial elephant in the parlor, cannot be ignored. But their stories have traditionally been told from the perspective of the ruling elites; I reexamine them from a different point of view.

SOCIAL HISTORY AND PEOPLE'S HISTORY

What distinguishes a "people's history" from the more general category of social history? They are overlapping but not identical approaches to understanding the past. Social historians have done an admirable job of describing the social context in which the traditional Heroes of Science worked. Explaining

Boyle's and Newton's activities in light of the English Civil Wars and the Glorious Revolution, for example, is a valuable corrective to idealized history of science.[45] A people's history, however, aims at deepening the understanding of science as a social activity by emphasizing the collective nature of the production of scientific knowledge.

Social historians for many years now have been presenting their pictures of the past from "the bottom up" rather than from "the top down." Some have described the activities of the common people or have in other ways broadened the social context in which historic events have been understood, but without abandoning the point of view of the dominant social classes.

Also, some social historians have purposefully directed their attention toward the privileged classes. Steven Shapin's *A Social History of Truth* is an excellent example of social history of science, but his account of what he calls "the gentlemanly constitution of scientific truth" in seventeenth-century England is explicitly grounded in an elite perspective. "I am both well aware of, and deeply sympathetic towards, the new cultural history of the disenfranchised and the voiceless," Shapin wrote. "To the extent that I focus on the society of gentlemen," however, "this is a story told 'from their point of view.'"[46]

Shapin's focus was on the role of seventeenth-century gentlemen in the production of scientific knowledge. That role was essentially an epistemological one—that is, the certification or legitimation of knowledge. The question Shapin examined is not how gentlemen came to know new things about nature, but how they came to agree among themselves that they knew those things. Those who were actually making new discoveries about nature were artisans—people who worked with hands as well as brains and who were motivated not primarily by curiosity but by material necessity: the need to make a living. In a nutshell, the birth of modern science occurred when gentlemen began to appropriate artisans' knowledge and to systematize it. This theme is developed more fully in chapter 5.

Artisans, it should be noted, needed no formal means of

legitimizing what they knew and were generally unconcerned with whether others agreed with it. Their knowledge of nature was tested, confirmed, and continuously reconfirmed in their daily practice. If what they knew "worked" for them, that was legitimacy aplenty. Robert Boyle observed that empirical data discovered by artisans, "if they really serve the craftsman's turn, must be true" and are therefore "fit to be admitted into the history of nature."[47] Sailors and fishermen once again provide a cogent example. Their precise knowledge of the relation between the moon's position and the tides, which they recorded in accurate tables, had helped them safely reach port for many centuries before Galileo erroneously denied the moon's agency and instead attributed the tides to the earth's rotation.[48]

While acknowledging that his book is about "a small group of powerful and vocal actors," Shapin challenged other historians to look in other social directions as well: "If there are past voices—of women, of servants, of savages—in the practice to be attended to and made audible, then there is every reason why historians should, if they choose, concern themselves with them." On the other hand, he added, "if there are no such voices, or if they are almost inaudible," then historians should pay attention to the "*practices of inclusion and exclusion* through which some speak and others are spoken for, and some act and others are acted upon."[49]

I have attempted to heed Shapin's advice. Although there are some faint voices of knowledge-seeking members of the subordinate classes that can be amplified and made audible, an alternative approach to uncovering their place in the history of science is to analyze carefully the words and deeds of those who "spoke for" and "acted upon" them. It is by the testimony of Bacon and Boyle, of Gilbert and Galileo, that a strong case for the significance of the "illiterate mechanicks" can be made.

WHOSE KNOWLEDGE WAS IT, ANYWAY?

ALMOST ALL AUTHORS who have commented on the subject of craft secrecy have condemned it and deplored the backwardness of benighted craftsmen who attempted to hoard their knowledge, while praising elite *virtuosi* such as Robert Boyle for making new knowledge about nature public. Those who freely shared the results of their investigations by publishing them have been hailed almost universally for their unselfish contributions to human enlightenment and progress. That, of course, is the light in which Boyle, Bacon, and their like-minded colleagues saw themselves, and most subsequent authors simply echoed their point of view, perhaps without thinking about it very deeply. But although Baconian propaganda claims "mankind" in general as its concern and the "bettering of man's life" as its motive, Bacon's own program for progress clearly included maintaining and strengthening the social status quo—to increase the power of the dominant elite over the popular masses.[50]

Artisans saw the practice of craft secrecy in quite a different light. It was not motivated by evil intent on their part; they looked on it as a necessary condition for their economic survival: "Technical knowledge was the craftsman's most valuable property, even more valuable than his materials or his labor."[51] Their knowledge of natural processes had been gained by hard work and years of apprenticeship; it was the source of their income, the basis of their ability to make a living and support themselves and their families. When well-off gentlemen who could afford to be magnanimous exposed the lore of the craftsmen to public view, it was, from the standpoint of the artisans' interests, an act of robbery.

The *virtuosi's* posture as champions of the free exchange of scientific ideas may seem warranted because they did not hoard the stolen knowledge for themselves but broadcast it to the world. But their liberality mirrored the hypocrisy of laissez-faire economics; those countries that are the staunchest advocates of free markets are invariably those most able to dominate them: Great Britain in the nineteenth century and the United States today. In the "marketplace of ideas," the intellectual elite was likewise in a position to control the valuable knowledge it "liberated."

Robert Boyle attempted to justify appropriating the

craftsmen's knowledge by arguing that it would be repaid eventually with interest: "as the naturalist may . . . derive much knowledge from an inspection into the trades, so by virtue of the knowledge thus acquired . . . he may be as able to contribute to the improvement of the trades."[52] There is no reason to doubt Boyle's sincerity in assuming that "improvement of the trades" would result in bettering the tradesman's condition, but that was not what happened. In the context of a nascent capitalist economy, the benefits of increased productivity went not to the producers but to a privileged few whose access to capital allowed them to gain control of the productive process. The artisans who forfeited their knowledge were for the most part eventually forced into dependency as wageworkers. Boyle's "improvement of the trades" was a forerunner of the nineteenth-century introduction of so-called labor-saving machinery, which did not serve to ease the labor of the laborers but to reduce the *labor costs* of their employers. No matter how socially progressive it may have been in the long run, the immediate effect was that displaced workers lost their livelihoods while factory owners enriched themselves.

The general themes and arguments of this book should by now be sufficiently clear. It is time, then, to back up and begin our examination of people's science at the beginning—with the earliest people: the hunter-gatherers in all parts of the globe.

NOTES

1. Walter Burkert, *Lore and Science in Ancient Pythagoreanism*, p. 217. The case of Pythagoras is discussed further in chapter 3.
2. If ever a final word has been written on any subject, surely it is Robert Merton's *On the Shoulders of Giants* ("OTSOG" for short) with regard to the famous aphorism commonly attributed to Newton. Newton used it in a letter to Robert Hooke, February 5, 1675/76 (Newton, *Correspondence*, vol. 1, p. 416).
3. Margaret Rossiter has made this point particularly well. See her *Women Scientists in America*.

4. Lynn White Jr., *Medieval Technology and Social Change*, p. 39.

5. T. S. Ashton, *The Industrial Revolution, 1760–1830*, p. 27–28, 62.

6. The contributions of seamen are explored more fully in chapter 4.

7. In this book the abbreviations "C.E." and "B.C.E.," for "Common Era" and "Before the Common Era" are used instead of "A.D." and "B.C.," respectively.

8. The status of the better-off members of the incipient capitalist class as an urban elite was relative; in the context of aristocratic society as a whole, they were a "subelite," or "middle," class.

9. William Eamon, *Science and the Secrets of Nature*, p. 80.

10. *Egyptian Hieratic Papyri in the British Museum*, second series (London, 1923); quoted by J. D. Bernal, *Science in History*, vol. 1, p. 130.

11. Xenophon, *The Economist*, chapter 4, pp. 22-23.

12. Eamon, *Science and the Secrets of Nature*, p. 80.

13. "The millennial separation of . . . thinker from craftsman," a leading historian of technology comments, "is not vanished even in our own time." Lynn White, Jr., "Pumps and Pendula," p. 110.

14. Eamon, *Science and the Secrets of Nature*, p. 37.

15. "In the late Middle Ages," Eamon wrote, "as literacy spread, craftsmen began more frequently to record their technical secrets in writing. They composed handbooks to train other artisans and to stake claims to their inventions." He cited "dozens of examples." Eamon, *Science and the Secrets of Nature*, p. 83. This artisanal literature is discussed further in chapter 5.

16. A book, Dava Sobel's *Longitude*, and a four-hour television documentary based on that book attempt to win entry for Harrison into the pantheon of great scientists.

17. Clifford Dobell, ed., *Antony van Leeuwenhoek and His "Little Animals."* The quoted phrase appears on the cover of the book.

18. See, e.g., Edgar Zilsel, "The Origins of Gilbert's Scientific Method," p. 91.

19. "Physicists Seek Definition of 'Science,'" *Nature*, April 30, 1998.

20. J. D. Bernal, *Science in History*, vol. 1, pp. 3, 34. The author of this important four-volume study of the history of science was not primarily a historian. J. D. Bernal's contributions to the science of crystallography place him among the leading twentieth-century physicists.

21. Benjamin Farrington, *Science in Antiquity*, p. 3.

22. On the other hand, some historians of science may find this definition of science not inclusionary enough, on the grounds that it begs the question of exactly what is meant by "knowledge." Does it include all of what people of the past considered to be knowledge, or is it restricted to what we today, with benefit of hindsight, accept as knowledge? (See, e.g., the introductory essays in Michael H. Shank, ed., *The Scientific Enterprise in Antiquity and the Middle Ages*.) Although I agree that retrospectively discredited knowledge should not be excluded from the history of science, in this book the emphasis is on the origins of knowledge that most readers would accept as still valid today.

23. The "stamp-collecting" insult originated a century ago with the physicist Ernest Rutherford, but it has been repeated many times since.

24. I have borrowed this phrase from Richard Creath, "The Unity of Science," p. 168.

25. Daniel S. Greenberg, *Science, Money, and Politics*, pp. 451-453.

26. Feminist philosophers of science have been among the most insightful critics of the notion of "value-free" science. See Sandra Harding, *The Science Question in Feminism*, pp. 43-44, 227-228, 232-233; Helen Longino, "Can There Be a Feminist Science?" and Helen Longino, *Science as Social Knowledge*, chapters 4 and 5.

27. Harding, *The Science Question in Feminism*, p. 47.

28. See the section "Feminism versus Medical Science" in chapter 8.

29. Derek J. de Solla Price, "Of Sealing Wax and String," p. 239.

30. Jacques Barzun distinguished between the practical arts, *techne*, and "their ology." Barzun, *From Dawn to Decadence*, p. 205. I stick with the more familiar term *technology*.

31. V. Gordon Childe, *Man Makes Himself*, p. 171.

32. Claude Lévi-Strauss, *The Savage Mind*, pp. 13-15.

33. Cyril Stanley Smith, "Preface" to Denise Schmandt-Besserat, ed., *Early Technologies*, p. 4.

34. On mass-produced chemical dyes as the first significant science-based technology, see David Landes, *The Unbound Prometheus*, pp. 274–276.

35. Ironically, the best presentation of the case for the ontological priority of technology over (theoretical) science I have encountered was made by authors who would, I strongly suspect, reject my definition of science: James McClellan and Harold Dorn, *Science and Technology in World History*. In my view, their definitions minimize the empirical side of science and focus only on the theoretical side. In so doing, they draw an unreasonably sharp line between technology and science, which led them to conclude that "science and technology followed separate trajectories during 2,000 millennia of prehistory" (p. 5). Harold Dorn, *The Geography of Science*, argued even more forcefully against "conflating technology and science" (pp. 17-21). Nonetheless, McClellan and Dorn did an excellent job of demonstrating that technology was the ground from which science grew, and that is the main point I am trying to make here.

36. Richard Westfall, "Science and Technology during the Scientific Revolution," p. 69. Westfall was echoing the sentiments of John Flamsteed (1646-1719), England's first royal astronomer, who wrote: "All our great attainments in science . . . have come . . . from the fire-sides of thinking men . . . and not from Tarpawlins, tho' of never so great experience." Quoted in E. G. R. Taylor, *The Mathematical Practitioners of Tudor and Stuart England*, p. 4.

37. See the section "Isaac Newton and the Hessen Thesis" in chapter 6.

38. Strabo, *The Geography of Strabo,* vol. VII, p. 269.

39. Christoph J. Scriba, ed., "The Autobiography of John Wallis, F.R.S.," p. 27 (emphasis in original).

40. Derek J. de Solla Price, *Science since Babylon,* p. 47.

41. "Leaders of science," Daniel Greenberg observed, "are not the most reliable commentators on the historical, political, and financial realities of their profession." Greenberg, *Science, Money, and Politics,* p. 77.

42. To avoid being charged with hypocrisy, I must again acknowledge that this book is not entirely free of the tunnel-vision syndrome. (See note 22 above.) My defense is that this is a brief survey of a very large subject, not intended to be comprehensive, focusing on continuities of knowledge that will be recognizable to nonspecialists in the history of science.

43. Alexander Koyré and A. Rupert Hall were among the most influential idealizers of the history of science. See chapter 5. As for the "very capable historians of science," I have identified a representative sample in the acknowledgments to this book.

44. See Mott T. Greene, "History of Geology," p. 97.

45. See James Jacob, *Robert Boyle and the English Revolution;* and Margaret Jacob, *The Newtonians and the English Revolution, 1689-1720.*

46. Steven Shapin, *A Social History of Truth,* p. xxi.

47. Robert Boyle, *That the Goods of Mankind May Be Much Increased by the Naturalist's Insight into Trades,* p. 444.

48. See the section "The Tides" in chapter 4.

49. Shapin, *Social History of Truth,* p. xxii (emphasis in original).

50. See the epigraph to chapter 6. Bacon's ideas regarding the social uses of science are discussed further in that chapter.

51. Eamon, *Science and the Secrets of Nature,* p. 81.

52. Boyle, *That the Goods of Mankind May Be Much Increased by the Naturalist's Insight into Trades,* p. 446.

2

PREHISTORY: WERE HUNTER-GATHERERS STUPID?

IN SUCH CONDITION, there is no place for Industry; because the fruit thereof is uncertain: and consequently no Culture of the Earth; no navigation, nor use of the commodities that may be imported by Sea; no commodious Building, no instruments of Moving, and removing such things as require much force; no Knowledge of the face of the Earth; no account of Time; no Arts; no Letters; no Society; and which is worst of all, continuall feare, and danger of violent death; And the life of man, solitary, poor, nasty, brutish and short.

—THOMAS HOBBES, *Leviathan* (1651)

THOMAS HOBBES, WRITING in the seventeenth century, had a low opinion of the knowledge possessed by prehistoric humans. In his eyes, their condition was scarcely different from that of animals. His estimation was not based on evidence; it was simply commonsense conjecture. He imagined what life had been like in the distant past without such benefits of civilization as the rule of law and assumed that it must have been, in his oft-quoted phrase, "nasty, brutish and short."

In the following century, an opposite but no less abstract appreciation of human prehistory was advanced by another social theorist, Jean-Jacques Rousseau. According to Rousseau's theory of the social contract, prehistoric humans were noble savages. In that original primitive state, Rousseau maintained, people "were as free, healthy, good, and happy as their nature permitted them to be." The rise of civilization, however, "brought on the downfall of the human race" by creating property, inequality, slavery, and poverty.[1] But noble though they may have been, early humans were still essentially savages who, to borrow the biblical metaphor, had not yet tasted the fruit of the tree of knowledge. Rousseau's noble savages were characterized by "stupidity and obtuseness"; they were no more knowledgeable or intelligent than Hobbes's nasty brutes.[2]

Precisely what prehistoric humans knew and did not know is not easily determined, but it is certain that Hobbes, Rousseau, and the Bible all seriously underestimated their intellectual capacity and accomplishments. Unlike other species, early humans did not simply survive in limited ecological niches to which they were able to adapt, but spread throughout the globe, shaping their surroundings wherever they went to meet their own needs. It is reasonable to assume that they could have done so only because of their uniquely human ability to gain and apply an immense body of knowledge of nature.

A people's *prehistory* of science is an account of the impressive extent of the nature-knowledge possessed by humans during the many millennia before the advent of literacy, for which no documentary evidence is possible. The difficulty of finding and interpreting solid evidence of prehistoric knowledge hampers the inquiry. On the other hand, it is made simpler by not having to distinguish between elite and popular knowledge, because the era under consideration mostly preceded the rise of social differentiation into dominant and dominated classes. In other words, whatever science can be said to have existed then was *by definition* people's science.

Dating the origin of the human species is a matter of ongoing discovery and scholarship, but the most recent fossil finds suggest that the first humanoids distinguishable from apes may have appeared in Africa more than seven million years ago. The fossil record reveals subsequent variants of humanoids progressively more like ourselves until—*voilà*—a species appears, sometime between forty and ninety thousand years ago, that is anatomically indistinguishable from us. As the old cliché goes, if you were to put one of those human beings from fifty thousand years past on a New York subway with a shave, a haircut, and a new suit of clothes, the other strapholders probably would not raise an eyebrow.[3] A more important supposition is that if you could send one of them to Harvard, he or she would prove to be as educable as a person born in our era. Although that is an untestable hypothesis, there is no good reason to suspect that it is false. Anthropologist Sally McBrearty asserted that "the earliest Homo sapiens probably had the cognitive capability to invent Sputnik."[4]

In any case, the evidence of paleontology indicates that close members of our own "human family" have inhabited the earth for many tens of thousands of years. Until the end of the last Ice Age, about thirteen thousand years ago, they all depended for their subsistence entirely on hunting and gathering. The post–Ice Age origins of agriculture and domestication of animals began to lessen human adherence to the hunting-gathering lifestyle, to the point where today hunter-gatherers—perhaps better identified as "foragers"—constitute only a very small proportion of the world's population.[5] Nonetheless, it is a safe estimate that more than 99 percent of all people who have ever lived were foragers.[6]

During the nineteenth century and most of the twentieth, the way our foraging forebears were perceived owed more to Hobbes than to Rousseau. Their lifestyle was almost universally assumed—by scholars and laypeople alike—to have been one of unrelieved poverty, endless labor, and abysmal ignorance. Small wonder, then, that the "Neolithic revolution" that initiated agriculture and the domestication of animals was thought of as liberating humans from the miserable existence of hunting and gathering. In accord with the heroic view of the history of science, that great act of liberation was assumed to be the innovation of a few superior humans whose intelligence allowed them to perceive the advantages of settling down to produce a regular food supply.

The radicalism of the 1960s, however, prompted a rethinking of many traditional ideas about society, and prehistoric societies were not left out of account. A 1966 conference entitled "Man the Hunter" witnessed a turning point in perceptions of the foraging lifestyle. Anthropologists R. B. Lee and I. DeVore made a startling proclamation: "To date," they said, "the hunting way of life has been the most successful and persistent adaptation man has ever achieved."[7] Marshall Sahlins stated the case even more provocatively: the prehistoric foragers, he declared, constituted "the original affluent society."[8] Sahlins contended that the hunters and gatherers typically needed only a few hours of work a day to satisfy their material needs, leaving them with plenty of leisure time and a relatively relaxed existence. Sahlins called it a "Zen economy": the foragers had everything their hearts desired because their hearts did not desire very much in the way of material goods. Some critics charged that he overstated the case for a prehistoric paradise, but he succeeded in fundamentally altering the way anthropologists and archaeologists interpret the cultures and artifacts they study.

It had traditionally been assumed that "primitive" people failed to advance technologically because their desperate struggle for survival left them no time for deep thoughts and innovative experimentation. Sahlins's claim, however, that foragers did not lack for free time has been amply confirmed by subsequent studies. If foragers "failed" to make "progress," it was neither because they were too busy or too stupid, but because what appears in retrospect to be progress simply held little attraction for them.

Whereas modern scholars considered the transformation to agriculture a liberating event, the foragers themselves may well have perceived it as expulsion from the Garden of Eden. Instead of their being able to pick up their means of sustenance in the course of a leisurely day, the obligations of agriculture would henceforth sentence them to hard labor from sunup to sundown. They would make that change only as a last resort, when, under the relentless pressure of population growth, they would be forced to wrest their means of survival from ever-diminishing areas of land. Furthermore, it does not stand to reason that the "pioneers" of the Neolithic revolution were necessarily the most intelligent of the foragers; they were simply those who were first confronted with the choice between producing food or going hungry.

The Garden of Eden is appropriate as a metaphor for the foraging lifestyle in only the most relative sense. Hunting and gathering may have required less work than agriculture, but it would be misleading to suggest that food and shelter came so easily to prehistoric humans that very little effort or knowledge on their part was necessary. The growth of their knowledge of nature was driven by the need to maintain access to natural resources that varied in more or less predictable ways. Survival required that they learn as much as they could about migratory habits of animals, seasonal changes in water supply, and fruit-bearing cycles of plants, among other things, over large territorial expanses. In order to keep open a variety of resource options, many presently existing foragers "feel compelled to maintain knowledge of enormous areas. The Nunamiut [of northern Alaska] maintained knowledge of nearly 250,000 square kilometers; the Australian Pintupi have knowledge of over 52,000 square kilometers."[9]

As radical as the "Man the Hunter" conference was, its very title exposed its limitations. One commentator observed, "With few exceptions, twentieth century anthropology has treated women as at best peripheral members of society and at worst as nonexistent."[10] But the 1960s also saw a rebirth of feminist thought, so it was not long before the obvious questions were raised:

> If men were hunting and creating culture, where were *women*, and what were they doing? Shivering naked in the cave with the children? It didn't make ethnographic sense. From these seeds the Woman-the-Gatherer model developed further, along with its corollary, Woman-the-Inventor-of-Agriculture.[11]

The idea of Man the Hunter was thus complemented by that of Woman the Gatherer.[12] Feminist scholars made a good case that women's activities accounted for considerably more than half of the food intake of foragers. The implications for a people's history of science are obvious: If women were the principal gatherers of plants and small animals, then a major share of foragers' knowledge of nature must be attributed to them—especially the intimate knowledge of plant characteristics that were vital to human survival.[13]

Regardless of how the earliest humans have been perceived by their modern progeny, it is undeniable that hunting and gathering was the way all people made a living during the first many thousands of years of human existence on Earth. A people's history should therefore begin with the foragers. And the primary consideration must be how people came in the first place to possess the intelligence that is prerequisite to knowledge of nature.

THE ORIGINS OF HUMAN INTELLIGENCE

WHICH CAME FIRST: the brain or the hand? The unquestioned assumption among evolutionary thinkers, even before Darwin published *Origin of Species* and *Descent of Man,* was that intelligence had led the way in human evolution. The process was believed to have been primarily driven, every step of the way, by increasing brain size. The accompanying increase in intelligence supposedly provided the selection advantage that gradually transformed apes into humanoids, and humanoids into *Homo sapiens.*

Before the first discoveries of early hominid fossils in the late nineteenth century, it was taken for granted that if a "missing link" between ape and man were to be found in the fossil record, it would have a brain size intermediate between ape and man but would not have the upright posture characteristic of humans. Only after attaining a sufficient intellect would a creature be expected to display its humanity by standing erect. This "dogma of cerebral primacy," as Stephen Jay Gould called it, was dealt a death blow in the 1920s when the remains of "smallbrained australopithecines" who "walked as erect as you or I" were discovered in Africa.[14] It was henceforth undeniable that the transition from apelike creature to humanoid was well advanced before any appreciable growth in brain capacity occurred.

Why, Gould wondered, had "Western science" been "so hung up on the a priori assumption of cerebral primacy"? He found the answer in an essay written by Frederick Engels in 1876 but not published until twenty years later, after Engels's death.[15] Engels argued in *The Part Played by Labor in the Transition from Ape to Man* that "the decisive step" occurred when apes ceased to use their arms for locomotion and began to "adopt a more and more erect gait." That freed their hands for tool use, or *labor*: "Only by labor, by adaptation to ever new operations," Engels wrote, "has the human hand attained the high degree of perfection that has enabled it to conjure into being the pictures of Raphael, the statues of Thorwaldsen, the music of Paganini."[16]

Engels's theory held that the development of the hand as a product of labor propelled the acquisition of intelligence, and hence the augmentation of brain size, in early humans. Although his hand-first hypothesis was no more based on evidence than was the brain-first assumption, his conjecture was confirmed as soon as fossil evidence became available.

Engels's successful insight was possible because he had not been hampered by the ideological disposition favoring intellectual endeavors over manual labor. Gould noted that scholars traditionally tended to look down upon laborers, and to them "cerebral primacy seemed so obvious and natural that it was accepted as given, rather than recognized as a deep-seated social prejudice related to the class position of professional thinkers and their patrons."[17] It was that ancient bias, Engels said, that had prevented "even the most materialistic natural scientists of the Darwinian school" of his day from understanding human origins and recognizing "the part that has been played therein by labor."[18]

If human evolution had been brain-driven, then the leading role in the process would have to be ascribed to individuals of above-average intelligence—an early version of an intellectual elite—whose natural superiority gave them and their genes a preferential edge in the struggle for survival. But the fossil evidence indicates otherwise. It was the toolmaking and tool-using activities of entire humanoid populations—"working people"—from which human intelligence emerged, together with the capacity for language, prescientific knowledge, and eventually science.

SEARCHING FOR EVIDENCE

THERE ARE TWO general kinds of evidence that might help us determine what prehistoric people knew about nature, but each is unsatisfactory in its own way. The first are the human-made tools and other objects unearthed by archaeologists, on which inferences can be based about how they were used. But whereas "a rich legacy of material artifacts" testifies to the existence of extensive *technologies* in the Paleolithic and Neolithic eras, "only a feeble record exists of any *scientific* interests in those preliterate societies, mainly in the form of astronomically oriented structures."[19] A familiar mantra, however, among archaeologists and paleontologists is that "absence of evidence is not evidence of absence." In other words, although the material artifacts provide few clues as to the explicit scientific interests of their owners, that does not prove that these people had no scientific interests.

What is more important, the Paleolithic and Neolithic technologies could not have existed without the prior acquisition of "an imposing body of scientific knowledge—topographical, geological, astronomical, chemical, zoological, and botanical—of practical craftlore on agriculture, mechanics, metallurgy, and architecture, and of magical beliefs that might also enshrine scientific truth."[20] Those early technologies therefore constitute an indispensable part of the history of science. "Because the essential character of natural science is its concern with the effective manipulations and transformations of matter," J. D. Bernal explained, "the main stream of science flows from the techniques of primitive man."[21]

The second kind of evidence comes from the field reports of anthropologists who live among and observe the daily lives of still-existing groups of foragers in the Kalahari Desert, in the Australian outback, above the Arctic Circle, and elsewhere. This is a much more fruitful source of data regarding the knowledge of nature possessed by hunter-gatherers, but it poses a significant methodological problem: To what extent are we justified in assuming that what foraging Bushmen, Aborigines, or Eskimos know today is at all similar to what Paleolithic or Neolithic foragers knew? Archaeologists and anthropologists alike warn us against "falling prey to the temptation to use a modern hunter-gatherer people . . . as an analogy for reconstructing the past."[22]

For one thing, conclusions drawn from studying one group of foragers cannot be automatically universalized to apply to other groups, especially when comparing a present-day people to one that lived tens of thousands of years ago. Second, no "pure stone-age cultures" exist anywhere in the world today: "All ethnographically known hunter-gatherers are tied into the world economic system in one way or another."[23] Furthermore, the foragers that anthropologists study today live in deserts and rainforests—the most marginal ecosystems on Earth for humans—whereas prehistoric foragers inhabited the far more fertile lands now occupied by farms and cities. But although modern ethnographic evidence cannot give us absolute assurance about prehistory, it can certainly provide a reasonable indication of the kinds of knowledge preliterate, preagricultural people were capable of developing.

First of all, anthropological data can serve to counteract prejudicial notions of the intrinsic intellectual inferiority of "primitive" peoples. Biologist Jared Diamond's "33 years of working with New Guineans in their own intact societies" led him to conclude that "modern 'Stone Age' peoples are on the average probably more intelligent, not less intelligent, than industrialized peoples." In mental abilities, he added, "New Guineans are probably genetically superior to Westerners, and they surely are superior in escaping the devastating developmental disadvantages under which most children in industrialized societies now grow up."[24] This is, admittedly, a subjective judgment, but it is supported by recognition of the intellectual accomplishments of many preliterate societies. Because early Pacific Islanders, for example, could not refer to written star catalogs or charts, the immense knowledge of the star positions by which they navigated had to be entirely committed to memory—an awesome mental feat indeed.

FORAGER SCIENCE

CULTURAL ANTHROPOLOGIST PETER Worsley studied the foragers of an Aboriginal tribe on Groote Eylandt, an island off the coast of northern Australia, for many years. "Because they lived by hunting and collecting and hadn't developed agriculture, the Aborigines are sometimes considered devoid of scientific thought," Worsley wrote. "Yet they depended for their very survival on observing plants and animals accurately, on coming to correct conclusions about the world and on reaching an understanding of cause and effect." They have, he added, "developed categories remarkably similar to those of Western biologists, zoologists and botanists," and "in the process, they use similar intellectual procedures."[25]

The Aborigines he studied "recognize, and name, no fewer than 643 different species." Because that is almost double the number of edible species they distinguish, it belies "the common belief that their knowledge is limited simply to the utilitarian." Worsley stressed that "it is not just the sheer *amount* of Aboriginal knowledge that is impressive. It is that . . . everything is classified within a *taxonomy*, the basic division being between plants *(amarda)* and animals *(akwalya)*, which are then subdivided into lower-level sets."[26] Another biologist and anthropologist, Donald Thompson, examined the biological knowledge of a different group of Aborigines, the Wik Monkan of Cape York Peninsula, and concluded that their system bears "some resemblance to a simple Linnaean classification."[27]

In the Groote Eylandters' taxonomy, the first division among plants separates the woody-stemmed from the non–woody-stemmed: "Altogether they distinguish no fewer than 114 kinds of woody plants and eighty-four kinds of non-woody ones."[28] In one classification system, the woody plants were then put into eight and the non-woody ones into three subcategories, "partly on the basis of similarity in form, and partly on the basis of shared habitat."[29] As for classification of aquatic animals, they

> are first divided into three named subdivisions: fish (137 kinds), shellfish (sixty-five kinds) and marine turtles (six); while cartilaginous fish *(aranjarra)* are distinguished from bony ones *(akwalya)*. The twenty-three kinds of cartilaginous fish are then subdivided into sharks (nine kinds), with a second subdivision (with no overall name) made up of stingrays (eleven kinds), shovel-nosed rays and sawfish (three kinds) and suckerfish. The 113 kinds of bony fish they distinguish are divided into twelve categories.[30]

Worsley called attention to the "remarkably high degree of correspondence" between the Aborigines' basic categories and the genera and species used by Western biologists. "For four-footed land mammals and reptiles, it is as high as 86 percent," he reported. "For animals as a whole, it is 69 percent; for plants it is 74 percent."[31] It is not my intention to suggest that Groote Eylandt taxonomy is equivalent in content or sophistication to that of modern biological science, which recognizes millions rather than hundreds of species. These samples should, however, sufficiently demonstrate that foragers are capable of creating a systematized body of knowledge that is no less deserving of being called *science*.

The quantity and quality of Aboriginal geographical knowledge are also worth noting:

> To the eyes of white men, Groote Eylandt, like northern Australia in general, is featureless. To the Aborigines it is anything but. . . . Waddy has recorded no fewer than 600 named places on the coast of Groote and its offshore islands, while David Turner, who, in 1969 and 1971, studied

Bickerton Island, off the coast of Groote—only two or three miles across from north to south and from east to west—recorded ninety-three named spots on the coast alone.[32]

The islanders "recognize no fewer than sixteen different ecological zones: eight for the land and eight for the sea." Most important to them is the coastal area, where they "distinguish between the deep sea, the shallow sea, the coral reefs, the inter-tidal rock platforms and outcrops, and the shallower intertidal zone close inshore, as well as the sand and mud flats, and the beach itself."[33]

Worsley noted that some of the foragers' most important foods are, in their natural state, highly toxic to humans, which means that they had to develop technologies of food processing. The burrawang, for example,

> is a singularly problematical food, since the nut inside the fruit contains a poison for which there is no known antidote. To remove it, the Aborigines treat the nuts by heating them, using hot stones and ashes, then pounding or grinding the nuts to make "flour" . . . using bent-over fronds in running water as a strainer, which allows the poison to leach out without the food being washed away.

Such processes pose questions of how prehistoric peoples originally discovered them. How did they come to find out that removing the poison was possible in the first place? How did they work out the complicated procedures that often require several days to complete? "Whatever the answers," Worsley said, "the development of the necessary treatments must have involved a good deal of abstract reasoning and a lot of experimentation."[34]

Another practical technology from which Aboriginal biological knowledge emerged was medicine. The details of the empirical methods of discovery are unknown, but the results of the Groote Eylandters' healing arts have been documented:

Medicines were made from leaves, vines, roots, bulbs, berries, sapwood, bark, fruit pulp, the various parts of bee-hives, . . . young shoots, seeds, salt, seawater, powdered cut-tlefish "bone" and even dingo manure. For generalized aches and pains, vines and leaves were crushed, heated or soaked in water, then applied to the body. But there were special medicines for illnesses of the chest and of the ears; for headaches; . . . for toothaches; for snake-bite and for bites from spiders, sea-centipedes, cup-moth caterpillars, sea-wasps, stonefish and stingrays; for minor cuts and for more serious wounds; for boils, burns and sores; and for broken bones; . . . to close up cicatrices; to cure constipation, coughs and colds; for diarrhoea; for difficulties in passing urine; for the eyes; for leprosy; for skin troubles; and for swellings. They also used contraceptives so powerful that they inhibit-ed conception not just in the short term but forever.[35]

There is no class of full-time, professional intellectuals among these Aborigines, which implies that maintaining the culture's body of knowledge is to some degree the responsibil-ity of every individual member. That is not to suggest that all members are equally knowledgeable or equally involved in passing knowledge on to the younger generation. On Groote Eylandt, some individuals "spend more time than most in thinking." Some "apply their minds to practical concerns," while others, *though they have to go out hunting like everyone else*, spend a lot of time thinking about thinking":

> In Aboriginal society, the experience of generations has been distilled, elaborated, codified and handed down from gener-ation to generation by these part-time "popular" intellectu-als, working within an oral culture (until recently) in which knowledge is not stored or available to be studied in books and libraries.[36]

After many years of interactions with anthropologists and

others from the outside world, the Groote Eylandters no longer constitute a purely traditional society of foragers. The younger generation is literate, and some of the women of the island have accomplished "a dramatic reappropriation of their own culture" by creating a 350-page encyclopedia of plants and animals arranged according to their traditional system of classification.[37] The fact that it could be written by committee with the general agreement of the society as a whole confirms that the Groote Eylandt taxonomy was a coherent, broadly held body of knowledge of long standing rather than the recent creation of a few exceptionally brilliant individuals.

TRACKING: "THE ORIGIN OF SCIENCE"?

THE GROOTE EYLANDTERS and other Aborigines discussed previously are not unique among foragers with respect to their knowledge of nature. Anthropologists studying the San people, or "Bushmen," of the Kalahari Desert have shown that their subjects are not only able to recognize and categorize hundreds of species of plants and animals but, more important, possess a deep understanding of animal behavior. Hunting is not merely a matter of seeing an animal and killing it; in most cases, prey is elusive and must be tracked by knowing its habits and reading its spoor.

The most important spoor is an animal's footprints, but there is much more to tracking than simply following a visible trail of animal tracks. The ability to read spoor requires a hunter to draw inferences from a wide range of barely perceptible clues, including traces of animal feces, urine, saliva, or blood; fur or feathers; broken twigs, branches, and blades of grass; odors and sounds; and various indicators of animal feeding and other behavior. "Hunters of the Kalahari desert," one author explained, "are able to discern, even in loose sand, the spoor of numerous creatures, ranging from beetles and millipedes . . . to the snake and the mongoose. They are even able to distinguish different species of mongoose by the spoor alone."[38] And, it may be added, the subtlest of these clues often suffice to allow them to determine an animal's sex, approximate age, and how recently it passed by.

Historian Carlo Ginzburg contends that this "may be the oldest act in the intellectual history of the human race: the hunter squatting on the ground, studying the tracks of his quarry."[39] Anthropologist Louis Liebenberg has extended this idea, arguing at book length that the sophisticated tracking ability of foragers constitutes "the origin of science."[40] According to his thesis, tracking "is based on hypothetico-deductive reasoning," and "is a science that requires fundamentally the same intellectual abilities as modern physics and mathematics." The knowledge of the hunter-gatherers he studied in the Kalahari "contains fairly detailed information on the feeding, breeding and hibernating habits" of many species; they "appear to know more about many aspects of animal behaviour than European scientists." That knowledge underpins "a process of creative problem-solving in which hypotheses are continuously tested against spoor evidence, rejecting those which do not stand up and replacing them with better hypotheses."[41]

Two other anthropologists who studied !Kung San hunters reached similar conclusions. "Such an intellective process," they wrote, is evidently "a basic feature of human mental life":

> It would be surprising indeed if repeated activation of hypotheses, trying them out against new data, integrating them with previously known facts, and rejecting ones which do not stand up, were habits of mind peculiar to western scientists and detectives. !Kung behavior indicates that, on the contrary, the very way of life for which the human brain evolved required them. . . . Man is the only hunting mammal with so rudimentary a sense of smell, that he could only have come to successful hunting through intellectual evolution.[42]

Whether or not we accept Liebenberg's assertion that Kalahari trackers are scientifically equivalent to modern physicists who "track" subatomic particles,[43] we cannot help but be impressed by the sophistication of the logical techniques these foragers utilize to analyze and exploit their natural environment. No less impressive is the wealth of knowledge about the sea and the stars possessed by the people who populated the islands of the Pacific thousands of years ago.

PACIFIC OCEAN PIONEERS

THE LARGEST EXPANSE of water on earth is the Pacific Ocean. Schoolchildren everywhere are taught to revere the name of Ferdinand Magellan for discovering it and for being the first navigator to explore it. The magnitude of the feat attributed to Magellan is indicated by the fact that it took the better part of a century for any other Europeans to duplicate such a journey.[44] But the island-dwellers of the Pacific could hardly have been impressed; their ancestors had learned to navigate that ocean many thousands of years earlier. To them, going back and forth across its vast expanses had been a routine matter for a long, long time. Thomas Hobbes could not have been more mistaken in his assumption that prehistoric humans had "no navigation."[45]

How did the preliterate people of the Pacific—with no charts, magnetic compasses, or other navigational instruments, or even the use of metals—succeed in traversing the open seas? As Captain James Cook, the leader of the first group of Europeans to reach the Hawaiian Islands, exclaimed in wonderment on encountering the Polynesians there, "How shall we account for this Nation spreading itself over this Vast ocean? We find them from New Zealand to the South, to these islands to the North (Hawaii) and from Easter Island to the Hebrides."[46]

Human beings managed to reach Australia from Southeast Asia some forty to sixty thousand years ago, and although the distances over water were then shorter than they are now, the crossings certainly required being out of sight of land for significant periods of time. Archaeologists originally thought that the peopling of Australia and New Guinea must have occurred accidentally—by offshore fisherpeople being caught in storms and blown across seas to unknown lands—but evidence of extensive two-way travel makes it clear that there were many intentional voyages of colonization very early on.[47] The growth of populations in the colonized areas confirms that the ancient voyagers included people of both sexes.

As significant as the navigational skills of the early Australians and New Guineans were, they must have been limited, because it would be at least another thirty thousand years before the next big wave of population expansion was to take place in the Pacific. But a thousand years before Magellan entered that ocean, indigenous Austronesians had succeeded in colonizing virtually all of the inhabitable islands throughout the immense expanses of Melanesia, Micronesia, and Polynesia—an accomplishment that presupposes the existence of a highly sophisticated system of navigation based on extensive astronomical, geographic, and oceanographic knowledge. The original Pacific Islanders were driven to gain knowledge of their oceanic environment by the same imperatives that compelled Aborigines and the !Kung San to master their natural surroundings on the land.

Dating the Austronesian expansion remains an open-ended question; approximations are possible, but they are subject to revision as new archaeological data come to light. The best estimates now available indicate that it began no later than five thousand years ago, that the Solomon Archipelago was occupied no later than 1600 B.C.E., and that over the next four hundred years, so were the Santa Cruz Islands, the Gilberts, Carolines, Marshalls, Fiji, Tonga, and Samoa. Within another thousand years—by the time of the birth of Christ—the Cook Islands, Tahiti, the Marquesas, and Hawaii were all occupied. By 500 C.E. the expansion had reached as far east as Easter Island and as far west as Madagascar, off the coast of Africa.

INDIGENOUS ASTRONOMY AND GEOGRAPHY

WHEN MAGELLAN AND his successors arrived in the Pacific, they were amazed to find that the islanders, whom they regarded as savages, knew how to navigate and were adept at astronomy. In 1769 Joseph Banks, a naturalist accompanying Captain Cook, expressed surprise that native Tahitians knew

> a very large part [of the stars] by their Names and the clever ones among them will tell in what part of the heavens they are to be seen in any month when they are above the horizon; they know also the time of their annual appearing and disappearing to a great nicety, far greater than would be easily believed by an European astronomer.[48]

French and Spanish explorers, in Tahiti at about the same time as Cook, made similar observations. Louis Antoine de Bougainville was shocked to discover that the Pacific Islanders were capable of traveling back and forth over great distances. Bougainville took aboard a Tahitian navigator named Aotourou, who, after "attentively observing" the night stars,

> pointed at the bright star in Orion's shoulder, saying, we should direct our course upon it; and that in two days time we should find an abundant country. . . . He had likewise told us that night, without any hesitation, all the names which the bright stars that we pointed at, bear in his language.[49]

The Spaniard Andia y Varela described the Tahitians' direction-finding methods in more detail:

> When the night is a clear one they steer by the stars. . . . not only do they note by them the bearings on which the several islands with which they are in touch lie, but also the harbours in them, so that they make straight for the entrance by following the rhumb of the particular star that rises or sets over it; and they hit it off with as much precision as the most expert navigator of civilized nations could achieve.[50]

Because the Europeans already had their own highly developed navigational sciences, they did not bother to investigate that knowledge in any depth. On the other hand, the *geographical* knowledge of the indigenous navigators was something they did need. Hobbes had taken for granted that such people could have "no Knowledge of the face of the Earth," but once again he was in error.[51]

Captain Cook, like Bougainville, was able to gain the cooperation of a native navigator, Tupaia. He provided Cook with information on the "existence and approximate bearing of every major island group in Polynesia and Fiji, with the exception of Hawaii and New Zealand." Cook had a chart of seventy-four islands drawn up under Tupaia's guidance, and Tupaia personally directed Cook's ship to Rurutu, an island 300 miles south of Tahiti that had previously been unknown to Europeans. Tupaia "had a most impressive geographical horizon" that "extended for 2,600 miles from the Marquesas in the east to Rotuma and Fiji in the west, equivalent to the span of the Atlantic or nearly the width of the United States."[52]

An earlier encounter in 1696 between Spaniards in the Philippines and indigenous navigators from the Caroline Islands is also instructive. "What is significant," David Lewis comments,

> is that it was these Carolinians who were eagerly questioned about their islands by the Spaniards, and not the other way around. They listed thirty-two islands including Saypen (Saipan) in the Marianas, and a map was drawn from their statements that depicted even more islands.

The geographical range of the Carolinians, "extending as it did 2,000 miles east of the Philippines, and embracing Saipan 500 miles to the north, far surpassed the sketchy knowledge of the Spaniards."[53]

Even as late as the nineteenth century, Pacific Islanders continued "to instruct the geographically uninformed European explorers."[54] In 1817 Otto von Kotzebue, on Ailuk Atoll in the Marshall Islands, was informed by a chief named Langemui of a second chain of islands in the Marshall group, 130 miles to the west. Langemui placed stones on a mat to indicate the positions of islands in both chains, "Radeck" and "Ralick." Von Kotzebue, who had known of the existence of only part of the Radeck chain, later wrote:

> As the groups, as far as we were acquainted with them, were accurately laid down; his information respecting the Ralick chain deserves equal credit . . . . The chart of the Ralick chain, which, I hope, will be pretty correct, I drew according to Langemui's information; and have added it to my atlas.[55]

The appropriation of indigenous geographical and seafaring knowledge was most nakedly exhibited in the routine kidnapping of local navigators who were forced to serve as pilots, a practice that was initiated by Columbus in the Atlantic and Magellan in the Pacific and which thereafter became standard operating procedure for "explorers." Antonio Pigafetta, the eyewitness chronicler of the Magellan voyage, casually described how that expedition found its way to the source of the European merchants' most profitable commodities, exotic spices: "We one day took two pilots by force that we might learn from them of Molucca."[56]

HOW DID THEY NAVIGATE?

THERE ARE NO records of how the earliest seafarers accomplished their voyages, but some very strong evidence exists with regard to the navigational methods they employed. In the second half of the twentieth century, a small number of anthropologists became aware of the significance of Polynesian and Micronesian navigation and realized that although the traditional seafaring techniques were still utilized on a few isolated islands, these practices were rapidly disappearing under the relentless pressure of Westernization. Fortunately, some very competent researchers went to those islands, placed themselves under the tutelage of indigenous navigators, and preserved their knowledge in a number of excellent books.[57]

One investigator, David Lewis, reports that in the late 1960s encounters with traditional navigators in Tonga and Papua alerted him to "the realization that parts of the sea lore of the ancient voyagers remained alive." He learned "that there still existed, scattered among the islands, a mosaic of fragments of a former Pacific-wide system, or systems, of navigational learning only waiting to be put together." Accordingly, he made a methodical search for indigenous seafarers who could instruct him in the ancient knowledge of the seas. Anthropologists customarily call such people their "informants," but Lewis says that in this case such a designation "hardly seems appropriate, since so many of the accomplished men who helped were our teachers, who instructed us, mainly by demonstration, both ashore and afloat." Although "the majority were illiterate," that made their prodigious knowledge of nature all the more remarkable.[58]

Lewis's primary instructors were Hipour, a Micronesian from Puluwat Atoll in the Carolines, and Tevake, from Pileni Atoll in the Reef Islands of the Santa Cruz group. Thomas Gladwin, another researcher who went to Puluwat to study under Hipour, declared in his published account of what he learned, "I am one of those fortunate anthropologists whose native instructor . . . almost wrote the book for him."[59] Another anthropologist, Richard Feinberg, went to the tiny island of Anuta in the Solomons in the 1970s and '80s and produced a book about Polynesian navigation based on information provided to him by Pu Nukumanaia, Pu Koroatu, and Pu Maevatau, among other Anutan mariners.[60] In the 1980s, a young American sailor and author named Steve Thomas apprenticed himself to Mau Piailug, a master navigator on Satawal in the Caroline group. Piailug, he says, "took me into his family, assumed responsibility for my material and political well-being, and taught me his navigation without reserve. The knowledge he gave me about navigation is considered priceless in his culture."[61]

But how much about ancient seafaring can really be inferred from these relatively recent researches? A great deal, it seems. First of all, the traditional navigational techniques of the Pacific are so completely incompatible with those of Western origin that there is little possibility of the surviving indigenous knowledge having been contaminated by the introduction of external ideas or innovations. And although standard notions of progress might lead to the assumption that Polynesian navigation had improved steadily over the past thousands of years—which would make living practitioners like Hipour and Tevake qualitatively more knowledgeable than their ancestors—the evidence points in the opposite direction. As island societies became more complex, conflicts developed that tended to inhibit contact among them. The loss of ability to navigate, for example, between the Carolines and Hawaii reveals that a significant *deterioration* of seafaring ability had occurred throughout the Pacific even before the arrival of Western imperialists, whose trade and prohibitions against native sailing accelerated the trend and almost entirely wiped out the traditional practices. The navigational knowledge recorded by recent anthropologists, as impressive as it is, must be considered vestigial—a pale reflection of that possessed by the pioneers of the Austronesian expansion.

The vessels the Pacific Islanders used are customarily referred to as canoes, but with the following qualification noted by David Lewis:

> The word "canoe" is rather misleading in the present context, conjuring up as it does a picture of some tiny craft hollowed out from a tree trunk. The vessels with which we are here concerned . . . deserve the appellation "ship," rather than "canoe." As an indication of their size, some were longer than Cook's *Endeavour.*[62]

Polynesian and Micronesian voyaging canoes, usually from fifty to seventy-five feet in length, were designed for long-distance blue-water sailing. Linguistic evidence makes it clear that the earliest of these seagoing craft were powered by sails rather than paddles: "The 5,000-year antiquity of the proto-Austronesian word for sail, *lay(r)*, along with those for mast, outrigger, and outrigger boom, leave little doubt as to the nature of their craft and the manner in which they were propelled."[63]

In the past, some scholars contended that the Austronesian expansion could have occurred entirely by accidental one-way voyages—by boatsmen caught unawares by gales and drifting to previously unknown islands.[64] That argument is no longer tenable. Computer simulations of the winds, gales, and currents of the Pacific have disproved the drift hypothesis by showing that "drifts could not account for certain crucial stages of contact":

> The probability of drifts occurring was negligible, or zero, across the seaways between Western Melanesia and Fiji; between Eastern Polynesia and Hawaii, New Zealand, or Easter Island; and Eastern Polynesian contact, in either direction, with the Americas. The probability of their having been drifts from Western to Eastern Polynesia and from Western Polynesia to the Marquesas zone was very low.[65]

That is not to deny that some accidental voyages of discovery occurred, but they would not have led to colonization had they not been followed up with deliberate two-way communication between islands. There is no reason to believe that unintentional discoveries were the norm. Adventurous navigators, confident in their seamanship, can easily be imagined setting out on deliberate voyages of exploration, sailing forth into the unknown against the wind in order to be assured of a quick and easy return trip home with the wind at their backs.

THE SIDEREAL COMPASS

THE FIRST REQUIREMENT of any navigational system is direction-finding. To navigate the open seas you must be able to set a course (specify the direction of the place you want to reach) and maintain that course (keep your vessel heading in that direction). A navigator with a magnetic compass and a destination known to lie on a north-northeast course, for example, has simply to steer along that compass heading to landfall. Although the Pacific Islanders had no magnetic compass, they developed a technique every bit as accurate based on a comprehensive knowledge of star positions; anthropologists have dubbed it the "sidereal compass." Even when Europeans introduced the magnetic compass to indigenous navigators, the latter used it only to supplement their own system, which is no less capable of providing accurate bearings.

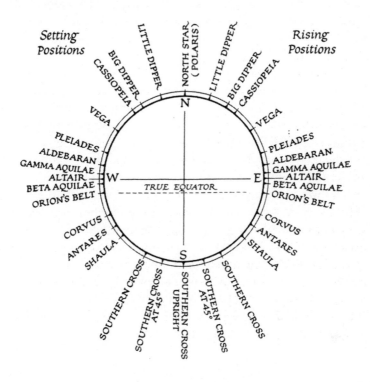

STAR COMPASS
after Goodenough (1953)

The Polynesian sidereal compass, like the European magnetic compass, was divided into thirty-two segments. This may seem like an improbable coincidence and perhaps even an indication that they did not develop independently of each other, but it is less coincidental than it appears at first glance.

Both undoubtedly began with the four cardinal directions North, South, East, and West, which were subsequently subdivided as more accuracy was desired. The first subdivision would yield eight directions, the next, sixteen, and the next, thirty-two. It is not difficult to believe that sailors in both the Mediterranean and the Pacific could have independently settled on thirty-two compass headings as optimal for their needs. In any event, it can be stated with certainty that the sidereal compass predated the introduction of the magnetic compass into the Pacific.[66]

Here is how the sidereal compass works. The night sky is filled with a pattern of stars that never varies from night to night. The earth's rotation causes the stars to appear to move in arcs from east to west, but their motion is perfectly regular. Prehistoric navigators knew that a star that rose at a given point on the eastern horizon would rise at exactly the same point night after night after night, and would set at a specific point on the western horizon with the same dependable regularity. Their system was therefore based on a number of recognizable stars or small star groups that were approximately equally spaced around the 360-degree perimeter of the horizon, analogous to the thirty-two points of the magnetic compass rose. A sailor in the Caroline Islands, for example, wanting to head north-northeast would sail toward the point on the horizon where the star we know as Vega rises. For due east, one would head toward where Altair rises, and for due west, toward where Altair sets.

Altair does not, of course, remain on either horizon all night long; it rises in the east and proceeds through its arc to its setting point in the west. As a directional guide, it is useful only during the times when it is not very high above the horizon. But fortunately the sky is full of stars, and there happen to be two bright ones, Procyon and Bellatrix, that follow an arc very close to Altair's. Their spacing on the arc is such that "when one is down another is up. Between the three of them they can provide a rising or setting bearing at almost any season or time of night."[67]

Compass rose. From a surveying compass made by David King, Salem, Massachusetts, dated 1744.

Because horizon points at which stars rise and set vary with an observer's latitude, the ancient navigators had to take into account how far north or south they were and adjust their sidereal compass calculations accordingly. That they were familiar with the stars at latitudes far removed from those of their home islands is evidenced by the importance of Polaris, the star very near the north celestial pole, in the cosmogony of Tahiti, which is a thousand miles south of where Polaris can first be seen above the horizon.[68]

Another significant complication: Each day, a given star rises four minutes earlier than it did the day before. The *sequence* of the cycle of stars rising at a particular point of the sidereal compass is always the same, but the first star visible after sundown changes as the weeks and months pass. A complete cycle therefore must cover twenty-four hours. Because the same stars can serve when they rise in the east and when they set in the west, the total number of stars necessary to make up a sidereal compass is on the order of only thirty to forty. Nonetheless, it is obviously a highly sophisticated system requiring knowledge of hundreds of star positions, which to the modern mind would seem practicable only if the star sequences for each of the thirty-two compass points were arranged on a written list for ready reference. The Polynesian navigators, however, had no such list at their disposal and were therefore compelled to commit the entire system of star positions to memory.

The education of a traditional navigator was the work of a lifetime, and learning the sidereal compass was but a small part of it. The star positions were usually taught by master seamen to their sons and nephews, using both classroom techniques ashore as well as on-the-job training at sea. A typical classroom session might present the students with a model canoe surrounded by a circle of thirty-two stones representing the direction-finding star cycles, which they would be required to name when pointed to at random. (One star in each cycle can serve to designate the entire cycle; "Altair," for example, could be considered a shorthand way of referring to the Altair-Procyon-Bellatrix sequence.) They would further be expected to name all the stars of each cycle in order of their appearance and, as a more advanced exercise, to name the *reciprocal* of each star—that is, the star that lies on the exact opposite compass point. The reciprocals are important because they represent the directions of return voyages.

The novice navigators were also expected to memorize the star directions that would point them to all of the other islands to which they might be called on to sail. Anthropologists discovered that many of the navigators with whom they studied had committed to memory accurate star courses to faraway islands they never expected to visit and which had not been visited for many generations past. In other words, although neither they, nor their fathers, nor even their grandfathers had ever made the voyage from, say, Tahiti to Hawaii, the sidereal compass directions for that voyage were preserved in the education that had passed down to them through the generations.

The anthropologists found a remarkable degree of consistency from island to island in the information thus preserved. Because some of the island groups had long been isolated from each other (by imperialist compulsion or otherwise), the correspondence of their sea knowledge points to a common origin and suggests that it had not appreciably degraded over time, as might be expected of an orally transmitted body of knowledge. The master navigators embedded the star sequences in chants as a means of solidly fixing them in the minds of their apprentices. "This is a body of knowledge," Gladwin explains, that can be learned only "through the most painstaking and lengthy instruction" and therefore "is taught and memorized through endless reiteration and testing." It is not, however, "a litany memorized by rote." The information "is learned so that each item is discreetly available, as it were floating on the surface of the navigator's mind rather than embedded in a long mnemonic chain."[69]

Through use of the sidereal compass, Polynesian navigators were at least as competent at direction-finding as were their counterparts in the Mediterranean, and probably much more so until the latter gained use of the magnetic compass in the thirteenth century C.E. But direction-finding is only one element of a navigational system; knowledge of one's position at sea is almost as important. Mariners want to know, as accurately as possible, how far they have gone along their course and how far they have left to go. But until the solution of the longitude problem in the eighteenth century,[70] no navigators anywhere were able to pinpoint their position at sea with total assurance. Before that, in the Pacific as well as in the Mediterranean and Atlantic, sailors could estimate their east-west position only by means of dead-reckoning techniques.

Their north-south position, on the other hand, was readily determined by astronomical means. Calculating latitude at sea by measuring the height of stars above the horizon—especially Polaris, "the star that never moves"—was the primary key to position-finding for European navigators. Polynesian navigators utilized the same technique, but only secondarily—as one among many clues to their whereabouts. Perhaps it was less important in the Polynesian system because a great deal of their voyaging took place below the equator, where Polaris is not visible. But Jacob Bronowski's assumption that the people of the Southern Hemisphere could not have developed "a sense of the movement of the stars by which to find their way" because "there is no Pole Star in the southern sky" is patently false.[71] He seems to have been completely unaware of Polynesian navigational knowledge.

In order to keep track of their east-west position, European sailors practiced dead reckoning by estimating their speed through the water and multiplying by the time they had been traveling. But the speed estimates had to take current flows into consideration, which more often than not introduced a considerable degree of imprecision. Pacific Islanders used similar estimates, but the primary technique by which they maintained knowledge of their position was one that anthropologists call the *etak* system. This, too, was based on knowledge of the sidereal compass. The navigators visualized the path between the island from which they departed and their target island, and then also visualized a third island off to the side (the *etak* island) that was chosen to serve as a reference point. By keeping track (in the mind's eye, because the *etak* island was always beyond the horizon and no maps or charts were used) of the relationship between their canoe, the *etak* island, and the sidereal compass, navigators could estimate how far they had traveled and how far still remained before landfall.

Although the *etak* system, like any form of dead reckoning, is by nature imprecise, the anthropologists who tested it with electronic position-finding technology found that the indigenous navigators were generally fairly close to where they thought they were. Their "mental map" of the islands among which they travel is evidently highly accurate. Learning the *etak* system raises the degree of difficulty considerably for apprentice navigators, who must memorize not only the static star positions for target islands but also the changing star positions for *etak* islands, which are visualized as moving from star to star along the horizon as a voyage progresses.

KNOWLEDGE OF OCEAN SWELLS

Although the sidereal compass is no less accurate in principle than the magnetic compass, the latter has the great advantage of being useable in daylight hours and in overcast conditions. Indigenous Pacific navigators needed secondary direction-finding techniques for times when the stars were not visible. In the daytime, they could steer by the sun's position, but only just after sunrise, just before sunset, and at high noon. When the sun was low on the horizon, its direction could easily be matched with that of a compass-point star, and at noon the shadow of the canoe's mast was known to lie exactly in the north-south direction. At other times of day, however, the sun was less useful as a guide. Numerous other clues to direction were exploited, but the most important were ocean swells.

Direction-finding by reading and analyzing ocean swells was a technique that was utterly foreign to Western navigation, but the Pacific Islanders had developed it into a high art. An ocean swell might be thought of as an "old wave." It does not crest; it has a longer, slower undulation. Waves are caused directly by winds, but the waters thus set into motion continue to rise and fall for hundreds of miles beyond the region of the blowing winds. Because the winds of the Pacific follow regular, predictable patterns, the ocean swells that those winds create are also predictable over large areas of the open sea. An experienced navigator can detect the direction of the swells by the way the canoe pitches and rolls and can determine sailing direction by relating it to that of the known direction of the swells.

The principle of reading the swells is simple, but the practice is anything but. Although the winds and swells of the Pacific are predictable, they vary with the seasons. More problematic is the fact that swells in the open sea rarely come from only one direction at a time. There are often three and sometimes four separate swell systems operating simultaneously, and the navigator must analyze the subtle pitching and rolling of the craft into its component motions. Although this increases the difficulty of the task immensely, each additional swell that can be identified also provides additional information as to direction. The technique has been described succinctly by Steve Thomas:

> In the tropical Pacific, where the steady tradewinds blow from easterly quadrants for most of the year, the wind pushes up long, low groundswells that march across the sea in steady

lines. The vector of the swells' march remains steady, enabling the skilled *palu* [navigator] to maintain his direction by keeping a constant angle between the swells and his canoe. Where two or three swell systems interact, the navigator will steer by what are called "knots," the peaks the swells make as they come together, like the converging wakes of two motorboats. At dawn and dusk, he must check the swells' vector against the stars. At night, if the sky is overcast and there is no moon to light the swells on the ocean, he must steer by the pitch and roll of his canoe in the seaway. This technique . . . is the ultimate test of the navigator's skill.[72]

When visual clues are of no use at all, the navigator may lie face up on the deck of the canoe with eyes closed and sort out the swell patterns by feel. One Western seaman reported:

> I have heard from several sources, that the most sensitive balance was a man's testicles, and that when at night or when the horizon was obscured, or inside the cabin this was the method used to find the focus of the swells off an island.[73]

Unlike glancing at a magnetic compass, reading the ocean swells is a technique that requires hours of patient observation and immense knowledge of the natural environment, but it served the needs of Pacific Island sailors well. In addition to direction-finding on the open seas, it also provided a means of detecting the presence of land before it came into view. When a navigator begins to notice the characteristic patterns of reflected or refracted swells, it is an indication that land is nearby.

LAND-FINDING

PACIFIC ISLAND NAVIGATION was unlike that which developed in other parts of the ancient world in that its target landfalls were tiny specks of land surrounded by immense stretches of open sea. Where the Polynesians and Micronesians live, "the proportion of dry land, exclusive of New Zealand, is on the order of two units of land for every thousand of water."[74] If making landfall required directly sighting land, a small, low-lying island would present a target only about twenty miles wide, and even a very small angular error in heading could cause a navigator to miss it. But the ancient seafarers' knowledge of nature enabled them effectively to expand a twenty-mile-wide target to as much as a hundred miles, making it much more difficult to miss. One set of land-finding clues was the reflected or refracted swells mentioned previously. Another was the distinctive shapes and movements of clouds above islands that lay beyond the horizon. Most useful of all, however, was knowledge of the habits of birds.

The navigator first had to distinguish between the species that wander all over the ocean and those that are land-based, such as terns, noddies, boobies, and frigatebirds. Because each species has a characteristic flight range, identifying the birds provides an estimate of maximum distance from land. But even more valuable is their land-finding function. Just after dawn, when they fly to their fishing grounds, and just before sunset, when they make their return flight,

> their flight paths indicate the direction of land. Toward evening the frigatebirds, for example, will be seen to abandon their leisurely patrolling, climb even higher, and set off in one direction, probably homing by sight. About the same time the boobies will tire of their inquisitive inspections and fly low and arrow-straight for the horizon. As the noddies depart they will weave slightly in and out between the crests of the larger waves, while the terns will be flying a little above them, but all will be following a very exact path toward their home island.[75]

The navigator then has only to set his course in the same direction to be assured of making landfall.

This description of traditional navigation in the Pacific barely scratches the surface of the subject. It should suffice, however, to establish that it constitutes not simply a group of techniques or a miscellaneous assemblage of practical lore, but a sophisticated body of theory empirically derived from an extensive knowledge of nature. Only the most unreconstructed modernist could exclude it from the realm of science.

GEOGRAPHY AND CARTOGRAPHY OF "SAVAGES"

IT HAS BEEN noted that European explorers tapped the geographic knowledge of indigenous peoples of the Pacific in order to "discover" unknown islands and construct maps of them. That process occurred on land as well as on water:

> It is clear from the narratives and journals of scores of explorers, from Columbus onward, that Amerindian cartographers and guides in every region of the continent contributed significantly to the outlining and filling of the North American map. Columbus, for example, relied on Indian geographic information from the outset of his landings in the New World and, when available, Indian drawn maps as well.[76]

Unfortunately, "the manifold contributions of Amerindians to the exploration and mapping of North America have been largely ignored in the literature of the history of cartography."[77] In an attempt to begin to rectify that injustice, one historian examined the process by which the Chesapeake Bay area of Virginia was initially mapped. Captain John Smith, who was governor of Virginia in 1608 and 1609,

> was to become known as the colony's first historian and author of the first detailed printed map to show the area with any degree of accuracy in 1612. Cartographic scholars have been unstinting in their praise of this map and it is universally acclaimed as one of the most influential pieces of cartography in the history of the United States.[78]

Smith's map explicitly acknowledges "the significant input of

Amerindians who guided, instructed, and informed him and his associates as they painstakingly explored the wilderness during the first months and years of Virginia's founding."[79] Describing the comprehensiveness of his map, Smith wrote that it presents "the way of the mountaines and current of the rivers, with their severall turnings, bays, shoules, isles, inlets, and creeks, the breadth of the waters, the distances of places and such like." He marked with little crosses the places that he or other white men had actually seen with their own eyes; "the rest," he stated, "was had by information of the *Savages*, and are set down according to their instructions."[80]

Smith's dependence on Amerindians was typical of how the explorers gained their geographical knowledge. American history textbooks credit Henry Rowe Schoolcraft with discovering the source of the Mississippi River in 1832, but "Schoolcraft 'discovered' the source of the Mississippi only because an Ojibwa chief [Ozawindib] guided him and his small expedition to the site."[81] Samuel Champlain wrote of his interaction with natives during the exploration of the St. Lawrence Valley region:

> I had much conversation with them regarding the source of the great river and regarding their country, about which they told me many things, both of the rivers, falls, lakes, and lands, and of the tribes living there, and whatever is found in those parts. . . . In short they spoke to me of these things in great detail, showing me by drawings all the places they had visited.

"Some things," he added, "were cleared up about which I had been in doubt until they enlightened me about them."[82]

One Amerindian has received posthumous recognition for her contribution to the exploration of North America: Sacajawea, the Shoshone Indian who helped guide Lewis and Clark on their historic expedition and whose likeness now appears on the United States dollar coin. A scholar who examined the published journals of that expedition found "at least thirty direct references to Indian-made maps and some ninety-one descriptive statements of a geographical nature by Indians." Lewis and

Clark, he further noted, "faithfully copied most of the maps the Indians traced out for them on a swept-sand surface, on a smooth bark and on a leathern chart."[83]

An account written in 1709 by a traveler in previously uncharted parts of North Carolina described the assistance he received from the natives of the area:

> They will draw Maps very exactly of all the Rivers, Towns, Mountains and Roads, or what you shall enquire of them. . . . These Maps they will draw in the Ashes of the Fire, and sometimes upon a Mat or Piece of Bark. I have put a Pen and Ink into a savage's Hand, and as he has drawn me the Rivers, Bays, and other Parts of a Country, afterwards I have found to agree with a great deal of Nicety.[84]

William Gerard De Brahm, Britain's surveyor general of America's southern colonies, encountered Creek Indians while exploring Florida and praised their "natural knowledge in Geometry."[85] Another adventurous voyager in early North America, Baron de Lahontan, commented that the Amerindians

> draw the most exact Maps imaginable of the Countries they're acquainted with, for there's nothing wanting in them but the Longitude and Latitude of Places: They set down the True North according to the Pole Star; the Ports, Harbours, Rivers, Creeks and Coasts of the Lakes; the Roads, Mountains, Wood, Marshes, Meadows, etc.[86]

Some of the geographic knowledge thus acquired by Europeans was given voluntarily, but not all. In 1502, Columbus, on his fourth voyage to the New World, inaugurated a tradition of capturing Amerindians and forcing them to serve as pilots: "He found an old man among the Indians, whom he kept as a guide, since the savage could draw a sort of chart of the coast."[87] In 1534, Jacques Cartier abducted two native Americans named Taignoagny and Dom Agaya, transported them to France, where they were taught to speak French, and then brought them back to guide him up the St. Lawrence River.[88] "In yet other cases, Indians were kidnapped and sent back to England to be debriefed in depth about their geographical knowledge."[89] In 1576, Martin Frobisher seized Inuit fishermen and made them pilot his Arctic expedition. These were not isolated incidents: "Kidnapping Indians and forcing them to become interpreters, guides, or slaves grew into a well-established cultural pattern among explorers. Such practices became a standard part of the business."[90]

COLONIALIST CARTOGRAPHY

THE GEOGRAPHICAL KNOWLEDGE thus stolen from Amerindians was then used to rob them of their ancestral lands, as the science of cartography became "an epiphenomenon of imperial control."[91] J. B. Harley described "the tragedy in American history" that "saw the map as an instrument through which power was exercised to destroy an indigenous society. The maps of seventeenth-century New England provide a text for studying the territorial processes by which the Indians were progressively edged off the land." Those maps were a "two-edged weapon" that served "the double function in colonialism of both opening up and later closing a territory."[92]

The earliest of the settlers' maps were "full of empty spaces that are ready for taking by Englishmen." The lands represented on paper by blank areas were not empty, of course, but by helping to "render Indian peoples invisible in their own land," the maps served a crucial ideological purpose. They "were read as graphical articulations of the widely held doctrine that colonial expansion was justified when it occurred in 'empty' or 'unoccupied' land." The cartographers thus "contrived to promote a durable myth of an empty frontier . . . that allowed the English conveniently to ignore the realities of the Indian societies that they encountered in the New World."[93]

Later, as the blank spaces were filled in, the function of the maps shifted to make them "practical documents on which the subdivision and bounding of Indian territories would occur":

By the mid-seventeenth century, maps were becoming a necessary device for juridical control of territory. . . . As early as 1641, the General Court of Massachusetts Bay Colony enacted a law requiring every new town within its jurisdiction to have its boundaries surveyed and recorded in a plan. The authority of the map was thus added to the authority of legal treatises, written histories, and the sacred books in sanctioning the taking of the lands of the Indians.[94]

By the end of the seventeenth century,

maps were being made to describe all tiers of territory, from the individual estate to the colonies as a whole. Even at a local level, they often tended to stress boundaries rather than other features. This is characteristic of a type of European colonial mapping that focussed on private property but failed to make legible the usufructuary rights of conquered peoples. Such maps are more than an image of the landscapes of English colonization in New England. They are a discourse of the acquisition and dispossession that lie at the heart of colonialism.[95]

Maps are not simply graphic depictions of geographical entities; they inevitably reflect the society that produced them. Early American cartography amply demonstrates the hollowness of claims for the neutrality and disinterestedness of science when conflicting material interests are at stake. But this subject has carried us deep into the historic era, and now we must return to prehistory to consider the origins of another branch of science.

ARCHAEOASTRONOMY

THE REFERENCE TO Amerindian use of the Pole Star indicates that the navigators of the Pacific were not the only ancient people who studied the skies. The celestially aligned megaliths of Stonehenge may be the most famous artifacts of prehistoric science, but "the oldest astronomically oriented monument in the world" is Newgrange in Ireland.[96] Newgrange is much larger than Stonehenge and was built circa 3200 B.C.E., about four hundred years before the earliest phase of Stonehenge was constructed.[97] Hundreds of lesser-known megalithic sites in Europe and Africa, the pyramids of the Aztecs and Mayans, the palaces of the Incas, the temples of Angkor Wat, the Great Pyramid at Giza, and many other structures oriented to the sun and stars testify to ancient traditions of astronomy among diverse peoples in all parts of the world. Analyzing the astronomical functions of these monuments has in recent years developed into a flourishing scientific discipline called archaeoastronomy. Its serious practitioners have been at pains to rescue their subject from the imaginative exaggerations of popular authors who have, for example, described Stonehenge as a "Neolithic computer" for predicting eclipses.[98]

The great astronomically oriented monuments represent not the origins but the culmination of sky-watching traditions already thousands of years old. The production of monumental architecture required forms of social organization more complex than foraging. As one leading student of archaeoastronomy observed with regard to Newgrange,

> In the last half of the fourth millennium B.C., cultivation of wheat and barley and pasturing livestock must have created surpluses that could subsidize these larger monuments. Knowledge of the sun's seasonal shifts along the horizon is not especially esoteric, but the ability to incorporate it into monumental architecture implies some specialization.[99]

Those who constructed Stonehenge and Newgrange were not foragers but settled agricultural peoples, and those who produced the Mayan and Egyptian pyramids had achieved literacy. For agriculturalists, astronomy was necessary for calendar-making, which established the annual cycles of plowing, planting, and harvesting. In Egypt at the time of the Pharaohs, for example, each new year began on the day when the bright star Sirius was first detected rising just before sunrise—

a "heliacal rising" that was known to occur at about the same time as the annual flooding of the Nile, on which Egyptian agriculture depended.

But knowledge of the heavenly bodies and their cyclical motions far predated the megaliths and pyramids. The evidence is not as dramatic, but it is clear that foragers were aware of the basic astronomical principles that were later built into the solstitial and equinoctial alignments of monumental architecture: "It began, of course, in ancient observations of celestial cycles. Long before we became farmers or built civilizations, our brains must have focused on the rhythmic changes in the sky and measured the behavior of the world in terms of them." Foragers studied the heavens "to orient themselves in time and space."

> From the sky they gained—and we, their descendants, have inherited—a profound sense of cyclic time, of order and symmetry, and of the predictability of nature. In this awareness lie not only the foundations of science but of our view of the universe and our place in it.[100]

The most obvious celestial cycle is the daily rising and setting of the sun, but tracking its path with the use of a gnomon— a vertical shadow-casting pole—reveals much more than the passage of days:

> The basic facts concerning the sun's behavior came from the common man, the shepherd and the farmer, the fisherman and the camel driver of primitive civilization. . . . The origin of the gnomon was humble indeed. In its most primitive form it was a shepherd's staff, a tent stake stuck in the ground, or any other kind of rod, tree or vertical shaft, which, by casting a shadow in the sunlight indicated the sun's position. The slant of the shadow told the herdsman and the moneylender how the day was progressing, while the length of the shadow indicated the passage of the seasons.[101]

Using gnomons, tribesmen of Borneo and other preagricultural peoples "established the length of the year and the time of the solstice by measuring the length of the shadow at, say, noon."[102]

Solstices and equinoxes were highly portentous events with significant implications for people directly dependent on nature's bounty. As a means of anticipating seasonal change, determining the winter and summer solstices in the annual solar cycle by identifying the northernmost and southernmost points of sunset along the horizon was as important to foragers as it was to agriculturalists. For Desana Indians who live in the Amazon rainforest,

> the equinoxes are important, for they signal the start of each rainy season. One begins in March, the other in September. At the equinoxes, when the rivers rise, fish head upstream to spawn and so become relatively scarce. Likewise, game is less available. The rainy seasons are regarded as periods of gestation.[103]

Keeping track of the regular changes in the shape of the moon must also have been among the earliest of astronomical observations. Alexander Marshack has cited credible evidence "of lunar observation in notational sequences and markings dating from the Upper Paleolithic period; these extend backward in an unbroken line from the Mesolithic Azilian to the Magdalenian and Aurignacian cultures, a span before history of some 30,000 to 35,000 years."[104] Tabulations of lunar cycles can also be seen in the petroglyphs left by North American foragers at Presa de la Mula, near Monterrey, Mexico.[105] The archaeological evidence is not

Notched bone. Ksar Akil, Lebanon.

completely unambiguous, but anthropological data support the astronomical interpretation: "Wands marked in similar ways by inhabitants of the Nicobar Islands in the Indian Ocean are known to be calendar sticks with lunar tallies. . . . They parallel what we see from the Upper Paleolithic."[106]

So-called Amerindian "medicine wheels," dozens of which have been discovered on the Northwest Plains of North America, have also been cited as evidence of hunter-gatherer astronomy. John A. Eddy analyzed one of the best-known of these large-scale stone arrangements, the Bighorn Medicine Wheel in Wyoming, and concluded that four of its "spokes" were aligned with the summer solstice sunset and the heliacal risings of three bright stars, Aldebaran, Rigel, and Sirius. Aldebaran's appearance heralded the summer solstice, and the other two marked two successive twenty-eight-day periods (the equivalent of lunar cycles), which might have made them useful for calendrical purposes. "The probability that the Bighorn Medicine Wheel's astronomical alignments are intentional and not coincidental," observes archaeoastronomer E. C. Krupp, "was increased when Eddy verified that another ruin, 425 miles north in Saskatchewan, had the same basic plan as the Wyoming wheel."[107]

Not all researchers agree that the medicine wheels are astronomical artifacts, but the proposition that foraging Amerindians studied the skies is nonetheless undeniable.[108] Anthropological evidence demonstrates, for example, that the Chumash Indians of what is now southern California "counted the cycles of the moon, established the times of the solstices, [and] observed the seasonal appearances of stars."[109]

Even if the medicine wheels' celestial alignments could be conclusively proved, that would not mean they were "observatories" in any modern sense of the word or that the foragers who built them shared the scientific motives of modern astronomers. The earliest sky-watchers may have studied the motions of celestial bodies for ceremonial, or religious, or magical, or who-knows-what purposes. But regardless of their motivations, they were engaged in a process of probing nature's secrets and creating a foundation of astronomical knowledge. "If careful observation of the world around us counts as science," Krupp commented, "then there is no doubt that our ancient and prehistoric ancestors were scientists."[110]

Should astronomically oriented ancient structures such as the Great Pyramid and the Mayan temple of Chichén Itzá be thought of as monuments to people's science? Quite the opposite: They represent the earliest examples of the domination of nature-knowledge by social elites and, as a corollary, the creation of the first scientific elites. First of all, those at the top of the earliest class-divided societies mobilized slaves and other forms of involuntary labor to construct the monumental architecture. And when the work was finished, they used the structures to monopolize astronomical knowledge as a primary component of political power. The rulers "consolidated the power of the state with the help of an astronomical bureaucracy" made up of "specialists in esoteric knowledge of the sky" whom they subsidized. Astronomy thus became "the prerogative of trained experts." When a society

> gets complex enough to install a king and centralizes regional power . . . the power invested in the ruler must be explained and justified . . . and so ideology is enlisted to legitimize political power. . . . The sacred king and the divine emperor trace their lineage to the celestial realm. In this way, they justify their monopoly of power and institutionalize astronomy in the bargain.[111]

The codices and cuneiform tablets left by the astronomers of Mesoamerica and Mesopotamia demonstrate that with the introduction of written record-keeping and mathematics, the accuracy of observation and the precision of calendars was increased and important innovations such as eclipse prediction were accomplished, but in essence the subsidized specialists were building on and refining the fundamental knowledge inherited from the sky-watchers of many millennia past.

THE ORIGINS OF READIN', WRITIN', AND 'RITHMETIC

THE WRITTEN RECORD-KEEPING of the royal astronomers evolved from the Paleolithic tally marks on pieces of bone.[112] Such tally marks constitute the earliest evidence of human number awareness and point to its origins in foragers' enumeration of the animals they hunted: "Notched sticks—tally sticks—were first used at least forty thousand years ago. . . . Notch-marks found on numerous prehistoric cave-wall paintings alongside outlines of animals leave no doubt about the accounting function of the notches."[113]

The further development of numerical reasoning into mathematics, which also stimulated the development of the first writing systems, emerged from the routine economic activities of farmers, artisans, and traders. Because literacy and numeracy are each a *sine qua non* of scientific advance, their origins are crucial milestones in the history of science.

Literacy appeared independently in several parts of the prehistoric world, but the earliest evidence of writing is the cuneiform Sumerian script on the clay tablets of ancient Mesopotamia.[114] The roots of that writing system in commercial activity were revealed by archaeological detective work on the part of Denise Schmandt-Besserat.[115] Schmandt-Besserat demonstrated that preliterate people, to keep track of the goods they produced and exchanged, created a system of accounting using clay tokens as symbolic representations of their products. Over many thousands of years, the symbols evolved through several stages of abstraction until they became wedge-shaped signs on clay tablets, recognizable as writing.

Plain tokens.
Seh Gabi, Iran.

The original tokens (c. 8500 B.C.E.) were three-dimensional solid shapes—tiny spheres, cones, disks, and cylinders. A debt of six units of grain and eight head of livestock, for example, might have been represented by six conical and eight cylindrical tokens. To keep batches of tokens together, an innovation was introduced (c. 3250 B.C.E.) whereby they were sealed inside clay envelopes that could be broken open and counted when it came time for a debt to be repaid. But because the contents of the envelopes could easily be forgotten, two-dimensional representations of the three-dimensional tokens were impressed into the surface of the envelopes before they were sealed. Eventually, having two sets of equivalent symbols—the internal tokens and external markings—came to seem redundant, so the tokens were eliminated (c. 3250–3100 B.C.E.), and only solid clay tablets with two-dimensional symbols were retained. Over time, the symbols became more numerous, varied, and abstract and came to represent more than trade commodities, evolving eventually into cuneiform writing.[116]

The evolution of the symbolism is reflected in the archaeological record first of all by the increasing complexity of the tokens themselves. The earliest tokens, dating from about ten thousand to six thousand years ago, were of only the simplest geometric shapes, such as spheres and cylinders—"the shapes which emerge spontaneously when doodling with clay."[117] But about 3500 B.C.E., more complex tokens came into common usage, including many "naturalistic forms" shaped like miniature tools, furniture, fruit, and humans.[118] The earlier, plain tokens were counters for agricultural products, whereas the complex ones "stood for finished products, such as bread, oil, perfume, wool, and rope, and for items produced in workshops, such as metal, bracelets, types of cloth, garments, mats, pieces of furniture, tools, and a variety of stone and pottery vessels."[119] The signs marked on clay tablets likewise evolved from simple wedges, circles, ovals, and triangles based on the plain tokens to pictographs derived from the complex tokens.

Before this evidence came to light, the inventors of writing were—as usual—assumed to have been an intellectual elite.

Scholars, Schmandt-Besserat said, had previously believed that "writing emerged as the rational decision of a group of enlightened individuals." V. Gordon Childe, for example, hypothesized that writing emerged when members of the priestly caste agreed among themselves "upon a conventional method of recording receipts and expenditures in written signs that shall be intelligible to all their colleagues and successors."[120] But the association of the plain tokens with the first cultivators and of the complex tokens with the first artisans—and the fact that the token-and-envelope accounting system invariably represented only small-scale transactions—testifies to the relatively modest social status of the creators of literacy.

And not only literacy, but numeracy as well. The evidence of the tokens provides further confirmation that mathematics "originated in man's desire to keep a record of his flocks and other goods."[121] Another immensely significant step occurred around 3100 B.C.E., when Sumerian accountants extended the token-based signs to include the first real numerals—that is, "signs encoding the concept of oneness, twoness, threeness, etc., abstracted from any particular entity."[122] Previously, eight units of grain had been represented by direct one-to-one correspondence—by repeating the token or symbol for a unit of grain eight times. The accountants, however, devised numeral signs distinct from commodity signs, so that eight units of grain could be indicated by preceding a single grain symbol with a symbol denoting "8." Their invention of abstract numerals and abstract counting was one of the two most revolutionary advances in the history of mathematics.[123]

What was the social status of the anonymous accountants who produced this breakthrough? The immense volume of clay tablets unearthed in the ruins of the Sumerian temples where the accounts were kept suggests a social differentiation within the scribal class, with a virtual army of lower-ranking tabulators performing the monotonous job of tallying commodities. One archaeologist visualized the temple counting-room as containing "rows of scribes, either sitting or squatting down beside little mounds of clay, busy checking and adding long accounts."[124]

We can only speculate as to how high or low the inventors of true numerals were in the scribal hierarchy, but it stands to reason that this labor-saving innovation would have been the brainchild of the plebeian types whose drudgery it eased.

Although literacy and numeracy arose from the activities of farmers, artisans, and traders, as political power centralized in ancient empires and imperial bureaucracies took command of commerce, writing and reckoning came under the control of educated elites—astronomer-priests and high-ranking court scribes. "The appearance of writing," Claude Lévi-Strauss declared, was always and everywhere linked to "the establishment of hierarchical societies, consisting of masters and slaves, and where one part of the population is made to work for the other part."[125]

Nonetheless, subordinate social classes—merchants, in particular—continued to play a major role in mathematical progress. Almost a century ago, Karl Kautsky gave a cogent account of how mercantile practices encouraged the further development of mathematics. "The merchant," he wrote,

> is interested in the last analysis only in price conditions, in other words . . . in abstract numerical relations. As trade develops more and more . . . the more varied the money conditions with which the merchant must deal . . . and the more advanced the stage of development of the credit system and interest payment, the more complicated and varied do these numerical relations become. Therefore trade must stimulate *mathematical* thought, and simultaneously *abstract* thought.[126]

Complex tokens.
Susa, Iran.

POSITIONAL NUMERATION

THE PROGRESS OF early mathematics, however, was long hampered by the lack of an adequate system of writing numbers. The fundamental operations of arithmetic were extremely complex before the adoption of positional numeration—the place-value system—which makes a digit's value dependent on its position in a number. The digit "9," for example, means "nine hundred" in the number 2,945 but "ninety" in 2,495. To appreciate the difficulty of "simple" operations in previous systems, try performing this addition using Roman numerals: MMCMXLV plus MMCDXCV. If you can do that without cheating (i.e., mentally translating them into the numbers you are familiar with), try multiplying them.

Before positional numeration, everyday calculations were done on the fingers or with the help of a reckoning device such as an abacus, but learning how to actually add, subtract, multiply, and divide required advanced education. A historian of mathematics, writing about medieval Europe, noted: "Computations which a child can now perform required then the services of a specialist, and what is now only a matter of a few minutes meant in the twelfth century days of elaborate work."[127] For thousands of years, then, beginning with the first civilizations, arithmetic ("the foundation of all mathematics, pure or applied")[128] was the province of royal astronomers and other elite types. "A man skilled in the art was regarded as endowed with almost supernatural powers. This may explain why arithmetic from time immemorial was so assiduously cultivated by the priesthood."[129]

The introduction of the place-value system together with the zero symbol to hold "empty" columns—a watershed event in the history of mathematics—is particularly relevant to a people's history for three reasons. First of all, it "democratized" arithmetic, making its operations accessible and useful to people at all levels of society, including merchants, sailors, and artisans. Second, it did not originate in the brilliance of elite mathematicians of Athens or Alexandria but in the activities of anonymous people—perhaps ordinary accounting clerks—in India between the third and fifth centuries C.E. As the celebrated mathematician Laplace exclaimed,

> It is India that gave us the ingenious method of expressing all numbers by means of ten symbols, each symbol receiving a value of position as well as an absolute value; a profound and important idea which appears so simple to us now that we ignore its true merit. But its very simplicity and the great ease which it has lent to all computations put our arithmetic in the first rank of useful inventions; and we shall appreciate the grandeur of this achievement the more when we remember that it escaped the genius of Archimedes and Apollonius, two of the greatest men produced by antiquity.[130]

And third, this revolutionary innovation was not transmitted by means of mathematics journals or other organs of scholarly discourse, but by merchants pursuing their occupation along the trade routes between India and the rest of the world. By the tenth century C.E., the new, improved way of calculating had been adopted by Arab traders, and they in turn introduced it into Europe in the early thirteenth century. "The first notable work in mathematics" of that era was accomplished by Leonardo Fibonacci, "a merchant by vocation" who "had traveled considerably in the Near East and had absorbed the Arabic knowledge of the period."[131]

There is more to the story. Not only did elite mathematicians *not create* this decisive innovation, many of them—perceiving it as a threat to their status as guardians of esoteric knowledge—*resisted* it and delayed its acceptance in Europe for several centuries:

> The struggle between the *Abacists*, who defended the old traditions, and the *Algorists*, who advocated the reform, lasted from the eleventh to the fifteenth century and went through all the usual stages of obscurantism and reaction. In some places, Arabic numerals were banned from official documents;

in others, the art was prohibited altogether. And, as usual, *prohibition* did not succeed in abolishing, but merely served to spread *bootlegging,* ample evidence of which is found in the thirteenth century archives of Italy, where, it appears, merchants were using the Arabic numerals as a sort of secret code.[132]

It was the use of the place-value system by the rising mercantile class that assured its triumph and opened the way for further mathematical progress.

As to who the original innovators were, no positive identification is possible. Lancelot Hogben suggested that they were probably clerical workers "in the counting houses of India."[133] The first *documented use* of the place-value system and zero are in a work, reliably dated to 458 c.e., titled the *Lokavibhâga,* attributed to an Indian holy man named Sarvanandin who may have lived a century or more earlier.[134] But, as Georges Ifrah observed, the names of "the inventors themselves are irretrievably lost, perhaps because . . . these brilliant inventions belonged to relatively humble men whose names were not deemed worthy of recording."[135]

ALPHABETIC WRITING

Just as a new style of numerical notation democratized arithmetic, the introduction of the alphabet likewise democratized writing. The many hundreds of pictograms, ideograms, and other abstract symbols that constituted cuneiform, hieroglyphic, or Chinese characters could be mastered only by many years of education, which meant that in the earliest civilizations, writing was practiced exclusively by court officials, priests, and professional scribes. But writing with an alphabet of a few dozen letters can be learned much more easily and quickly and therefore creates the possibility of mass literacy.

Again, this momentous innovation emerged not from the top of the social hierarchy but from the bottom. The first alphabetic writing in the archaeological record (c. 1800 b.c.e.) was of a Semitic language, but its elements—the first "letters"—were derived from Egyptian hieroglyphs. Because these earliest examples were found in mines in the Sinai peninsula, a case can be made that the creators of the original alphabet were Semitic slaves who "began to adapt a small battery of Egyptian hiero-glyphics to represent the sounds of a tongue alien to that of their masters and overseers."[136]

The democratization of writing paralleled the democratiza-tion of arithmetic in yet another way: The new Semitic alpha-bet was not spread by learned discourse but by Phoenician sailors and traders who disseminated it throughout the Mediterranean world. Although writing was invented inde-pendently in various places, *alphabetic* writing was invented only once. All of the existing alphabets of the world today derive from that original Semitic script found in the Egyptian mines.

Before the democratization of writing and arithmetic, their monopolization in the earliest civilizations by priestly castes placed astronomy in the hands of scientific elites. But astronomy was not the only scientific field that was developing during the lat-ter part of the Neolithic era. More important was what the first arti-sans and farmers were contributing to knowledge about nature and natural processes. It must once again be conceded that archaeo-logical evidence does not provide us with much information about the specific knowledge or scientific practices of the people of whom we treat. Nonetheless, let us inquire, insofar as possible, into two crucial bodies of Neolithic knowledge: the materials sci-ences (including ceramics and metallurgy) and the agronomic sci-ences (plant and animal domestication).

STONE, CLAY, METAL, AND FIRE

STONE TOOLS SHOW up in the archaeological record some two million years before anatomically modern human beings came onto the scene, so it seems that *Homo sapiens* inherited the rudiments of materials science—knowledge of the properties of materials—from ancestor hominids. Although tools made of stone are by far the oldest that have survived, that does not prove that stone was the first or the only material of choice for

Paleolithic toolmakers. It may simply mean that other tools made of more perishable materials—wood, shells, bones, horns, antlers, leather, and so forth—deteriorated over time.

"A finished tool," V. Gordon Childe explained, "does really reflect, albeit rather imperfectly, the science at the disposal of its makers." The creation of stone tools implies a certain amount of knowledge about the properties of various kinds of stone as well as of the materials on which the tools were intended to be used. Early hominids "had to discover by experiment the best stones for making tools and where such occurred," thus building up "a considerable body" of geological and other kinds of knowledge. "In acquiring and transmitting this our forerunners were laying the foundations of science."[137]

Evidence that hominids had learned to control fire at least a million years ago suggests that experimentation with materials may have included transforming them with heat—in cooking, for example—but when that first occurred can only be a matter of speculation. Small stone lamps that probably burned animal fat as fuel, however, bear witness that by 15,000 B.C.E. humans had begun to gain empirical knowledge of fire-induced chemical processes. An innovation of "great significance for human thought and for the beginning of science" was the conscious use of fire to modify clay in the manufacture of pottery.[138]

The earliest pottery that has been found by archaeologists (c. 8000 B.C.E.) shows evidence of a long history of prior development:

> When the full-scale baking of earths for ceramic, bonding, vitric, or metallurgic purposes began some 8,000 to 10,000 years ago, artisans were already launched full-scale into the chemics and physics of materials . . . they came to learn temperatures of melting, forms of reduction of oxides, chemical combinations of elements (iron oxide with silicon, iron with sulphur, etc.), the electropotential of certain elements, as well as the complex relationships of carbon with iron or of lime and soda with clays.[139]

"Like so many early inventions," a historian of technology wrote, "it seems unlikely that we shall ever be able to retrace the steps by which pottery came into being and it is probable that at the back of the invention lay a long period of experimentation with vessels made of unfired clay."[140] Nevertheless, two important inferences can be made about who invented pottery. First, because kilns do not appear in the archaeological record until much later (c. 6000 B.C.E.), the earliest known pots were apparently fired in the domestic hearth. Given the gender division of labor in Neolithic societies wherein women tended the home fires and cooked and stored the food (which was what pottery was used for), there can be little doubt that the pioneer ceramicists who achieved the momentous discovery of baked clay's properties were women.[141] And second, it is unlikely that it was the accomplishment of a female "lone genius" but rather of communal groups of women: "The craft traditions are not individual, but collective traditions. The experience and wisdom of all the community's members are constantly being pooled."[142]

Weaving was another noteworthy Neolithic technology: "The invention of the loom was one of the great triumphs of human ingenuity. Its inventors are nameless, but they made an essential contribution to the capital stock of human knowledge, an application of science that only to the unthinking seems too trivial to deserve the name."[143] Because weaving, like pottery, was a domestic occupation, the innovators were likewise probably groups of women.[144]

By definition, the Neolithic era ended when people began using metal tools on a large scale, but it is evident that they had begun to accumulate knowledge of the properties of metals much earlier. Unfortunately, this is a blank page in the history of science. Not only are the first metallurgists nameless; virtually nothing is known about the empirical processes by which they gained the knowledge underlying the sophisticated smelting processes necessary to produce metals.

The Paleolithic toolmakers who experimented with various kinds of rocks were undoubtedly drawn to the colorful ones that contained metallic ores. In fact, the colors of the famous pre-historic cave paintings of Lascaux and Altamira were made from dyes produced from metallic ores. But extracting metals from the ores is a far more complex matter.

Most metal exists in nature in chemical combination with other elements. A common form of copper ore, for example, is malachite, in which copper, oxygen, and carbon are joined by strong chemical bonds. Some metals, however, including copper and iron, can also be found in their "native," uncombined state.[145] Prehistoric people discovered and used small amounts of native copper and iron, an experience that no doubt prepared them to recognize the metals when they later succeeded in separating them from their respective ores. The advent of that process—smelting—marked the true beginning of the age of metals.

Because there is no direct evidence as to how smelting was discovered, it has long been a subject of scholarly speculation. Smelting requires, among other things, heating a metallic ore to a high temperature, so an obvious conjecture was that it first occurred when someone inadvertently dropped a piece of copper ore into a fire and noticed the results. This campfire theory is not very convincing, however, because open wood or charcoal fires rarely exceed a temperature of 700°C, whereas copper smelting requires reaching 1084°C, the melting point of copper. Furthermore, simply heating malachite to 1084°C will not produce metallic copper; it must be heated for hours in a "reducing atmosphere" (an oxygen-poor, carbon-rich atmosphere), which an open fire does not provide.

More persuasive is the pottery kiln theory, which suggests that metallurgy began when potters using malachite to color their pots may have found small bits of smelted copper in their kilns, prompting them to undertake deliberate experiments. The latter theory is more plausible for several reasons: first, because the fires in the closed kilns could reach the requisite temperature; second, because they produced a reducing atmosphere; and third, because copper smelting and high-temperature kilns make their initial appearance in the archaeological record at approximately the same time. Both the campfire and pottery-kiln theories, it should be noted, would suggest that the pioneer metallurgists may well have been women.

Although the kilns were adequate to the task of accidentally producing small bits of metallic copper, they were far from ideal for the job of smelting. That suggests that a great deal of experimentation was necessary before copper smelting became a viable technology in its own right. And the experimentation continued with the aim of improving copper as a material for tool and weapon production.

BRONZE, "BEGRIMED MINERS," AND "SWEATING BLACKSMITHS"

COPPER ORES USUALLY contain small amounts of other metals, such as lead, silver, and iron. The smelters discovered by trial and error that by varying the proportions of these "impurities," they could change the properties of the metal—making it harder or softer, more or less malleable, and so forth. Their most momentous result was the recognition that combining about 88 percent copper and 12 percent tin would produce an alloy—bronze—that is considerably more durable and yet easier to work than pure copper. That discovery (somewhere in the Middle East, c. 3300 B.C.E.) initiated an era of profound technological revolution, retrospectively canonized as the Bronze Age.

That the discovery of bronze was a result of deliberate experiment rather than a happy accident is suggested by the fact that copper ores only rarely contain tin. Because copper and tin hardly ever coexist in nature, it can safely be assumed that the anonymous metallurgists who first tried to alloy them brought them together intentionally to test a hypothesis. These experimenters were not, however, professional specialists working in laboratories. As one historian of metallurgy said of them, "We are beginning to glimpse, through the dust and fumes and smoke of ten thousand years of mining and metal working, the contributions to human comfort and material progress that were made by the begrimed miner and the sweating blacksmith."[146]

The precise geographical location of the earliest Middle Eastern bronze-makers is likewise unknown, but some reasonable inferences about where they were also gives us more clues as to who they were:

> It is most unlikely that this discovery was first made in Mesopotamia. In all probability it was made much nearer the sources of the metals, in the mountainous areas of Syria or eastern Turkey, for example, but the Mesopotamians had the wealth to purchase the new metal and the wealth to employ craftsmen to fashion it. Thus, it is from the tombs of the early Sumerian kings that we find the first examples of bronze used in any quantity.[147]

This suggests that the innovators were "begrimed miners" of the mountainous regions and that their knowledge was bought or otherwise appropriated by Sumerian imperialists.

A traditional assumption—the "diffusionist" model—held that the knowledge of bronze-making spread from a single Middle Eastern point of origin to all other parts of the ancient world. That supposition was based on the notion that the high level of scientific knowledge required to smelt copper and produce bronze could only have been the product of rare genius and was therefore unlikely to have arisen more than once. Archaeologists, however, have uncovered evidence that copper and bronze technology did indeed emerge independently in several other places: the Balkans, China, India, Nigeria, Peru, and, most surprising of all, Southeast Asia.[148] Evidence of an independent bronze-making tradition in northern Thailand as early as 3000 B.C.E. suggests that the genius necessary to produce bronze may even have been within the capacity of "the hunter-gatherer Stone Age culture, called the Hoabinhian, which at that time existed throughout South-east Asia."[149] In general, those who defend the diffusionist model of technological transmission tend to underestimate the creative intellectual abilities of ordinary human beings working collectively.

FROM BRONZE TO IRON

ALTHOUGH THE DISCOVERY of bronze represented a major technological advance, it "was very much an elitist metal, useful to the ruling classes for war and ornament, and little applied to agriculture or peasant life. The so-called Bronze Age was really a minor graft onto the still dominant era of stone, wood, and bone." Iron was the first metal really to "dent the world of the farmer and housewife . . . through its application to knives and axes and plows."[150]

The Bronze Age in the Mediterranean world came to a sudden end and the Iron Age began about 1000 B.C.E. It might be assumed that the transition occurred as a result of the discovery of iron smelting, simply because iron is both more plentiful than bronze and superior as a material for making tools and weapons. Not so. The Bronze Age apparently came to a halt because previously available supplies of the tin needed to make bronze were suddenly cut off by widespread political turmoil. Where the tin had come from in the first place is one of the great mysteries of archaeology. All that is known is that vast amounts of tin were being imported into the Middle East and eastern Mediterranean between 3000 and 1000 B.C.E.:

> A far-flung and fairly reliable network of trading contacts had been established across the ancient world. Extending in the west as far as Spain and perhaps Cornwall, and in the east as far as India (and possibly further . . .), these routes were the major arteries of supply for the tin which was in such demand in those early centres of civilisation.[151]

That the Bronze Age was predicated on this mercantile activity is another indication of the social, collective nature of scientific history. Without the arduous labors of many traders transporting heavy loads of tin over long distances, the experimentally created bronze would have been simply an interesting curiosity.

Techniques for producing iron from iron ore were already known well before the sudden collapse of the Bronze Age, but the methods were crude, and the iron they produced was clearly inferior to bronze; iron was considered a second-rate material. When metalworkers could not obtain bronze, however, they had to make do with iron, which stimulated them to experiment to find ways to improve the quality of their iron. The occupational identity of the experimenters who led the way into the Iron Age is not in question: They were the "sweating blacksmiths."

The original discovery of smelted iron most likely occurred during copper smelting in which iron ore was used as a catalytic agent. In one documented instance, malachite and iron oxide were heated together. The malachite "contained silica sand. The iron combined with the silica to form slag, thus freeing the copper." At the same time, "some of the iron in the hottest part of the furnace was occasionally reduced, along with the copper."[152] Calling this discovery accidental undervalues the knowledge of the discoverers. If it was indeed made during a planned metallurgical procedure, it would be better described as serendipitous rather than accidental.

This iron's inferiority to bronze derived from the much higher melting point of iron (1537°C), which Bronze Age furnaces could not reach. As a result, "iron could be reduced from its ores, but only to a spongy mass mixed with slag, called a 'bloom.' This bloom became the raw material of the blacksmith." By repeated hammering, a blacksmith could drive the slag out of the bloom, producing what is known as wrought iron. Tools and weapons made of wrought iron were softer than those of bronze, so their cutting edges dulled more quickly with use. "But if iron was so inferior to bronze," Robert Raymond asked,

> how was the ancient world able to turn to it so widely and emphatically, to such a degree that within a few centuries iron had become an almost universal replacement for bronze in virtually every aspect of daily life? The answer lies in the discoveries made by blacksmiths in the treatment of iron which transformed its very character.[153]

According to ancient testimony, those innovators were of the Hittite people, who occupied what is now southern Turkey and northern Syria. Whether that is true or not, it is certain that blacksmiths of that general area made three key discoveries in the late second millennium B.C.E. First, they empirically determined that certain methods of heating the iron bloom in a charcoal furnace resulted in a higher-quality iron; in fact, they had turned some of their iron into the alloy of iron and carbon that we call steel. Second, they discovered that the quality of their iron could be further improved by taking it directly from the fire and quenching it in cold water. And third, they found that if, after quenching, they reheated it briefly and let it cool—a process called tempering—it would become less brittle.

The experimentally derived procedures the blacksmiths used to transform and improve metals with the help of fire and the orally transmitted "recipes" by which they passed their craft secrets to succeeding generations of apprentices represented a great wealth of knowledge about the properties of materials and constituted the basis for all later advances in the materials sciences. The roots of alchemy and then chemistry in the early metal crafts are clearly evident. "On the evidence available," one historian reported, alchemy "seems to have sprung up among the skilled metallurgists and metal-workers of the Middle East, possibly in Mesopotamia, whence it spread westwards to Egypt and Greece, and eastwards along the caravan routes to India and China." Few details of this transmission were recorded for posterity, so one suggestive example will have to represent the entire process:

> It is known that as far back as the sixth century B.C., there was a great intermingling of the natural philosophy of Persia, Syria, and Greece in the ancient and long-forgotten city of Harran, in Syria. The Sabian craftsmen of Harran were skilled in metallurgy and in many other operations calling for a knowledge of the materials of primitive chemistry.[154]

AGRICULTURE

As IMPORTANT AS the advent of literacy, numeracy, cities, and metal use were, they were subordinate aspects of the Neolithic revolution. The most revolutionary innovations—from which the others followed—were the domestication of plants and animals. Systematic food production led to food surpluses, which were the economic basis of craft specialization that generated new technologies and their associated sciences. Food surpluses also permitted the rise of social and scientific elites.

Agriculture is not simply a matter of growing plants found in nature and harvesting them; it requires domesticating species of wild plants and transforming them into *crops* designed to serve human needs. In spite of the "natural" labels that proliferate in our supermarkets today, almost none of the food we eat is truly natural. Domesticated crops are plants that were created by human beings and that cannot exist apart from human beings. They have been "genetically engineered" in such a way that they cannot propagate themselves. An ear of corn is an *artifact;* "corn can survive only if man removes the kernels from the cob and plants them."[155]

Although the transition from foraging to agriculture is called a "revolution" because of the dramatic transformation it wrought in human social life, it can be thought of as a sudden burst of change only if viewed on a geological timescale. On the scale of human lifetimes, it was a gradual process that undoubtedly occupied many hundreds, if not thousands, of years. It began when foragers started to manage the environment of wild plants, a transitional step toward domestication. Hunter-gatherers "should not be seen as passive participants in the ecosystem who simply conform their lives to a rigid, unyielding natural environment. These societies have actively and continually experimented with manipulation of plant and animal communities."[156]

As an example of how preagricultural people could "domesticate" a landscape, consider the case of the Kumeyaay Indians of California:

From coastal sandbars and marshes up through floodplains, valleys, and foothills, to high mountain deserts, the Kumeyaay had made experimental plantings of a variety of food and medicinal plants. They created groves of wild oaks and pines producing edible nuts at higher elevations and established plantings of high-desert species such as desert palm and mesquite along the coast. They planted agave, yucca, and wild grapes in various micro-habitats. They also planted cuttings of cacti and other succulents near their villages. They carefully burned many of the groves and other plantings of wild species to keep yields high, and by regularly burning off chaparral they improved the browse for deer. In early summer they harvested large stands of a wild grain-grass, now extinct, by hand stripping seeds from the stalk. Then they burned off the stands and broadcast a portion of the harvested seed across the burned areas.[157]

Even after true domestication appeared, its earliest practitioners continued for a long time also to depend on gathering wild plants and hunting wild animals. It is obvious, then, that the knowledge underlying the domestication of plants and animals originated in foragers' experience with the wild species they were familiar with. "Such peoples are walking encyclopedias of natural history," Jared Diamond said, "with individual names (in their local language) for as many as a thousand or more plant and animal species, and with detailed knowledge of those species' biological characteristics, distribution, and potential uses."[158] The profound knowledge of nature they accumulated—with "Woman the Gatherer" in the vanguard—was the direct antecedent of the modern agricultural sciences.

That knowledge was not confined to a single group of superior prehistoric people. Agriculture did not originate only once and then diffuse throughout the rest of the world. According to the archaeological record, plant domestication first took place in the Middle East about ten thousand years ago, and animal domestication (not counting dogs, which were a special case) followed about a thousand years later. But similar processes also occurred independently in China, the Americas, and sub-Saharan Africa—at least seven different places in all. The highlands of New Guinea have been added to the list, because evidence has recently been found of crop cultivation there also occurring as early as ten thousand years ago.[159]

I placed the phrase "genetically engineered" in quotation marks because those words could not have been used before the gene was identified as the biological unit of inheritance. But the domestication of plants and animals was essentially a matter of manipulating their genetic material; in that sense it is not illegitimate to say that the foragers who created agriculture were practicing de facto genetic engineering.

How conscious a process was it? Did people create crops like wheat and corn on purpose, or was it the outcome of a series of fortuitous discoveries? The first steps, no doubt, were taken by the plants themselves, evolving in symbiosis with humans according to the principles of Darwinian natural selection. But plants could never have naturally forfeited their ability to survive in the wild; that required *artificial* selection: "When human beings took control of the reproductive cycles of some populations of certain species by harvesting, storing, and planting their seeds in prepared areas, they effectively created a separate and parallel world for these plants."[160]

Animal domestication was a far more conscious process right from the start. Humans were surrounded by wild plant species and were "unintentionally drawn into long-term relations with them . . . but wild mammals do not tolerate the presence of humans. It requires deliberate, intentional human action (such as capturing the young and raising them in captivity) to alter the normal avoidance behavior of adult wild animals."[161] Whereas humans had previously concentrated on killing animals, domesticating them required the opposite: keeping them alive. That generated "complex new areas of knowledge . . . the knowledge of how to sustain, manage, and regenerate animal populations."[162]

That plant domestication by humans was at first inadvertent, occurring in the course of their daily quest for subsistence, illustrates once again that in the beginning was the *deed*. But eventually foragers paved the way for cultivation by intentionally experimenting with all of the species that were available to them. By so doing, they discovered—from among the hundreds of thousands of wild species of flowering plants—the very few that could be altered to serve human purposes better. And they did an admirably thorough job of it. As Jared Diamond pointed out, the first farmers, "knowing far more about local plants than all but a handful of modern professional botanists . . . would hardly have failed to cultivate any useful wild plant species." Furthermore, "our failure to domesticate even a single major new food plant in modern times suggests that ancient peoples really may have explored virtually all useful wild plants and domesticated all the ones worth domesticating."[163] Another author listed the food crops that Amerindian foragers (on their way to becoming farmers) domesticated—

> tomatoes and potatoes; all the squashes and pumpkins; almost all the kinds of beans; peanuts, pecans, hickory nuts, black walnuts, sunflower seeds, cranberries, blueberries, strawberries, maple syrup, and Jerusalem artichokes; all the peppers, prickly pears, chocolate, vanilla, allspice, sassafras, avocados, wild rice, and sweet potatoes

—and observed that "in the four hundred years since the European settlers began coming to North America, they have not found a single American plant suitable for domestication that the Indians had not already cultivated."[164] Important non-food plant products that could be added to the list include rubber, cotton, and tobacco.

Once the Amerindians had identified those plants, they proceeded to "improve" them. "Without question," Jack Weatherford declared, "the Indians were the world's greatest plant breeders."[165] Their development of corn "remains man's most remarkable plant-breeding achievement"; pre-Columbian peoples "bred most of the major varieties of corn that exist today, including red corn, yellow corn, field corn, sweet corn, dent corn, flint corn, flour corn, pod corn and popcorn."[166]

Amerindian genetic engineering differed in important ways from that of the prehistoric Middle East and elsewhere:

> Most of the traditional Old World grains had very small seeds that the farmer broadcast by the handful onto the prepared ground. American Indians knew that corn could be planted only by placing the kernels firmly in the ground. The Indians selected each seed to be planted rather than merely grabbing a random handful of seeds from a bag and throwing them. This process of selecting the seeds allowed the Indians to develop the hundreds of varieties of each plant that they cultivated. . . . This diversity developed through the Indian farmers' profound understanding of practical genetics. To make the corn grow the farmers had to fertilize each plant by putting corn pollen on its silk. They knew that by taking the pollen from one variety of corn and fertilizing the silk of another variety, they created corn with the combined characteristics of the two parent stalks. Today, this process is known as hybridization and scientists understand the genetic reasons behind this process; the Indian farmers developed it through generations of trial and error.

"Without the treasure of diversity created by the trial-and-error methods of early Indian farmers," Weatherford concluded, "modern science would have lacked the resources with which to start."[167]

Even more important is the impact Amerindian plant-breeding has had on the world's demographic growth: "Much of the post-Columbian explosion of European populations was driven by the introduction of two New World crops: potatoes and maize."[168] The European colonists' dependence on Amerindian foods is symbolized "by the traditional corn-beans-squash-cranberry-turkey dinner that marks the Thanksgiving holiday commemorating European settlement in North America."[169] Furthermore, Amerindian cultivars "also fueled much of the population explosion of China, for more than a third of China's current food supply is provided by crops of New World origin."[170] Ask yourself which had a more important impact on the food you eat: modern plant geneticists or the anonymous foragers, including Amerindians, who created domesticated crops? It is no contest—the foragers win a thousand to one.[171]

Amerindian use of fertilizers such as guano (the feces of seabirds) to revitalize their land is also deserving of notice. It took hundreds of years before the conquerors from Europe appreciated its value, but eventually they did: "The 'discovery' of guano by European agriculture in the nineteenth century initiated modern farming in Europe. . . . The guano age marked the beginning of modern agriculture and eventually led to artificial fertilizers made from other resources."[172]

AFRICAN AGRONOMY:
"KNOWLEDGE TRANSFER" BY ENSLAVEMENT

AFRICA PROVIDED ANOTHER crucial source of agronomic knowledge for European colonists in the Americas, and in this case the process often benignly referred to as "knowledge transfer" was accomplished through the medium of slavery. "Rather than importing African crops, planters more often discovered them in the gardens of their slaves. For these crops, blacks were the true experimenters":

> Through the Atlantic slave trade, blacks had gradually transferred African plants (like sesame, Guinea corn, okra) and American crops transplanted in Africa (peanuts and capsicum peppers) to lands where they were enslaved. Whites discovered uses for slaves' products only when they learned of external markets for them. This was clearly the case with peanuts. Blacks had often grown and marketed peanuts, but whites paid little heed until European chocolate manufacturers wanted the product for its bland oil.[173]

The presence of African crops in the Americas has usually been described as a matter of "seed transfer," as if there were nothing more to it than dropping seeds in furrows in the ground and harvesting what they produce. The inadequacy of that picture can be seen most clearly in the case history of rice production in the United States. "African knowledge of rice farming," Judith Ann Carney explained, "established . . . the basis for the Carolina economy":

> Slaves with knowledge of growing rice had to submit to the ultimate irony of seeing their traditional agriculture emerge as the first food commodity traded across oceans on a large scale by capitalists who then took complete credit for discovering such an "ingenious" crop for the Carolina and Georgia floodplains.[174]

Rice is commonly thought of as a crop of Asian rather than African origin, but it was in fact brought into being separately on both continents. Of the more than twenty species of rice that exist in nature, only two were domesticated: *Oryza sativa* in Asia and *Oryza glaberrima* in Africa. Thousands of years ago people of the Sahel discovered the latter, found that it thrived in swamplands, and went on to create "one of the world's most ingenious cultivation systems." *Oryza glaberrima* was cultivated "over a broad region of West Africa from Senegal southward to Liberia and inland for more than one thousand miles to the shores of Lake Chad."[175]

The key to growing African rice in the Americas was not seeds but "the sophisticated knowledge that is emblematic of a fully evolved wet rice culture." In South Carolina, where rice exports were exceeding sixty million pounds annually by the time of the American Revolution, "slaves from West Africa's rice region tutored planters in growing the crop." But it was not a body of knowledge that slave owners could easily learn from Africans and then apply themselves. It was an "indigenous knowledge system" that required the presence of human beings already familiar with rice culture, the knowledge to grow the crop in wetland environments and the means to mill the rice once it had been harvested. The only people in South Carolina possessing this familiarity were Carolina slaves who originated in the rice region of West Africa. To find the origins of rice cultivation one must thus look to Africans.[176]

The understanding of nature required for rice cultivation "represented repositories of cultural knowledge built from generations of observation, trial, and error." West African rice growers

> displayed acute knowledge of landscape gradient, soil principles, moisture regimes, farming by submersion, hydrology, and tidal dynamics, and the mechanisms to impound water and to control its flow. The result was an array of rice production zones with a management portfolio more diverse than those occurring in Asia and more finely nuanced by microenvironmental soil and water parameters.[177]

Plantation owners in South Carolina understood that to create successful rice plantations, not just any African slaves would do. They actively sought to buy slaves of specific ethnic groups that were known to possess rice-cultivation knowledge. Ample documentary evidence is available in the form of newspaper advertisements:

> One ad in Charleston boasted of 250 slaves "from the Windward and Rice Coast, valued for their knowledge of rice culture"; another on July 11, 1785, announced the arrival of a Danish ship with "a choice cargo of windward and gold coast negroes, who have been accustomed to the planting of rice."[178]

It is no coincidence, then, that the proportion of slaves imported into South Carolina from rice cultivation areas of Senegal, Gambia, and Sierra Leone grew from 12 percent in the 1730s to 54 percent from 1749 to 1765 and then to 64 percent between 1769 and 1774. By the American Revolution slaves from Senegambia and Sierra Leone formed the majority of forced migrants into South Carolina.[179]

West African rice culture was a "gendered" practice. "From the earliest period of the Atlantic slave trade, Europeans had noted the crucial role of females in African rice cultivation." Seed selection was "the responsibility of females," which suggests "that females may have initiated the process of rice domestication. The large number of varieties selected just for milling and cooking attest to the traditional female role as plant breeders." Furthermore, the process of milling rice was a "female knowledge system" that "served as the linchpin for the entire development of the Carolina rice economy. For without a means to mill rice, the crop could not be exported." Because the slaveholders were aware of the women's crucial knowledge, a higher percentage of females were shipped to South Carolina than to the Caribbean, and "female slaves bound for South Carolina received a higher purchase price than in other plantation economies."[180]

This important body of African knowledge of nature has traditionally been ignored by historians. "African ingenuity in Carolina rice cultivation," Carney observed,

> would over time be attributed to Europeans. To the Portuguese who enslaved them in Africa and to the English and French slaveholders who relied on them to create Carolina's rice landscapes would go the credit for the inventive irrigated rice system Africans had developed under diverse wetland circumstances. Planter memoirs would celebrate the brilliant cultivation system invented by their forebears, while reducing the "savages" from the Guinea Coast to mere lackeys in their glorious schemes.[181]

"In an era of scientific racism and colonialism," she concluded, "the denial of African accomplishment in rice systems provides a stunning example of how power relations mediate the production of history."[182]

In spite of the Carolina rice growers' initial dependence on their slaves' knowledge, "by the mid-eighteenth century rice plantations had increasingly come to resemble those of sugar, imposing brutal demands on labor."[183] The slaves' knowledge of nature had thus been stolen and turned against them as a means of maintaining the social system that enslaved them.

MEDICINES, DRUGS, AND ETHNOBOTANY

ALTHOUGH THE SEARCH FOR food was the primary motive that drove early humans to accumulate botanical knowledge, it was not the only one. Natural selection has equipped plants with a broad range of strong chemicals to attract or repel animals, and foragers discovered the power of some of those chemicals to cure or alleviate physical ailments. The knowledge they gained is still directly benefiting us today. "From the herbs traditionally in use, modern medicine has derived such substances as salicylic acid, ipecac, quinine, cocaine, colchicines, ephedrine, digitalis, ergot, and other drugs besides."[184] About one-fourth of today's prescription medicines are plant-derived drugs, most of which "were originally discovered through the study of traditional cures and folk knowledge of indigenous peoples."[185]

Much of the indigenous peoples' medicinal knowledge has been documented by anthropological specialists who identify themselves as ethnobotanists. Their primary method of data collection is "interviewing healers, weavers, shipwrights, and other indigenous experts in the use of plants." Richard Evans Schultes, for example, spent fourteen years in the Amazon rain forest in the 1940s and '50s, where "he worked with many Amazonian tribes, identifying dozens of their hallucinogens and hundreds of their medicinal and toxic plants."[186] The work of the modern ethnobotanists is indispensable to a people's history of medicine, but it must be emphasized that the original creators of the scientific knowledge at issue were their subjects, the indigenous folk healers themselves.

The heroic narrative of the history of science traditionally identifies two first-century C.E. Roman writers, Pliny the Elder and Dioscorides, as "fathers" of ancient botany and the eighteenth-century Swede Carl Linnaeus as their modern counterpart. Pliny and Dioscorides, however, owed much of their knowledge to *rhizotomi*, people who made their living "preparing and selling of roots and herbs that were of repute in medicine."[187] Linnaeus also built on an ethnobotanical foundation by traveling to Lapland, north of the Arctic Circle, and tapping the Sami reindeer herders' knowledge of plants.[188] The point here is not to denigrate the scientific contributions of Pliny, Dioscorides, and Linnaeus but simply to note that their accomplishments built on the knowledge of others, and to identify those others insofar as possible.

Ethnobotanists report that "indigenous peoples experiment with the plants in their environment, often over hundreds of generations, and identify those that are bioactive." The botanical knowledge, for example, of Samoan healers (most of whom are women) "is formidable: a typical healer can identify over 200 species of plants by name, recognize over 180 disease categories, and compound more than 100 remedies."[189]

The contributions of Amerindians once again deserve special notice.[190] The "cornucopia of new pharmaceutical agents" they produced from plants "became the basis for modern medicine and pharmacology." Early in the post-Columbian era, "European doctors recognized that the Indians held the key to the world's most sophisticated pharmacy."[191] Nicholas Monardes, a sixteenth-century physician in Seville, wrote in praise of "all thynges that thei bryng from our Indias, whiche serveth for the arte and use of Medicine," and which "cure and make whole many infirmities, whiche if wee did lacke them, thei were incurable, and without any remedie."[192]

In general, however, Amerindian medicinal knowledge did

not make a direct leap into the Western medical mainstream; the transition was mediated by white folk healers. Whereas elite doctors initially resisted indigenous medicines, lay healers used them and demonstrated their therapeutic value, which led to their eventual acceptance. The popularity of traveling "Indian" medicine shows testified to early American popular culture's high regard for native remedies.[193]

THE HISTORY OF QUININE

AMONG THE EARLIEST and most important discoveries was that a traditional Quechua (Peruvian Indian) medicine for fevers was useful for treating malaria, a disease the Europeans had carried with them to the New World. The medicine, quinine, was made from the bark of the cinchona tree, which grew in the mountainous rainforests of Peru. Because of its bitterness, the ground-up bark was often dissolved in sugary water to produce "tonic water."

"The introduction of quinine," Weatherford says, "marks the beginning of modern pharmacology."[194] It also exemplifies the pattern of knowledge robbery: "Quinine quickly became too precious to the Europeans for them to allow the Indians to use it. The whites monopolized it to eradicate malaria in Europe while leaving the Indians to die from this disease that soon found a new permanent home in the American tropics."[195]

Later, as "the prime reason why Africa ceased to be the white man's grave," quinine was also instrumental in the European colonization of that continent. "In 1874 for instance 2,500 quinine-dosed British troops were marched from the Atlantic to the far reaches of the Asante empire in west Africa without serious loss of life; armed with quinine, the French began settling in Algeria in large numbers." It was quinine's efficacy that "gave colonists fresh opportunities to swarm into the Gold Coast, Nigeria and other parts of west Africa and seize fertile agricultural lands, introduce new livestock and crops, build roads and railways, drive natives into mines, and introduce all the disruptions to traditional lifestyles that cash economies brought."[196]

The importance of quinine is reflected in the history of international efforts to gain control of its source.[197] In the seventeenth century, Jesuit priests in Peru became familiar with the curative powers of what they called "Peruvian bark" and carried some of it to Rome, from whence its reputation spread throughout Europe. By the nineteenth century, millions of pounds of cinchona bark were being shipped to Europe every year. The governments of Peru and neighboring countries tried to protect their monopoly of this valuable resource by outlawing the exportation of cinchona seeds or saplings, which of course stimulated efforts to smuggle them out. In 1852 a Dutch colonialist in Java made a clandestine voyage to South America and, by bribing a corrupt official, obtained a quantity of cinchona seeds—an act of thievery for which the Dutch government knighted him. The trees his seeds produced, however, were found to be of a cinchona strain that was inferior in its ability to produce quinine, so smuggling efforts continued.

In 1861 an Australian illegally bought high-potency cinchona seeds from an Aymará Indian named Manuel Incra and spirited them out of Bolivia. When the crime was exposed, the Bolivian government arrested Incra and tortured him to death. Unfortunately for the Australian smuggler, he was unable to convince potential buyers that his contraband was genuine and finally sold a pound of seeds to the Dutch government—for twenty dollars! For the Dutch, "it was arguably the best $20 investment in history. . . . By 1930 the Dutch plantations in Java produced 22 million pounds of bark, yielding 97 percent of the world's quinine."[198]

At the beginning of World War II, the Allied powers were cut off from quinine supplies when the Germans overran Holland and the Japanese conquered Indonesia and the Philippines. Before the fall of the Philippines, however, the United States had managed to convey four million cinchona seeds to Maryland, where they were germinated and then transported to Costa Rica for planting. But it was too late:

More than 600,000 U.S. troops in Africa and the South Pacific had contracted malaria, and the average mortality rate was 10 percent. Since more U.S. soldiers were dying from malaria than from Japanese bullets, the lack of *Cinchona* bark immediately became a serious national security issue.[199]

The wartime emergency was resolved partially by American purchases of quinine on the black market and partially by the development of synthetic drugs such as chloroquinine, first synthesized in 1937. The synthetics proved useful for the treatment of malaria, but "the utility of quinine to treat certain heart arrhythmias . . . suggest that this bark . . . will remain an important botanical commodity for years to come."[200]

GOITER, SCURVY, CONSTIPATION, AND OTHER DISCONTENTS

QUININE WAS BUT one of many significant Amerindian discoveries that enriched the world's pharmacological knowledge. The Incas also used coca leaves, the source of cocaine, as an anesthetic, and dried seaweed, which is rich in iodine, as a preventive medicine against goiter.[201] By the time of European contact, Inca society had developed into a sophisticated civilization, but its medicinal knowledge originated in much earlier traditions and had continued to be the domain of "state-appointed herb collectors" and itinerant apothecaries who "went throughout the country with their stocks of mineral medicines and dried herbs."[202]

Although the Incas are considered to have been "the most highly developed medicinally of the Amerindian civilizations," the Aztecs and other Mexican Indians are known to have utilized some 1,200 medicinal herbs, and "North American Indian tribes had a similar though less extensive *materia medica*."[203]

The Indians of northern California and Oregon gave modern medicine the most commonly used laxative or cathartic.

They used the bark of the *Rhamnus purshiana* shrub as a cure for constipation. . . . it has spread to become the world's most commonly used laxative since its first introduction by the American pharmaceutical industry in 1878.[204]

The treatment of scurvy offers an instructive cross-cultural comparison. Scurvy is a devastating malady caused by vitamin C deficiency; it was often called the "sailor's disease" because the diet of seamen—typically devoid of fruits and vegetables—made them particularly vulnerable to it. When scurvy attacked Jacques Cartier's expedition in Canada in 1535, twenty-five members of his crew died, and another forty seemed on the brink of death. Cartier observed that a Huron Indian whom he had seen gravely ill with the same symptoms just ten or twelve days earlier had apparently been restored to full health. He asked the man, Dom Agaya (one of the Indians Cartier had earlier kidnapped and forced to serve as a guide), how he had been cured. Cartier's journal, in which he referred to himself in the third person as "the Captain," reported:

> Dom Agaya replied that he had been healed by the juice of the leaves of a tree and the dregs of these, and that this was the only way to cure sickness. Upon this the Captain asked him if there was not some of it thereabouts, and to show it to him. . . . Thereupon Dom Agaya sent two women with our Captain to gather some of it; and they brought back nine or ten branches. They showed us how to grind the bark and the leaves and to boil the whole in water. . . . The Captain at once ordered a drink to be prepared for the sick men. . . . As soon as they had drunk it they felt better . . . after drinking it two or three times they recovered health and strength and were cured of all the diseases they had ever had. . . . [It] produced such a result that had all the doctors of Louvain and Montpellier been there, with all the drugs of Alexandria, they could not have done so much in a year as did this tree in eight days; for it benefited us so much that all who were willing to use it recovered health and strength.[205]

The Eurocentric version of the history of medicine attributes discovery of the cure for scurvy to an eighteenth-century Scottish naval surgeon named James Lind. Lind, however, was well aware of the Huron treatment of two centuries earlier. "I am inclined to believe," he wrote,

> from the description given by Cartier of the *ameda* tree, with a decoction of the bark and leaves of which his crew was so speedily recovered, that it was the large swampy *American* spruce tree. . . . The pines and firs, of which there is a great variety . . . seem all to have analogous medicinal virtues, and great efficacy in this disease.[206]

Lind added that "Monsieur *Champlein*, who was then up the country, had orders to search for [the *ameda* tree] among the *Indians*, and to make provision of it for the preservation of their colony."[207]

FOLK HEALER SCIENCE

THE CONTRIBUTIONS OF folk traditions to medicinal knowledge is not a phenomenon unique to Amerindian culture. Aspirin is based on a chemical, salicylic acid, produced by plants discovered and used for treating aches and fevers independently by folk healers in widely separated parts of the world. North American Indians found the remedy in the bark of willow trees, but Europeans came across it in an herb growing in meadows. Arrow poisons used by Amazonian Indians and African hunters have yielded important drugs (curare, used as a muscle relaxant, and strophanthin, used to treat heart failure, respectively).[208]

For another example, the sedative reserpine originated in a plant in India that was used as a medication by local people. "How are we to characterize the discovery of reserpine?" asked two leading ethnobotanists:

> Does discovery of this important drug rest on "solid" science, such as structural chemistry and pharmacology, or is it

attributable to folklore and legend? Laboratory scientists may hail the invention of reserpine as serendipitous, but one fact is inescapable: a plant used by indigenous peoples eventually became the source of one of the world's most important pharmaceuticals.[209]

Furthermore, "every time a Shipibo hunter" in the Amazonian rainforest "fires a poison dart at an animal or a Tahitian healer administers a medicinal plant to a sick child, the efficacy of the indigenous tradition is empirically tested. It appears that indigenous traditions and science are epistemologically closer to each other than Westerners might assume."[210]

The discovery of digitalis, a cardiac drug of great historic importance, is usually credited to an eighteenth-century English doctor named William Withering. But in the 1785 publication in which he reported his findings, Withering stated that it was a folk healer who "first fixed my attention on the Foxglove":

> In the year 1775, my opinion was asked concerning a family receipt for the cure of the dropsy [edema]. I was told that it had long been kept a secret by an old woman in Shropshire, who had sometimes made cures after the more regular practitioners had failed.[211]

After talking to the woman and learning her "receipt" (i.e., recipe, or prescription), Withering wrote that among its twenty or more ingredients, "it was not very difficult for one conversant in these subjects, to perceive, that the active herb could be no other than the Foxglove."[212] His prior knowledge of foxglove (also known by its Latin name, digitalis) was likewise based on folk medicine; he was first alerted to the "vertues" of the plant by a compendium of English folk knowledge published by John Gerard almost two hundred years earlier, in 1597.[213] Gerard's herbal had suggested that foxglove could yield a medicine useful for treating dropsy.

It is also worth noting that "the American variety of foxglove was correctly used by [Amerindians] for its cardiac stimulant properties for hundreds of years before Withering discovered digitalis in England."[214] The drug produced from foxglove and the cardiac glycosides derived from it in the twentieth century are still used to treat congestive heart failure:

> More than 30 cardiac glycosides have been isolated from dried foxglove leaves, including digitoxin and digoxin. Neither of these drugs has ever been commercially synthesized; both are still extracted from dried foxglove leaves. Each year over 1500 kilograms of pure digoxin and 200 kilograms of digitoxin are prescribed to hundreds of thousands of heart patients throughout the world.[215]

VARIOLATION, INOCULATION, AND VACCINATION

A CONTEMPORARY OF Withering's, Edward Jenner, is traditionally hailed as the great doctor who rescued humankind from the scourge of smallpox by initiating the practice of vaccination. The history of smallpox prevention, however—"the one early striking instance of the conquest of disease"—is especially rich in folk tradition.[216] As one eighteenth-century commentator observed, "this Wonderful Invention was, first . . . found out, not by the *Learned* Sons of Erudition, but by a Mean, Course, Rude Sort of People. . . . It was rarely, if ever, used among *People of Quality*, until after the Beginning of the present Century."[217]

Healers in many parts of Africa and Asia had for centuries been practicing variolation or inoculation; that is, drawing pus from the sores of smallpox victims and introducing it into the bodies of healthy people.[218] The recipients of the attenuated *variola* virus would typically contract a relatively mild, nonlethal case of smallpox and would thereby gain lifetime immunity from the disease. The innovation attributed to Jenner was to inject fluids drawn from people infected with cowpox, a disease related to smallpox with a much milder effect on humans but which also confers smallpox immunity. (The root of *vaccination* is *vacca*, the Latin word for cow.)

The originators of smallpox prevention are anonymous, but the name of the African who introduced it into North America is known. The famous Puritan preacher Cotton Mather learned the technique of inoculation from a slave he owned named Onesimus. (A book titled *Cotton Mather: First Significant Figure in American Medicine* certainly does an injustice to Onesimus.)[219] In a letter of July 12, 1716, Mather wrote to friends in England:

> I do assure you, that many months before I mett with any Intimations of treating the *Small-Pox*, with the Method of Inoculation, any where in *Europe;* I had from a Servant of my own, an Account of its being practised in Africa. Enquiring of my Negro-man *Onesimus*, who is a pretty Intelligent Fellow, Whether he ever had the *Small-Pox*, he answered, both, *Yes*, and, *No;* and then told me, that he had undergone an Operation, which had given him something of the *Small-Pox*, & would forever preserve him from it; adding, That it was often used among the *Guramantese*, & whoever had the Courage to use it, was forever free of the fear of the Contagion. He described the Operation to me, and shew'd me in his Arm the Scar, which it had left upon him.[220]

In a later account, Mather added (interspersing his narrative with his version of African dialect):

> I have since mett with a Considerable Number of these *Africans*, who all agree in one Story; That in their Countrey *grandy-many* dy of the *Small-Pox:* But now they Learn This Way: People take Juice of *Small-Pox;* and cutty-skin, and putt in a Drop; then by'nd by a little *sicky, sicky:* then very few little things like *Small-Pox;* and no body dy of it; and no body have *Small-Pox* any more. Thus in *Africa*, where the poor Creatures dy of the *Small-Pox* like Rotten Sheep, a Merciful God has taught them an *Infallible Praeservative.* Tis a *Common Practice*, and is attended with a *Constant Success.*[221]

Mather launched a public campaign to convince his fellow citizens of the virtues of inoculation, but his efforts were met with fierce opposition. The resistance was not entirely irrational, but Mather's most vocal opponents appealed to racism by ridiculing him for adopting an idea from Africans: "There is not a Race of Men on Earth more *False Lyars*, &c."[222] Mather countered by reminding his critics of the proven value of indigenous medical knowledge: "I don't know why 'tis more unlawful to learn of *Africans*, how to help against the *Poison* of the *Small Pox*, than it is to learn of our *Indians*, how to help against the *Poison* of a *Rattlesnake*."[223]

The debt owed to Africans for introducing inoculation in North America was soon forgotten, but it was rediscovered by a scientific gadfly named Cadwallader Colden in 1753. "It seems probable," he wrote, "that the practice . . . came from Africa originally." Colden, who was at that time unaware of Mather's earlier revelation, continued:

> I have lately learned from my negroes, that it is a common practice in their country, so that seldom any old people have the disease. . . . It will be objected, how comes this not to have been sooner discovered, since so many negroes have been for near one hundred years past all over the colonies. But it is not to be wondered at, since we seldom converse with our negroes, especially those who are not born among us.[224]

The knowledge that was thus gained from African slaves was soon turned against them. Small-scale experimental trials designed to test the safety of inoculation had been performed in England on condemned prisoners, but the slave trade provided a much larger pool of involuntary subjects. Then, when it was found to be effective, slave dealers began routinely inoculating their chattels as a means of maximizing profits. Because immunized slaves were considered a safer investment, they brought a higher price.[225]

The knowledge of smallpox inoculation had reached Europe by another route. Rather than from African slaves, it had come from peasant women in Turkey. The traditional account makes an aristocratic European woman, Lady Mary Wortley Montagu, the star of the story, but a letter she wrote from Constantinople in 1717 shows where credit is due:

> The *smallpox*, so fatal and so general among us, is here entirely harmless by the invention of *engrafting*, which is the term they give it. There is a set of old women who make it their business to perform the operation every autumn in the month of September when the great heat is abated. . . . They make parties for the purpose . . . the old woman comes with a nutshell full of the matter of the best sort of smallpox, and asks what veins you please to have open'd. She immediately rips open . . . and puts into the vein as much [smallpox] matter as can lie upon the head of her needle.[226]

The evidence that inoculation preserved the lives of slaves and condemned felons led to its general acceptance as a legitimate medical procedure. In the hands of the medical elite, however, it became an expensive treatment that only the well-to-do could afford. When only a few people undergo inoculation, it actually puts the larger population of unimmunized people at greater risk of contracting smallpox. (Inoculees contract contagious smallpox; although their own case is relatively mild, the disease they can spread to others is the usual virulent form.) It is no wonder, then, that in America it generated a class-based controversy: The wealthy tended to support inoculation, whereas the less affluent opposed it. Benjamin Franklin was an advocate of inoculation on scientific grounds, but he recognized the social injustice it entailed. "The *expence* of having the operation perform'd by a Surgeon," he wrote, "has been pretty high in some parts of *America*." To immunize a typical workingman's family "amounts to more money than he can well spare." When smallpox claimed about three hundred lives in Philadelphia in 1774, Franklin was not surprised that "the chief of them were the children of poor people."[227]

Immunization was extended to the working classes in 1777

to 1778 when George Washington had the soldiers of the Continental Army inoculated in "the first large-scale, state-sponsored immunization campaign in American history."[228] Historian Elizabeth Fenn argues persuasively that Washington's decision to protect his troops from smallpox was essential to the victory of the American Revolution. To the extent that is true, the United States owes its very existence to the medical knowledge transmitted by Onesimus and other Africans.

Smallpox posed less of a threat to British troops in North America because widespread inoculation had begun earlier in England. Following Lady Montagu's reports of the Turkish peasants' method, "elite physicians predictably developed complex and costly techniques." A breakthrough occurred around 1750, however, when a family of "humble surgeons," Robert Sutton and his sons, "devised an easy, safe and cheap way, leading to mass inoculations."[229] The Suttons' main innovation was simply to eliminate extraneous (and medically worthless) procedures such as preliminary bleeding and purging that the upper-class doctors used to justify their inflated fees.

The later technique of vaccination proved superior to inoculation. Although most histories of medicine say that Edward Jenner invented it, the principles of vaccination were in fact deeply rooted in folk medicine. First of all, it was a simple variation of the inoculation procedure that, as we have seen, was created "not by the *Learned* Sons of Erudition, but by a Mean, Course, Rude Sort of People." Second, the people of rural England had long been aware that milkmaids rarely fell victim to smallpox. As Jenner himself explained, his interest in the subject

was *first excited* by observing that among those whom in the country I was frequently called upon to inoculate, many resisted every effort to give them the small-pox. These patients, I found, had undergone a disease they called cow-pox, contracted by milking cows affected with a peculiar eruption on their teats. On inquiry, it appeared that *it had been known among the dairies since time immemorial*, and that a vague opinion prevailed that it was a preventive of the small-pox.[230]

Modifying the inoculation process by injecting fluids drawn from a cowpox sore rather than from a smallpox pustule was an obvious experiment to try. Jenner was not the first to attempt it. In fact, the first documented vaccination was not performed by a doctor at all, but by someone who had spent his life in close proximity to cows. In 1774, a farmer at Yetminster in North Dorset named Benjamin Jesty injected cowpox matter into his wife Elizabeth and two of their children. Elizabeth became seriously ill but survived; the children showed no ill effects of the experience.[231] It was not until 1796, more than two decades later, that Jenner gave his first vaccination.

The Jenner myth has been dealt a further blow by evidence that some of those from whom he thought he was drawing cowpox pus may actually have been infected with smallpox. If so, his vaccine may have been inadvertently contaminated by the smallpox virus, which would mean that his patients had not been immunized by cowpox at all, but by an attenuated form of smallpox.[232] Although this cannot be conclusively proven, neither can it be disproven, which means that a sharp line cannot be drawn between the histories of inoculation and vaccination.

The disdain and ridicule with which the medical establishment greeted Jenner's ideas about cowpox and smallpox was typical of the way scientific elites often retarded science by closing their minds to the knowledge of ordinary people. Jenner stood apart from his peers in being willing to listen to farmers and milkmaids and to learn from them. In 1798, when he requested of the Royal Society in London that he be allowed to present his findings to them, he was warned by its president that he "ought not to risk his reputation by presenting to the learned body anything which appeared so much at variance with established knowledge, and withal so incredible."[233] To his credit, he took that risk.

CONCLUSION

This chapter does not constitute a comprehensive people's prehistory of science. That would require many, many volumes and the lifetimes of a number of scholars. Of the twenty-some-odd ingredients in the Shropshire healer's prescription for drop-sy, for example, foxglove was but one. What about the other nineteen or so?

By focusing on knowledge that has "stood the test of time," I have illuminated the rational side of the science of prehistoric people at the expense of superstitious, ritualistic, or accidental aspects that were no less a part of it. I do not, therefore, pretend to be pronouncing the final word on the subject. My more modest aim has been to demonstrate continuities, where they exist, between prehistoric knowledge and modern science. The hunter-gatherers' mastery of their natural environment had lasting consequences; their observation and experimentation laid the foundations of astronomy, botany, zoology, mineralogy, geography, oceanography, and many other sciences.

Finally, to bring this inquiry into the prehistory of science full circle, I call on two anthropologists for a direct answer to the question posed by the chapter's title:

> We have gained little or nothing in ability or intellectual brilliance since the Stone Age; our gains have all been in the accumulation of records of our intellectual achievements. We climb on each other's backs; we know more and understand more, but our intellects are no better. . . . Just as primitive life no longer can be characterised as nasty, brutish, and short, no longer can it be characterised as stupid, ignorant, or superstition-dominated.[234]

NOTES

1. Jean-Jacques Rousseau, *Discourse on Inequality among Men*, p. 180.
2. Ibid., p. 156.
3. This cliché seems to have originated with reference to Neanderthal man in a 1957 article by William L. Straus, Jr., and A. J. E. Cave, "Pathology and the Posture of Neanderthal Man."
4. Quoted in John Noble Wilford, "Debate Is Fueled on When Humans Became Human."
5. "In recent years the term *hunter-gatherer* has been discarded by some in favor of the more generic term *forager.* This term avoids privileging the hunting side of hunter-gatherer." Robert L. Kelly, *The Foraging Spectrum,* p. xiv.
6. "The proposition that human societies have spent over 99% of their cultural history as hunters and gatherers has become almost a truism." Geoff Bailey, ed., *Hunter-Gatherer Economy in Prehistory,* p. 1.
7. R. B. Lee and I. DeVore, *Man the Hunter,* p. 3.
8. Marshall Sahlins, "The Original Affluent Society."
9. Kelly, *The Foraging Spectrum,* p. 150. Kelly cites Binford, *In Pursuit of the Past* (1983), and J. Long, "Arid Region Aborigines: The Pintupi" (1971) with regard to the Nunamiut and the Pintupi, respectively.
10. Susan Carol Rogers, "Woman's Place," p. 126.
11. Sarah Milledge Nelson, *Gender in Archaeology,* p. 72 (emphasis in original). Although it can generally be said that in foraging societies, men hunt and women gather, Nelson stresses that the gender division of labor should not be considered absolute (p. 86).
12. Frances Dahlberg, ed., *Woman the Gatherer,* 1981.
13. Nelson, *Gender in Archaeology,* p. 101. Specific contributions of prehistoric women are discussed later in this chapter.
14. Stephen Jay Gould, "Posture Maketh the Man," pp. 208-210.
15. Ibid., p. 211.
16. Frederick Engels, *The Part Played by Labor in the Transition from Ape to Man.*
17. Gould, "Posture Maketh the Man," p. 212.
18. Engels, *Part Played by Labor in the Transition from Ape to Man.*
19. James McClellan and Harold Dorn, *Science and Technology in World History,* p. 5 (emphasis added). As was noted in chapter 1, these authors utilize a definition of science that (in my opinion) excessively separates it from technology.
20. V. Gordon Childe, *Man Makes Himself,* p. 106.
21. J. D. Bernal, *Science in History,* vol. 1, p. 61.
22. Kelly, *The Foraging Spectrum,* p. xiii.
23. Ibid., p. 26.

24. Jared Diamond, *Guns, Germs, and Steel*, pp. 19-21.

25. Peter Worsley, *Knowledges*, p. 14.

26. Ibid., p. 66 (emphasis in original).

27. Donald F. Thomson, "Names and Naming among the Wik Monkan Tribe," *Journal of the Royal Anthropological Institute*, vol. LXXVI (1946), quoted in Worsley, *Knowledges*, p. 66.

28. Worsley, *Knowledges*, pp. 66-67.

29. J. A. Waddy, *Classification of Plants and Animals from a Groote Eylandt Point of View* (1988), quoted in Worsley, *Knowledges*, p. 67.

30. Worsley, *Knowledges*, p. 69.

31. Ibid., pp. 71-72.

32. Ibid., p. 17. The studies he cites are J. A. Waddy, *Classification of Plants and Animals from a Groote Eylandt Point of View*, and David H. Turner, *Tradition and Transformation: A Study of Aborigines in the Groote Eylandt Area, Northern Australia* (1974).

33. Worsley, *Knowledges*, p. 20.

34. Ibid., p. 23.

35. Ibid., p. 56. His source is Dulcie Levitt, *Plants and People: Aboriginal Uses of Plants on Groote Eylandt* (1981). Among other sources, contraceptives were derived from the fruit and inner bark of the mistletoe tree.

36. Worsley, *Knowledges*, p. 123 (emphasis added).

37. Ibid., pp. 72-73.

38. Richard Rudgley, *Lost Civilisations of the Stone Age*, p. 110.

39. Carlo Ginzburg, "Clues."

40. L. W. Liebenberg, *The Art of Tracking*. See esp. chap. 6: "Scientific Knowledge of Spoor and Animal Behaviour."

41. Liebenberg, *Art of Tracking*, pp. 4, 29, 71, 87, 91.

42. Nicolas Blurton-Jones and Melvin J. Konner, "!Kung Knowledge of Animal Behavior," p. 343.

43. Liebenberg, *Art of Tracking*, pp. 156-157.

44. The second circumnavigation was accomplished by Sir Francis Drake in 1580.

45. See the epigraph at the head of this chapter.

46. Quoted in Steve Thomas, *The Last Navigator*, p. 5.

47. See Diamond, *Guns, Germs, and Steel*, pp. 41-42.

48. J. C. Beaglehole, ed., *The Endeavour Journal of Joseph Banks 1768–1771*, vol. 1, p. 368.

49. Louis-Antoine de Bougainville, *A Voyage Round the World*, pp. 275-276.

50. Bolton Glanvill Corney, ed., *The Quest and Occupation of Tahiti by Emissaries of Spain during the Years 1772–1776*, vol. 2, p. 286.

51. Again, see the epigraph at the head of this chapter. For another counterexample, see the section "Geography and Cartography of 'Savages'" in this chapter.

52. David Lewis, *We the Navigators*, pp. 9, 342-345.

53. Ibid., p. 306. Lewis's source is J. Burney, *A Chronological History of the Discoveries in the South Seas or Pacific Ocean*, vol. 5 (1967).
54. Lewis, *We the Navigators*, p. 248.
55. Otto von Kotzebue, *A Voyage of Discovery*, vol. 2 (1821), pp. 144-146.
56. Antonio Pigafetta, *Magellan's Voyage*, p. 110. As for Columbus, see the section "Geography and Cartography of 'Savages'" in this chapter.
57. Thomas, *The Last Navigator*; Lewis, *We the Navigators*; Thomas Gladwin, *East Is a Big Bird*; Richard Feinberg, *Polynesian Seafaring and Navigation*.
58. Lewis, *We the Navigators*, pp. 23-24, 30.
59. Gladwin, *East Is a Big Bird*; p. vi.
60. Feinberg, *Polynesian Seafaring and Navigation*.
61. Thomas, *Last Navigator*, p. viii.
62. Lewis, *We the Navigators*, p. 53.
63. Ibid., p. 7.
64. See, for example, Andrew Sharp, *Ancient Voyagers in the Pacific*, p. 153: "It cannot be too strongly emphasized that the people who settled both Polynesia and Micronesia were wanderers who had lost their way in the trackless wastes of the vast Pacific Ocean."
65. Lewis, *We the Navigators*, p. 16. Lewis was summarizing results of a study by M. Levison, R. G. Ward, and J. W. Webb, *The Settlement of Polynesia: A Computer Simulation* (1972).
66. Gladwin, *East Is a Big Bird*, p. 148.
67. Ibid., p. 154.
68. Lewis, *We the Navigators*, pp. 111, 284.
69. Gladwin, *East Is a Big Bird*, p. 131.
70. See chapter 4.
71. Jacob Bronowski, *The Ascent of Man*, p. 192.
72. Thomas, *Last Navigator*, p. 76.
73. Quoted in Lewis, *We the Navigators*, p. 127.
74. Lewis, *We the Navigators*, p. 3.
75. Ibid., pp. 206-207.
76. Louis De Vorsey, "Amerindian Contributions to the Mapping of North America," p. 211. I am indebted to De Vorsey for most of the information in this section.
77. Ibid., p. 211.
78. Ibid., p. 212.
79. Ibid., p. 212.
80. Captain John Smith, "The Description of Virginia," in *Travels and Works of Captain John Smith*, ed. Edward Arber, vol. I, p. 55. Quoted in De Vorsey, p. 211 (emphasis in original).
81. J. McIver Weatherford, *Native Roots*, p. 21.
82. Samuel de Champlain, *The Works of Samuel de Champlain*, ed. H. P. Biggar, vol. II, p. 191. Quoted in De Vorsey, p. 211.

83. Herman Friis, "Geographical and Cartographical Contributions of the American Indian to Exploration of the United States Prior to 1860" (unpublished paper), pp. 6-7. Quoted in De Vorsey, p. 211.

84. John Lawson, *A New Voyage to Carolina*, ed. Hugh Talmadge Lefler (Chapel Hill, 1967), p. 214. Quoted in De Vorsey, p. 215.

85. British Public Record Office, London, C.O. 700 Maps—Florida 3. Cited in De Vorsey, pp. 216-217.

86. Baron De Lahontan, *New Voyages to North America*, ed. Reuben Gold Thwartes (New York, 1900 [Reprint of 1703 English ed.]), vol. II, p. 427. Quoted in De Vorsey, p. 216.

87. Justin Winsor, *Christopher Columbus*, p. 442.

88. Jacques Cartier, *The Voyages of Jacques Cartier.*

89. J. B. Harley, "New England Cartography and the Native Americans," p. 174 and n. 16, p. 271.

90. Weatherford, *Native Roots*, pp. 23-24.

91. The phrase, from Lewis Pyenson, "Cultural Imperialism and Exact Science: German Expansion Overseas, 1900–1930," *History of Science*, vol. 20 (1982), is quoted in Harley, "New England Cartography and the Native Americans," p. 188.

92. Harley, "New England Cartography and the Native Americans," pp. 170, 187, 195.

93. Ibid., pp. 187–190.

94. Ibid., pp. 187–188, 191.

95. Ibid., p. 195.

96. E. C. Krupp, *Skywatchers, Shamans & Kings*, p. 136.

97. Ibid., p. 135.

98. Gerald Hawkins, "Stonehenge," *Nature*, January 27, 1964. See also Gerald Hawkins, *Stonehenge Decoded.*

99. Krupp, *Skywatchers, Shamans & Kings*, p. 140.

100. E. C. Krupp, *Echoes of the Ancient Skies*, pp. 1, 157.

101. Lloyd A. Brown, *Story of Maps*, pp. 35–37.

102. Krupp, *Echoes of the Ancient Skies*, p. 47.

103. Ibid., p. 165.

104. Alexander Marshack, "Lunar Notation on Upper Paleolithic Remains," p. 743. See also Marshack, *The Roots of Civilization.*

105. Anthony F. Aveni, *Ancient Astronomers*, pp. 32-33. Aveni cites the work of anthropologist William Breen Murray.

106. Krupp, *Echoes of the Ancient Skies*, p. 163.

107. Ibid., p. 145. Krupp cites John A. Eddy, "Medicine Wheels and Plains Indian Astronomy," in Kenneth Brecher and Michael Fiertag, eds., *Astronomy of the Ancients.*

108. For a dissenting view, see David Vogt, "Medicine Wheel Astronomy." Although Vogt concludes that "the evidence and analysis presented here

do not prove the case for Medicine Wheel astronomy," he adds: "all the indications are that Plains tribes were technologically capable of producing a practical lunisolar calendar."

109. Krupp, *Skywatchers, Shamans & Kings*, p. 156. See also Travis Hudson and Ernest Underhay, *Crystals in the Sky.*

110. E. C. Krupp, "As the World Turns," in E. C. Krupp, ed., *Archaeoastronomy and the Roots of Science.*

111. Krupp, *Skywatchers, Shamans & Kings*, pp. 149–150, 154, 228, 231, 234.

112. "The roots of science and of writing seem to be here." Marshack, *Roots of Civilization*, p. 57.

113. Georges Ifrah, *The Universal History of Numbers*, p. 64.

114. It is not certain whether the inventors of writing were the Sumerians or their Mesopotamian predecessors, the Subarians. See Denise Schmandt-Besserat, "On the Origins of Writing," p. 41.

115. Denise Schmandt-Besserat, *Before Writing*. Nonspecialists may prefer the abridged version: Denise Schmandt-Besserat, *How Writing Came About*. For a dissenting view, see S. J. Liebermann, "Of Clay Pebbles, Hollow Clay Balls, and Writing."

116. The chronology is from Schmandt-Besserat, "On the Origins of Writing," p. 42. My brief summary of her discovery cannot do it justice; interested readers can find the full story in the books cited in note 115.

117. Schmandt-Besserat, *How Writing Came About*, p. 17.

118. Ibid., p. 16.

119. Ibid., p. 83.

120. Ibid., p. 6; V. Gordon Childe, *What Happened in History*, p. 86.

121. Tobias Dantzig, *Number*, p. 21.

122. Schmandt-Besserat, *How Writing Came About*, p. 118.

123. The other, positional numeration, is discussed in the next section.

124. Edward Chiera, *They Wrote on Clay*, pp. 83–84.

125. Georges Charbonnier, *Conversations with Claude Lévi-Strauss*, pp. 29–30.

126. Karl Kautsky, *Foundations of Christianity*, p. 204 (emphasis in original).

127. Dantzig, *Number*, p. 27.

128. Ibid., p. 36.

129. Ibid., p. 25.

130. Quoted in Dantzig, *Number*, p. 19. Positional numeration was utilized much earlier in Babylonian mathematics, but in the context of a sexagesimal rather than a decimal system. As for what we call the Hindu-Arabic numeral system, there is a great deal of evidence for Indian origins, but it is not universally accepted. Lam Lay Yong and Ang Tian Se, *Fleeting Footsteps*, argue that our "Hindu-Arabic" system actually originated in China.

131. Dantzig, *Number*, p. 84. For more about Fibonacci, see chapter 5.

132. Ibid., p. 33 (emphasis in original).

133. Lancelot Hogben, *Mathematics in the Making*, p. 38.

134. Ifrah, *Universal History of Numbers*, pp. 416–418.

135. Ibid., p. 357. Ifrah himself attributes the place-value system to "Indian scholars" (p. 433) but concedes that there is no documentary or other evidence to support that conjecture.

136. Lancelot Hogben, *Astronomer Priest and Ancient Mariner*, p. 62.

137. Childe, *Man Makes Himself*, p. 40.

138. Ibid., p. 68.

139. Theodore A. Wertime, "Pyrotechnology," in Denise Schmandt-Besserat, ed., *Early Technologies*, p. 18.

140. Henry Hodges, *Technology in the Ancient World*, p. 41.

141. See Nelson, *Gender in Archaeology*, pp. 106–108, and 118–120. Although Nelson cautions that "nothing about the making of pots should be inherently gendered," she presents evidence from a number of cultures—"most native American groups," for example—in which pottery-making was clearly "women's work."

142. Childe, *Man Makes Himself*, p. 73.

143. Ibid., p. 72.

144. "Textiles are almost always attributed to women, though men weave in a number of societies." Nelson, *Gender in Archaeology*, pp. 109-111.

145. Most native iron is meteoric in origin.

146. Robert Raymond, *Out of the Fiery Furnace*, p. xi.

147. Hodges, *Technology in the Ancient World*, p. 92.

148. Raymond, *Out of the Fiery Furnace*, pp. 21-22, 36-49.

149. Ibid., p. 40.

150. Wertime, "Pyrotechnology," p. 22.

151. Raymond, *Out of the Fiery Furnace*, p. 35.

152. Ibid., p. 55.

153. Ibid., pp. 56-57.

154. John Read, *Through Alchemy to Chemistry*, pp. 12-13.

155. George W. Beadle, "The Ancestry of Corn," p. 125.

156. Bruce D. Smith, *The Emergence of Agriculture*, pp. 17-18.

157. Ibid., p. 17.

158. Diamond, *Guns, Germs, and Steel*, p. 143.

159. See T. P. Denham *et al.*, "Origins of Agriculture at Kuk Swamp in the Highlands of New Guinea," *Science* (July 2003).

160. Smith, *Emergence of Agriculture*, p. 23. It is worth noting that the first chapter of Darwin's *Origin of Species* was devoted not to natural but to artificial selection.

161. Patty Jo Watson, "Explaining the Transition to Agriculture," p. 35.

162. Smith, *Emergence of Agriculture*, p. 27.

163. Diamond, *Guns, Germs, and Steel*, pp. 132-133, 146.

164. Weatherford, *Native Roots*, p. 128.

165. Weatherford, *Indian Givers*, p. 88.

166. Beadle, "Ancestry of Corn," p. 125.

167. Weatherford, *Indian Givers*, pp. 84–85.
168. Michael J. Balick and Paul Alan Cox, *Plants, People and Culture*, p. 75.
169. Judith Ann Carney, *Black Rice*, p. 166.
170. Balick and Cox, *Plants, People and Culture*, pp. 75, 91.
171. Readers who want to pursue in more depth some of the issues I have touched on here are urged to consult Jared Diamond's fascinating and very readable *Guns, Germs, and Steel*. Another source that provides a wealth of technical data is Richard S. MacNeish, *The Origins of Agriculture and Settled Life*.
172. Weatherford, *Indian Givers*, p. 89.
173. Joyce Chaplin, *An Anxious Pursuit*, p. 156. For numerous instances of eighteenth-century European observers attributing the introduction of plants to African slaves, see William Grimé, *Botany of the Black Americans*, pp. 19-27.
174. Carney, *Black Rice*, pp. 140-141.
175. Ibid., pp. 44, 38.
176. Ibid., pp. 2, 81.
177. Ibid., pp. 136, 97.
178. Ibid., p. 90.
179. Ibid., p. 89.
180. Ibid., pp. 50, 117, 107.
181. Ibid., p. 97. For example: "Rice culture reached a development and a degree of perfection in the Carolina lowlands which had not been attained in any other rice-growing country in two thousand years. . . . The only labor at the disposal of the settlers who accomplished the feat was of the most unskilled character, African savages fresh from the Guinea coast. . . . The Southern planter who accomplished this result was a man who worked with his brains on an extended scale." Sass and Smith, *A Carolina Rice Plantation of the Fifties* (1936), p. 23; quoted in Carney, pp. 97–98.
182. Carney, *Black Rice*, p. 48.
183. Ibid., p. 141.
184. Roy Porter, *The Greatest Benefit to Mankind*, p. 35.
185. Balick and Cox, *Plants, People and Culture*, p. 25.
186. Ibid., pp. vii, 20–21.
187. Quoted in Balick and Cox, *Plants, People and Culture*, p. 14.
188. Balick and Cox, *Plants, People and Culture*, p. 18.
189. Ibid., pp. 38–39, 53–54.
190. An excellent source of information on this subject is Virgil J. Vogel, *American Indian Medicine*, which includes a 147-page appendix entitled "American Indian Contributions to Pharmacology."
191. Weatherford, *Indian Givers*, pp. 183–184.
192. Nicholas Monardes, *Joyfull Newes out of the Newe Founde Worlde*, vol. I, p. 10.
193. See Vogel, *American Indian Medicine*, pp. 263–265.
194. Weatherford, *Indian Givers*, p. 177.

195. Ibid., p. 195.
196. Porter, *Greatest Benefit to Mankind*, pp. 465–466.
197. See Balick and Cox, *Plants, People and Culture*, pp. 27–31.
198. Ibid., p. 29.
199. Ibid., p. 29. Italics in original.
200. Ibid., p. 31.
201. Weatherford, *Indian Givers*, pp. 183, 190.
202. Roderick E. McGrew, *Encyclopedia of Medical History*, p. 218.
203. Ibid., p. 218.
204. Weatherford, *Indian Givers*, p. 184.
205. Cartier, *Voyages of Jacques Cartier*, pp. 79–80.
206. James Lind, *A Treatise on the Scurvy*, pp. 177–178. For his knowledge of Cartier's experience, Lind cites "Hackluit's collection of voyages, vol. 3, p. 225."
207. Lind, *Treatise on Scurvy*, p. 303.
208. Balick and Cox, *Plants, People and Culture*, pp. 32, 118.
209. Ibid., p. 3.
210. Ibid., p. 3.
211. William Withering, *An Account of the Foxglove, and Some of Its Medical Uses*, p. 2. Edema, the modern term for dropsy, is the condition characterized by excessive fluid retention in body tissues or organs.
212. Ibid., p. 2.
213. John Gerard, *The Herball, or General Historie of Plantes*.
214. Vogel, *American Indian Medicine*, pp. 10–11. Vogel's source is Harlow Brooks, "The Medicine of the American Indian," *Journal of Laboratory and Clinical Medicine*, vol. XIX, no. 1 (October, 1933).
215. Balick and Cox, *Plants, People and Culture*, pp. 17–18.
216. "The introduction first of smallpox inoculation and then of vaccination . . . came not through 'science' but through embracing popular medical folklore." Porter, *Greatest Benefit to Mankind*, p. 11.
217. Attributed to Jacobus Pylarinus, "the Venetian Consul at Smyrna," by Cotton Mather in *The Angel of Bethesda*, p. 109 (emphasis in original). Pylarinus's letter appeared in the Royal Society's *Philosophical Transactions* in 1717.
218. For claims of a Chinese origin, see Joseph Needham, *The Grand Titration*, pp. 58–59.
219. Otho T. Beall and Richard Shryock, *Cotton Mather*.
220. George Lyman Kittredge, ed., "Lost Works of Cotton Mather," p. 422 (emphasis in original). I have taken the liberty of replacing "ye" in the original with "the" for ease of reading.
221. Mather, *Angel of Bethesda*, p. 107 (emphasis in original).
222. William Douglass, *Inoculation Consider'd* (1722), quoted in Kittredge, "Lost Works of Cotton Mather," p. 436 (emphasis in original). For an account of the inoculation controversy, see John B. Blake, "The Inoculation Controversy in Boston," pp. 489–506.

223. Kittredge, "Lost Works of Cotton Mather," p. 430 (emphasis in original).
224. Ibid., pp. 439–440.
225. See Larry Stewart, "The Edge of Utility."
226. Quoted in Porter, *Greatest Benefit to Mankind*, p. 275 (emphasis in original).
227. Quoted in Elizabeth Anne Fenn, *Pox Americana*, pp. 41–42 (emphasis in original).
228. Fenn, *Pox Americana*, p. 102.
229. Porter, *Greatest Benefit to Mankind*, pp. 275–276.
230. Edward Jenner, 1801. Quoted in Hervé Bazin, *The Eradication of Smallpox*, p. 180 (emphasis added).
231. Richard Horton, "Myths in Medicine," p. 62.
232. Peter Razzell, *Edward Jenner's Cowpox Vaccine.*
233. Quoted in C. N. B. Camac, ed., *Classics of Medicine and Surgery*, p. 211.
234. Blurton-Jones and Konner, "!Kung Knowledge of Animal Behavior," p. 348.

3

WHAT "GREEK MIRACLE"?

MUCH HAS BEEN written concerning the seemingly sudden emergence in the Greek Ionian colonies of the sixth century B.C. of a natural philosophy or science rational and surprisingly secular in character. In fact, historians . . . have called this phenomenon the "Greek Miracle."
—MARSHALL CLAGETT, *Greek Science in Antiquity*

IT WAS THE Greeks who invented science as we know it.
—A. C. CROMBIE, *Augustine to Galileo*

ULTIMATELY SCIENCE DERIVES from the legacy of Greek philosophy.
—C. C. GILLISPIE, *The Edge of Objectivity*

ANCIENT PHYSICAL SCIENCE begins with Thales of Miletus (about 600 B.C.).
—E. J. DIJKSTERHUIS, *The Mechanization of the World Picture*

THE CONCEPTION OF the infinity of the universe, like everything else or nearly everything else, originates, of course, with the Greeks.
—ALEXANDRE KOYRÉ, *From the Closed World to the Infinite Universe*

"OF COURSE," ALEXANDRE Koyré confidently asserted in 1957, as if no one would dream of disagreeing, "nearly everything" originated with the Greeks. According to the Greek Miracle doctrine, the Greeks of classical antiquity were the creators of philosophy, of science, of mathematics, of medicine, of politics, of theology—indeed, of everything of intellectual value. And they did it all on their own, without any significant external influences. Historian of ancient science

David Pingree labeled this attitude "Hellenophilia" and included in its definition the belief

> that one of several false propositions is true. The first is that the Greeks invented science; the second is that they discovered a way to truth, the scientific method, that we are now successfully following; the third is that the only real sciences are those that began in Greece.[1]

Pingree, writing in 1990, believed that this "thoroughly pernicious" mode of interpretation was on the rise among historians of science at that time. In my estimation, that is no longer true. Thanks to the labors of Pingree and like-minded scholars, it now seems beyond any question that Greek science was predicated on prior achievements in Mesopotamia and Egypt, and that other ancient cultures in China and India made significant contributions to the development of scientific knowledge as well. For a professional historian of science to express the blatant Hellenophilia exemplified by the epigraphs at the head of this chapter would now invite ridicule.

"The news, however," another commentator lamented, "has not yet reached many quarters. The disjunction between scholarly achievements and public consciousness stands out starkly."[2] A great deal of popular science writing continues to propagate Hellenophilia. But perhaps this, too, is changing for the better; major publishers have begun to produce books for a broad readership that challenge the Greek Miracle notion.[3]

One author in particular deserves special credit for exposing the fallacy of the Greek Miracle: Martin Bernal, whose *Black Athena* presents a powerful case for the Afroasiatic roots of Greek culture.[4] Bernal's work was not well received by Hellenophilic academics,[5] but its effectiveness can be discerned in the grudging admission of one of its critics that "*Black Athena* must be the most discussed book on the ancient history of the eastern Mediterranean world since the Bible."[6]

The Greek Miracle is incompatible with a people's history of science because it exalts individual Greek geniuses such as Thales and Pythagoras—and above all Plato and Aristotle—and awards them total credit for the creation of science. That is not to say, however, that all of the newly recognized antecedents of Greek science can be considered people's science. The scholars who have brought to light early scientific developments in Mesopotamia and Egypt, especially in astronomy and mathematics, have almost invariably attributed them to intellectual elites. Otto Neugebauer, considered by many to be "the greatest of all historians of ancient mathematics and astronomy," declared, "Ancient science was the product of *a very few men*."[7]

It would be a shame if future researchers were to treat Neugebauer's assertion as an unchallengeable truth. The earliest Mesopotamian and Egyptian astronomers and mathematicians were anonymous, and archaeological evidence has provided only vague insights into what social classes they represented. Were they very few in number? Were they all men? As Pingree pointed out, "We do not know by whom, when, or where any Babylonian lunar or planetary theory was invented; we do not know what observations were used, or where and why they were recorded."[8] Later Babylonian mathematics and astronomy were certainly under the hegemony of an educated elite, but perhaps further research will find that the social roots of the "exact sciences" in the ancient world were more complex than Neugebauer and others have heretofore imagined.

Be that as it may, allow me to reiterate two points that were made in the previous chapter: first, that the origins of astronomy and mathematics predated the rise of civilizations in Mesopotamia, Egypt, and elsewhere. Second, and more important, is that knowledge of nature entails much more than astronomy and mathematics. One historian who assumed that those two sciences were created by "the priestly scribes of Mesopotamia and Egypt" also noted that those scribes "rarely recorded a knowledge of the chemical arts, metallurgy, dyeing, and so on, which belonged to another tradition, that of the craftsmen who handed on their experience orally."[9] However, because we have no works on ancient chemical theory it does not follow that it did not exist. Though it may never have been formally expressed, the ancient chemists show in their products that they were acquainted with the general principles of oxidation and reduction and could introduce or remove non-metals, such as sulphur and chlorine.[10]

In preclassical antiquity as in other ages, illiterate artisans' contributions to the knowledge of nature were undoubtedly both crucial and undocumentable. "Techniques are a fertile seed-bed of science," Benjamin Farrington explained, "and progress from pure empiricism to scientific empiricism is so gradual as to be imperceptible." Between about 3000 and 2500 B.C.E., when the great pyramids were being constructed,

the Egyptians were also busy with agriculture, dairying, pottery, glass-making, weaving, ship-building, and carpentry of every sort. This technical activity rested upon a basis of empirical knowledge. . . . To deny it the name of science because it was, perhaps, handed down by tradition to apprentices instead of being written in a book is not wholly just. Technical problems also certainly clamoured for solution in connection with their gold-work, weaving, pottery, hunting, fishing, navigation, basket-work, culture of cereals, culture of flax, baking and brewing, vine-growing and wine-making, stone-cutting and stone-polishing, carpentry, joinery, boat-building, and the many other processes so accurately figured on the walls of the tombs of the nobles at Sakkara (2680 to 2540 [B.C.E.]). In all these techniques lay the germ of science.[11]

Documentary evidence does exist for one important aspect of ancient Egyptian science. The chance survival of a fragment of a single surgical treatise—known as the Edwin Smith papyrus after its discoverer—provides us a glimpse, at least, into "a traditional branch of knowledge which may quite possibly be as old as the fourth millennium [B.C.E.]." The document reveals anatomical knowledge that is "correct and considerable in amount," as well as "a beginning of physiological knowledge." All told, it constitutes "a body of knowledge which can only be regarded as the result of a long tradition of observation and reflection. As such it is a work of science in the modern sense."[12]

Before directly examining Greek science, let us inquire into the history of the Greek Miracle doctrine itself. One would think that the least scientific explanation for anything would be one that appeals to belief in miracles. The thesis that the classical Greeks suddenly burst onto the scene out of nowhere in the sixth century B.C.E. asks us to believe that in spite of the prior existence of high civilization in Egypt and in Mesopotamia for thousands of years, Greek culture developed *sui generis*, owing no debt whatsoever to its predecessors. Where and why did such an unlikely claim arise, and how could it have been sustained?

CAUCASIANS AND ARYANS

LINGUISTIC EVIDENCE INDICATES that the light-skinned people who populated Europe in historic times originated from some tribes that lived in the Caucasus mountains between the Black Sea and the Caspian Sea—hence the use of the word "Caucasian" to denote white people. These people, who spoke a language that linguists today call proto-Indo-European, spread out from the Caucasus in prehistoric times, and their language is now recognized as the ancestor of many of the tongues spoken in India, Iran, and Europe.

Those ancestral people of the Caucasus are also called Aryans, and the mystique that was built around them as a foundation myth for white supremacy was the basis on which Nazi ideologues developed their notion of the master race. But for more than a hundred years before Hitler's minions did so much to discredit the idea, European academic circles took for granted the proposition that virtually everything worthwhile in the human past was a product of Aryan genius. One of the pioneers of Egyptology, James H. Breasted, proclaimed as late as 1926: "The evolution of civilization has been the achievement of this Great White Race."[13]

Any notion that ancient Egyptians or Sumerians were Aryans was disproved when linguistic analysis made it clear that their languages were not members of the Indo-European family. Even before that, it had been obvious from ancient artwork that their skins were darker than Europeans' skins—that by modern standards they would be considered "people of color." One attempt to explain away that fact was the "suntan thesis": the proposition that these were really white people who appeared dark because they were exposed to the sun a lot.[14]

But the most important ploy that nineteenth-century European scholars devised to avoid acknowledging that the roots of civilization are Afroasiatic was to minimize the importance of Egyptian, Sumerian, and Semitic contributions and to focus instead almost entirely on the Greeks. According to this idea, the Egyptians, Sumerians, and Semites established rather static and uninteresting cultures, while the really worthwhile developments in the rise of civilization were the work of the dynamic and sophisticated Greeks, who were considered to be of Aryan stock because their language is part of the Indo-European family. Furthermore—and this is the crucial point—it was claimed that the Greeks developed their culture all on their own, with virtually no contribution from the earlier civilizations.

DID THE GREEKS BELIEVE IN
THE GREEK MIRACLE?

THIS NOTION DID not originate with the Greeks themselves. To the contrary, ancient Greek authors almost unanimously took for granted that their culture was rooted in the wisdom and accomplishments of earlier civilizations, and especially that of Egypt. It was a commonplace sentiment, not at all controversial. Herodotus, the fifth-century B.C.E. author who has been called "the father of history," acknowledged it, as did Hippocrates, the so-called "father of medicine," and even Plato and Aristotle. On the origins of Greek religious thought, Herodotus wrote,

Almost all the names of the gods came into Greece from
Egypt. My inquiries prove that they were all derived from a
foreign source, and my opinion is that Egypt furnished the
greater number. [Most] have been known from time immemorial in Egypt. . . . Besides these which have been here
mentioned, there are many other practices . . . which the
Greeks have borrowed from Egypt.[15]

As for philosophy, the orator Isocrates, a rival of Plato, wrote
that Pythagoras went to Egypt and on his return "was the first
to bring to the Greeks all philosophy."[16] As we shall see, attributions to Pythagoras cannot be taken at face value, but the
point here is that he was perceived in classical times as a *trans-
mitter* rather than as a *creator* of philosophy.

Thales of Miletus, hailed by Hellenophiles as the creator of
science (see the epigraphs at the head of this chapter), was
reputed to have spent many years abroad studying the ancient
wisdom of Egyptians, Babylonians, and Phoenicians; it was
even said by some that he was himself of Phoenician stock.[17]
According to no less an authority than Aristotle, "the mathematical arts were founded in Egypt."[18] Plato attributed to
Egyptian wisdom not only the invention of "arithmetic and calculation and geometry and astronomy," but also "the discovery
of the use of letters."[19]

"Geometry," Herodotus declared, "first came to be known in
Egypt, whence it passed into Greece."[20] In the first century
C.E., Strabo commented:

Geometry was invented, it is said, from the measurement of
lands which is made necessary by the Nile when it confounds
the boundaries at the time of its overflows. This science,
then, is believed to have come to the Greeks from the
Aegyptians; astronomy and arithmetic from the Phoenicians;
and at present by far the greatest store of knowledge in every
other branch of philosophy is to be had from these
[Phoenician] cities [Sidon and Tyre].[21]

This tradition was sustained by the authors of antiquity for more than a thousand years. In the fifth century C.E., Proclus repeated that it was "among the Egyptians that Geometry is generally held to have been discovered."[22]

And politics? Perhaps the best-known example of Greek political thought is Plato's *Republic*. Krantor, a commentator of the late fourth century B.C.E., reported that "Plato's contemporaries mocked him, saying that he was not the inventor of his republic, but that he had copied Egyptian institutions."[23]

These quotations do not prove that ancient Greece learned its theology, philosophy, mathematics, science, and politics from Egypt, but they certainly demonstrate that the Greeks themselves thought so. If the ancient Greeks did not ignore or deny their debt to their predecessors, from whence, then, did the notion of the Greek Miracle come?

THE GREEK MIRACLE AND RACIAL IDEOLOGY

IT WAS NOT until some two thousand years later, in the nineteenth century C.E., that a small but influential group of German scholars led by Karl Otfried Müller decided that the ancient Greek authors did not know what they were talking about—that their traditions of external influences were simply "myths." This school of thought got its start at the University of Göttingen and from there spread rapidly throughout Germany, to England, to France, and to the United States.[24] In this instance, it is not a science that has proved fruitful to later ages that is under consideration, but a body of ideas once thought of as science that has since been discredited. It is of the utmost relevance to a people's history of science because of the profoundly negative impact those false ideas had—and are still having—on the lives of people of color.

The key to understanding the Göttingen scholars' ideas about the relationship of Greece to the earlier civilizations is their conception of "scientific history." They were convinced that the primary scientific principle of historical explanation was race, and they believed they had discovered the "scientific laws of race." According to their laws of racial science, only the white race—the descendants of the Aryans—had the natural ability to create advanced civilizations. The black race, they maintained, was at the very bottom of the racial scale and had no aptitude for civilization whatsoever. The pre-Darwinian conception was not that white people represent a higher stage of evolution, but that the original human race created by God was pure Caucasian and that other races represent degenerate forms.

This "racial science" was a product of the age of triumphant European imperialism, and it served as a useful ideology to explain the "natural right" of white Europeans to dominate the darker peoples of the world. It must also be emphasized that the Göttingen scholars, for all of their glorification of science and their constant claim to be purely scientific in their investigation of history, did not think it necessary to present scientific proof that blacks were an inferior race. They simply treated black inferiority as self-evident.

Any evidence that black Africans were builders of civilizations was assumed to be false and had to be explained away, because it violated the fundamental axiom of black racial inferiority. For example, when German explorers came on the impressive ruins of Great Zimbabwe in 1871, at first they believed they had found King Solomon's lost mines. When that explanation proved untenable, they later attributed what they saw to other outsiders. The most obvious explanation—that these sophisticated structures had been built by the ancestors of the native peoples—was ruled out as ridiculous by the Europeans because they were convinced that black Africans were utterly incapable of such achievements.

Racial purity was a very important concept in this ideological program. The Greeks were considered to be the purest of Aryans and therefore the direct ancestors of the Germanic peoples. The progressive, creative, dynamic, and brilliant nature of the Greeks was attributed to the unadulterated quality of their Aryan blood. The ancient Egyptians, by contrast, were perceived as a mongrel race with a significant admixture of black blood. From these premises flowed the "scientific" conclusion that the Egyptians could not have contributed anything of value to Greek civilization. Again, any evidence to the contrary was summarily dismissed as impossible because it contradicted the inviolable axiom of "racial science."

Another aspect of the racial ideology of those late-eighteenth- to early-twentieth-century scholars was the rabid anti-Semitism that characterized the period. The Phoenicians, like the Jews, were a Semitic people (Hebrew and Phoenician are practically two dialects of the same language). The prevailing ideology of the racial purity of the ancient Greeks ruled out Phoenician influence as strictly as it did Egyptian influence. It took a great deal of ingenuity to explain away the undeniable fact that the Greeks adopted the Phoenician alphabet.

The University of Göttingen, Martin Bernal explains,

> in the period from 1775 to 1800, not only established many of the institutional forms of later universities, but its professors established much of the institutional framework within which later research and publication within the new professional disciplines was carried out. . . . [T]he center of the intellectual ferment was in Classical Philology, later to be given the more imposing and modern name . . . "Science of Antiquity."

This Science of Antiquity was "later transposed to Britain and America as the new discipline of 'Classics.'" The "chief unifying principle" of this new field of scholarship "was ethnicity and racism."[25]

One University of Göttingen professor of the late eighteenth century, Johann Friedrich Blumenbach, was the first to produce a scholarly work on the subject of human racial classification (*De Generis Humani Varietate Nativa*, 1775).[26] He coined the term "Caucasian" in 1795 to denote the white race, which he considered to be naturally superior to all others in beauty and intelligence. He believed other races to have devolved from the original Caucasian race of humans.[27]

Another Göttingen professor, Christoph Meiners, played a major role in the development of the new, allegedly scientific methodology of history. He insisted that historical studies should not focus on individuals but on "peoples," and he ranked various peoples on a hierarchical scale, with Germans and Celts at the top and Hottentots (a black African people) and chimpanzees on the bottom.[28]

The first major challenge to the tradition that the Greeks owed a substantial cultural debt to Egypt came from Göttingen scholar Karl Otfried Müller, whom Bernal characterizes as "ahead of his time in the intensity of his racialism and anti-Semitism."[29] These and other German scholars, including Barthold Niebuhr, Christian Gottlob Heyne, Friedrich Schlegel, and Friedrich August Wolf, were the creators of the doctrine of the Greek Miracle, which systematically sought to deny any creative role in the origins of civilization to people of color on the grounds that they simply lacked the necessary mental capacity.

THE SCIENTIFIC CONTEXT OF "RACIAL SCIENCE"

THIS AXIOM OF "racial science" was not a fringe notion, nor was it confined to historians and philologists. It was stated openly and repeatedly by the leading scientists of the nineteenth century. At the beginning of that century, the center of European science was Paris, and its primary institution was the Parisian Académie des Sciences. The Académie's leading spokesman was Georges Cuvier, the founder of comparative anatomy and the most prestigious scientist of his day. Cuvier saw blacks as "the most degraded of human races" and declared that their "form approaches that of the beast and [their] intelligence is nowhere great enough to arrive at regular government."[30] In what he considered to be a thoroughly scientific description of "the Negro race," Cuvier wrote, "The projection of the lower parts of the face, and the thick lips, evidently approximate it to the monkey tribe: the hordes of which it consists have always remained in the most complete state of barbarism."[31]

Another major nineteenth-century scientist, Charles Lyell, is frequently credited with founding the modern discipline of geology. Referring to an African people, Lyell wrote, "The brain of the Bushman . . . leads towards the brain of the Simiadae [monkeys]. This implies a connection between want of intelligence and structural assimilation. Each race of Man has its place, like the inferior animals."[32]

The most celebrated of all nineteenth-century scientists, Charles Darwin, was an outspoken opponent of slavery, but he nonetheless adhered to a hierarchical conception of human races that placed black Africans and Australian Aborigines in a position intermediate between Caucasians and chimpanzees. In his book *The Descent of Man*, he identified the dimensions of the gap separating humans from apes as the distance "between the negro or Australian and the gorilla."[33]

A disciple of Cuvier, Louis Agassiz, emigrated to the United States in the 1840s and became one of the most prominent American scientists of the day. Agassiz first encountered people of African descent when he came to America, and he was horrified by the experience. In 1846 he wrote to his mother in Europe of his extreme discomfort in the presence of black servants, whom he, too, perceived as members of a "degraded and degenerate race":

> And when they advanced that hideous hand towards my plate in order to serve me, I wished I were able to depart in order to eat a piece of bread elsewhere, rather than dine with such service. What unhappiness for the white race—to have tied their existence so closely with that of negroes in certain countries! God preserve us from such a contact![34]

Agassiz's feelings of revulsion toward blacks led him to the "scientific" conclusion that blacks and Caucasians were not merely different races but completely separate species. And here is his scientific conclusion regarding Africans and civilization:

> This compact continent of Africa exhibits a population which has been in constant intercourse with the white race, which

has enjoyed the benefit of the example of the Egyptian civilization, of the Phoenician civilization, of the Arab civilization . . . and nevertheless there has never been a regulated society of black men developed on that continent. Does not this indicate in this race a peculiar apathy, a peculiar indifference to the advantages afforded by civilized society?[35]

Another of the most prominent scientists of the nineteenth century was Paul Broca, a professor in the Parisian Faculty of Medicine. He saw it as his mission to raise the comparison of human races to a higher scientific level by means of quantification. If racial science were to become a real science, he believed, it would have to be based on numbers. Others before him had attempted to accomplish that by measuring and comparing the cranial capacity of skulls of various races of people. Broca followed the same program, but brought more sophisticated methods and a higher degree of precision to the measurements. Like his predecessors, he believed he had developed a purely objective way to demonstrate the superiority of the Caucasian race and the inferiority of black Africans. He concluded that "there is a remarkable relationship between the development of intelligence and the volume of the brain." His research, he claimed, showed that "the brain is larger," in general, "in men than in women" and "in superior races than in inferior races."[36]

Here is the bottom line, according to Broca: "A group with black skin, wooly hair and a prognathous face [one with a protruding jaw] has never been able to raise itself spontaneously to civilization."[37] As far as nineteenth-century elite science was concerned, the very idea of African civilization was an oxymoron. It just never could have happened.

When a few of Broca's contemporaries challenged his pronouncements on black inferiority, he responded by accusing them of allowing their political bias in favor of human equality to get in the way of the objective scientific truth: "The intervention of political and social considerations has not been less injurious to anthropology than the religious element."[38] But in retrospect, it is clear that it was Broca himself who was allowing his social prejudices to lead him to utterly worthless conclusions about brain size, race, and intelligence.

This is the context in which the scientific methods claimed by nineteenth-century historians of antiquity must be evaluated. To be fair, it must be acknowledged that little or no documentary evidence of ancient Egyptian or Mesopotamian science was available to these scholars. "Our whole knowledge of Egyptian written science depends on one discovery" made in the middle of the nineteenth century,[39] and the contents of the clay tablets of Mesopotamia were unknown until the twentieth century. That defense, however, is not available to those quoted in the epigraphs at the head of this chapter.

Nineteenth-century scientific discourse contained nonracist elements, as exemplified by Broca's critics, but the ideology of mainstream science was racist to the core. We should not, however, puff ourselves up with modernist pride and feelings of intellectual superiority over the men quoted above; after all, among their number were such Great Men of Science as Darwin, Lyell, and Cuvier! Clearly their profound error did not result from a lack of intelligence on their part. Instead, we should be humbled to see how easily social prejudice can insinuate itself into "science"—and ponder which of our own notions have been similarly distorted.

Furthermore, it should be noted that overt racism was not exorcised from science by procedures internal to science itself but by the transformation of the larger social context, which forced scientists to reexamine their racist premises. Although the generation of new scientific evidence eventually helped to refute racial theories, the final demise of racial science occurred only in the wake of the powerful wave of anticolonialism that swept the world after World War II. As postcolonial African, Asian, and Middle Eastern regimes began to establish their own universities, nonwhite scholars, freed for the first time from European domination, began to challenge the racist underpinnings of colonialist scholarship, which rapidly crumbled.

Finally, a people's history of science would be remiss if it failed to explicitly refute the insinuation of nineteenth-century "racial science" that the black people of sub-Saharan Africa were incapable of creating civilizations. The history of Kumbi Saleh—not to mention Gao, Jenne, Timbuktu, and any number of other African cities—is sufficient to counter that falsehood. More than a thousand years ago Kumbi Saleh was a thriving commercial city in West Africa, in the Kingdom of Ghana, with a population of fifteen to twenty thousand people. Neither London nor Paris was anywhere near that size until hundreds of years later.

It must also be emphasized that no sharp line can be drawn between the history of early Egypt and that of sub-Saharan Africa. Bernal maintained that "Egyptian civilization is clearly based on the rich Pre-dynastic cultures of Upper Egypt and Nubia, whose African origin is uncontested."[40] The prehistoric origins of Egyptian civilization came from far south along the Nile River, which is to say from the heart of the African continent. Sub-Saharan Africans constituted a significant part of the Egyptian population in the age of the pharaohs, and they frequently rose to the pinnacle of political power. Statues, wall paintings, and documents make it clear that there were black African Pharaohs—Pepy I, for example, circa 2360 B.C.E.—and that there were periods during which all of Egypt was ruled by the territories along the southern stretches of the Nile.

Herodotus, who traveled extensively in Egypt in the fifth century B.C.E., said of the Egyptian people that "they are black-skinned and have wooly hair."[41] Some scholars have argued that Herodotus's reports about Egypt are untrustworthy because he did not speak or read the Egyptian language and was therefore unable to evaluate critically the information he was given there.[42] That does not apply in this case, however, because in giving a physical description of the people in Egypt as he saw them, he was reporting what he had witnessed with his own eyes.[43]

WHAT EXACTLY WAS "GREEK SCIENCE"?

NOW THAT WE have inoculated ourselves against Hellenophilia, let us turn to a direct consideration of the science of antiquity. Although science did not originate with the Greeks, their contributions to its development were substantial. "By the end of the archaic period," M. I. Finley tells us,

> the Greeks had accumulated a very considerable body of empirical knowledge in agronomy, human anatomy and physiology, engineering, metallurgy, mineralogy, astronomy and navigation. We know next to nothing about the men who made the observations and transmitted the information, nor about the ways in which they worked, presumably because they were craftsmen who in their age-old manner learned and taught by doing, not by reading and writing. The practical consequences, however, are widely attested—in the pottery, the buildings, the sculptures, the range and variety of food products, the developments in navigation—and though much was inherited from older civilizations, much was surely new with the Greeks.[44]

When comparing Greek science to that of its predecessors, the underlying levels of material development must be taken into account. The earlier civilizations were of the Bronze Age; the Greeks of the eighth century B.C.E. had the great advantage of having entered the Iron Age. The accompanying cultural advances led to, among other things, much more thorough documentation of scientific endeavors from the sixth century B.C.E. on. The resulting imbalance in the historical record gives Greek science the appearance of having been a greater leap forward than it really was.

Greek science is not a unitary topic; it cannot be treated as a single undifferentiated historical subject. It was complex and multifaceted and evolved over more than a thousand years.

Arbitrary periodization schemes are always debatable, but for purposes of analysis it is useful to think of Greek science as divided into periods, and the most important watershed separates pre-Socratic and post-Socratic eras.

The period of pre-Socratic science occupies, roughly speaking, most of the sixth and fifth centuries B.C.E., ending, as the name implies, with the rise of Socrates's influence around the beginning of the fourth century. That turning point has often been called the "Socratic revolution" because it entailed a major shift in basic philosophical outlook from materialism to idealism. Whereas the main line of pre-Socratic thought was based on the primacy of matter, the post-Socratic era was dominated by an interpretation of nature in which mind took precedence over matter—a difference with immense implications for the development of science.

The pre-Socratic period is of particular importance to a people's history of science. Although it is usually presented in heroic-narrative form as a succession of great ideas of a few great thinkers, its real significance lay in the fact that it represented a major countertrend to the control of science by intellectual elites. As Benjamin Farrington observed, "The organized knowledge of Egypt and Babylon had been a tradition handed down from generation to generation by priestly colleges. But the scientific movement which began in the sixth century among the Greeks was entirely a lay movement."[45]

The thinkers who dominate the traditional narratives of pre-Socratic science[46] include Thales, Anaximenes, Anaximander, and Heraclitus, none of whom lived in Greece proper but in Greek colonies on the Ionian coast of Asia Minor, which is today part of Turkey. Three of them hailed from a single Ionian town, Miletus, and the fourth was from another one, Ephesus. Rather than focusing on these individuals, let us instead consider the social context that produced the materialist scientific tradition they represented. Thales, after all, was said to have "had many predecessors" on the Ionian coast.[47]

THE IONIAN SOCIAL CONTEXT

THE SOCIAL ENVIRONMENT of the Ionian Greeks was radically different from that of their Egyptian and Babylonian forebears. The older, agricultural-based civilizations were characterized by totalitarian forms of social organization in which learning, including science, came to be monopolized by conservative priestly castes that were in turn subordinate to absolute monarchs. Such a social climate nourishes traditionalism and discourages original, creative thought.

Very different forms of social organization began to develop in the Greek world, and especially on the Ionian coast, in about the eighth century B.C.E. There the economies were not based completely on agriculture; a substantial amount of mercantile activity, based on Phoenician precedents, began to develop. The increasing role of commerce led to the growth of nonagricultural social classes—merchants, manufacturers, artisans, shipbuilders, and sailors. These social classes, even in the cities, were only "a minor fraction of the population, but their very existence introduced a new dimension into the quality of the community and its structure."[48] According to Plutarch, "work was no disgrace" in the Greek world of the early sixth century B.C.E., "nor did the possession of a trade imply social inferiority."[49]

The new Greek settlements that grew up on the Ionian coast were trading centers. They exported oil, wine, weapons, pottery, jewelry, and clothing and imported grain, fish, wood, metals, and slaves. These port cities were populated by immigrants from all over the Greek world and beyond, as well as by the people indigenous to Asia Minor. They were people of diverse backgrounds who were away from their traditional settings and were exposed to a variety of "foreign" outlooks and customs. The existence of this multilingual, multiethnic population in a commercial economy during an economic boom created a situation conducive to intellectual ferment.

As the merchant and artisan classes grew in strength, new forms of government developed. First, the hereditary kings who had originally ruled the independent Ionian city-states were replaced by the rule of aristocracies of noble families. Then later, by the middle of the seventh century B.C.E., the aristocracies were overthrown by coalitions of merchants and manufacturers. Then, in the sixth century, these merchant oligarchies were replaced by the original "tyrants."

The words "tyrant" and "tyranny" have very bad connotations today, but that was not the case at first. Tyranny was a new form of government that reflected the development of class struggles between the rich merchants and the ordinary people, the plebeians. The plebeians became a potent political force. They fought for their interests by waging strikes and generally creating social turmoil. During the unrest, a prominent politician would typically step forward and claim to represent the interests of "the people." If he succeeded in winning the leadership of the plebeian masses, this politician would seize power and establish what was then called a tyranny. The original tyrants were the spiritual ancestors of the familiar populist demagogues of the modern world, such as Perón of Argentina or Nasser of Egypt. But within a generation or two, the tyrants would become what their name implies today—repressive and unpopular—and they, too, would be overthrown and replaced, in some cases by democratic republics.

On the Ionian coast, there was one city in particular that stood out as the most dynamic; that was Miletus. It had experienced an unprecedented maritime expansion. This one city had established ninety colonies all around the Black Sea and had gained a virtual monopoly of trade in that important area. Its colonization of the Black Sea area began about 650 B.C.E.; that is, only about fifty years before the first of the famous philosophers, Thales, appeared in Miletus.

Croesus, the king of Lydia, was the personification of great wealth in the ancient world. The thriving economy of Miletus enriched its merchant class to the point where King Croesus himself went to the bankers of Miletus when he needed to borrow money. But as the merchant princes grew richer in Miletus, the plebeians grew politically stronger. A tyranny was established in 604 B.C.E.; it was overthrown a few years later, after which followed two generations of political turmoil. Then a constitutional regime came to power, then a new tyranny, and finally a democratic government was established that ruled Miletus until it fell to the Persians in 546 B.C.E. These rapid shifts in government reveal a population that was politically active and difficult to suppress or intimidate. The social climate was one in which thought and speech were relatively uninhibited; it was a tumultuous "marketplace of ideas." We should think of Thales, Anaximander, Anaximenes, and Heraclitus not as isolated geniuses but as the leading representatives of a large and vigorous "people's science" movement that arose out of the class struggles of the ancient world.

The celebrated Ionian philosopher-scientists were either merchants themselves or greatly influenced by merchants. That is to say that they were not detached ivory-tower thinkers but were prominent and active citizens. Thales, for example, was reputed to have been a sagacious businessman. "According to the story," Aristotle tells us, Thales

> knew by his skill in the stars while it was yet winter that there would be a great harvest of olives in the coming year, so, having a little money, he gave deposits for the use of all the olive-presses in Chios and Miletus, which he hired at a low price because no one bid against him. When the harvest-time came ... [he] made a quantity of money.[50]

Whether this anecdote is based in fact or not, it illustrates the perceived connection between Ionian commerce and the origins of Ionian science.

The Ionian thinkers' materialistic interpretations of nature contributed something new and valuable to scientific understanding. Many of their ideas can be related directly to the economic activities in their social environment. In general, the mercantile economy of the Ionian Greeks shaped the way they tried to comprehend the workings of nature, because participating in commerce affects the way a person looks at the world and the things in it. As Karl Kautsky observed, the habit of seeing all commodities in terms of abstract price relations causes a merchant "to institute comparisons, enables him to discover the general element in the mass of particular details, the necessary element in the mass of accidentals, the recurring element which will result again and again from certain conditions."[51] That is precisely how Thales, Anaximenes, and Heraclitus approached the question of what the world is made of in putting forth their respective propositions that all matter derives from water, air, or fire.

Anaximenes attempted to provide a materialist account of natural phenomena by identifying air as the primordial element of the material world. He speculated that clouds are produced from air by a process he called "felting."[52] Felting was a word used to describe an important craft technique that involved subjecting woven materials to high pressure. It is evident that Anaximenes derived his ideas about nature from analogies drawn from the manufacturing processes of the time.

Heraclitus likewise sought a materialist understanding of nature, but he chose to recognize fire as the primordial material element. The metaphor he used to express his reasoning is revealing: "All things are an equal exchange for fire and fire for all things, as goods are for gold and gold for goods."[53] Although Heraclitus was an aristocrat—perhaps even a member of a royal family—the impact that commercial transactions and chemical reactions had on his thinking is obvious.

On the other hand, the fact that Anaximenes, Heraclitus, and other Ionian philosophers were thinkers rather than artisans limited their ability to contribute to science. Although they

> drew on the work of the craftsmen for the derivation of their ideas as to how Nature worked, they had little first-hand acquaintance with it, were not called on to improve it, and consequently they were unable to draw from it that wealth of problems and suggestions that was to create modern science in Renaissance times.[54]

PYTHAGORAS AND HIS THEOREM

Not all of the strands of pre-Socratic science were Ionian; Pythagoras and his followers were based in southern Italy. Although proponents of the Greek Miracle hail Thales as the originator of mathematics, they also credit a great deal of its early development to Pythagoras, whose name is forever linked with the proposition that in a right triangle, the square of the hypotenuse equals the sum of the squares of the other two sides.

There are two major problems with this latter attribution. First of all, "the 'Pythagorean' theorem was known [by Babylonian mathematicians] more than a thousand years before Pythagoras."[55]

Babylonian tablet showing that the Babylonian mathematician could determine the length of a square's diagonal from the length of its sides, which demonstrates knowledge of the "Pythagorean" theorem.

This mathematical idea was almost surely transmitted from the Babylonians to the Greeks, but even if the Greeks rediscovered it independently, it obviously cannot be considered primarily a Greek innovation.

More significantly, there is really no solid evidence to establish that "the semimythical founder of the Pythagorean sect" had any connection with mathematics at all. The idea of Pythagoras as mathematician first seems to have arisen toward the end of the fourth century B.C.E.[56] "The apparently ancient reports of the importance of Pythagoras and his pupils in laying the foundations of mathematics," Walter Burkert declared, "crumble on touch." The earliest writers to connect Pythagoras with mathematics were Hecataeus of Abdera and Anticlides, both of whom wrote about 300 B.C.E.—in other words, more than two centuries after Pythagoras's time. His mathematical reputation was solidified by much later writers such as Iamblichus (late third to early fourth century C.E.) and Proclus (fifth century C.E.). "Greek mathematics," Burkert concluded, "did not emerge from the revelation of a Wise Man, and not in the secret precinct of a sect founded for the purpose."[57] Claims that Pythagoras was a pioneer of medical science stand on even weaker foundations.[58]

The Pythagoreans sowed the seeds of a nonmaterialistic view of nature that would later culminate in the so-called Socratic revolution, which, from the standpoint of people's history, can be better understood as a *counter*revolution. Their heritage is the insistence that knowledge of nature is to be gained not by observation but by *a priori* arguments. "Subsequently," Farrington observed, "the Pythagorean method led to the most disastrous results. When it began to appear that nature is indifferent to Pythagorean mathematics . . . those who followed the Pythagorean tradition cast nature off and clung to mathematics. Mathematics became the master instead of the servant." Socrates's chief expositor, Plato, bears primary responsibility for solidifying this "mathematical idolatry" that "dominated European thought for centuries" and thereby exacted a heavy price from humankind.[59]

One commentator has claimed that although "these consequences of Pythagorean views are clearly reactionary they come from an age later than that of Pythagoras himself." The original Pythagoreans allegedly put forward "the first expression of *democratic* thought, that is of the rationalism of the merchant *middle* classes as against the traditionalism of the landed aristocracy," and as a result they suffered persecution.[60] If that is so, the later Pythagoreans are perhaps the earliest example of a political movement that survived by altering its ideology to placate the ruling powers.

Another pre-Socratic school of thought that arose in southern Italy carried the Pythagorean separation of reason and nature to its logical extreme. That was the Eleatic school founded by Parmenides, who propounded "a new philosophy which in the name of reason denied the reality of the whole world of the senses."[61] One of Parmenides's disciples, Zeno of Elea, devised four famous paradoxes to demonstrate how certain self-evident mathematical truths are incompatible with the material world of our sensory perceptions. Parmenides and Zeno were both partisans of the aristocratic and conservative party of their city; their attack on experimental science and their yearning for the absolute truths of pure mathematics "expresses the deep need for fixity that always recurs, usually on the losing side, in times of trouble."[62]

Zeno's best-known paradox has to do with a race between Achilles and a tortoise. The tortoise is given a head start. By the time Achilles reaches the point from which the tortoise started, the tortoise has moved ahead to point X. By the time Achilles reaches point X, the tortoise has moved ahead to point Y. By the time Achilles reaches point Y, the tortoise has moved ahead to point Z. If you repeat this procedure an infinite number of times, you find that Achilles will forever be in the wake of the tortoise. In the real world, of course, the much faster Achilles would very quickly leave the tortoise in his dust. How to reconcile this discrepancy between mathematical "truth" and the material world? Zeno and his followers, like the Pythagoreans, decided to keep the mathematics and discard the material world. The world that we know through our senses—all that we can see and touch—they declared to be only an illusion. Knowledge-seekers were therefore enjoined not to investigate nature but to despise it.

The effect of Zeno's paradoxes in the sphere of mathematics was "to banish number from geometry." A later mathematician who joined Plato's school, Eudoxus, devised a geometry "in which spatial relations can be symbolized quite independently of number and studied without reference to measurement." Plato adopted this kind of geometry as the basis of an "independent world created out of pure intelligence."[63] This was the culmination of the counterrevolution that overthrew the Ionian scientific tradition of using the senses to gain knowledge of nature through observation and experience. As Farrington explained, Plato "represents a political reaction against Ionian enlightenment, in the interest of an ideal of a slave-owning, class-divided, chauvinistic city-state which was already an anachronism."[64]

Meanwhile, what became of arithmetic, with its ordinary numbers and everyday calculations? Under the influence of Plato's disdain for anything "contaminated by practical uses," it fell into disrepute in the elite intellectual circles of the Greek world. It came to be looked down on as "a proper study for Phoenician traders, not for Greeks."[65]

Plato's social and political motives for preferring geometry to arithmetic are suggested by a Platonic dialogue written by Plutarch. One of the speakers, Florus, notes that Plato often compared Socrates to Lycurgus, the lawgiver of Sparta, and then goes on to say,

> Lycurgus is said to have banished the study of arithmetic from Sparta, as being democratic and popular in its effect, and to have introduced geometry, as being better suited to a sober oligarchy and constitutional monarchy. For arithmetic, by its employment of number, distributes things equally; geometry, by the employment of proportion, distributes things according to merit.

By way of explaining what Plato meant in his famous aphorism "God is always busy with geometry," Florus declares that geometry teaches us that social equality is unjust:

For what the many aim at is the greatest of all injustices, and
God has removed it out of the world as being unattainable;
but he protects and maintains the distribution of things
according to merit, determining it geometrically, that is in
accordance with proportion and law.[66]

Arithmetic, according to this, is the mathematics of what we
might call "affirmative action," whereas geometry upholds privilege in the name of "merit." Distributing "things . . . in accordance
with proportion" suggests that those already well off deserve a pro-
portionately large share of society's resources. If the logic of Florus's
discourse escapes you, do not despair; it is no fault of yours. But
it cogently exemplifies how Plato's scientific ideology cannot be
separated from his extreme elitism and the political philosophy
engendered by his disdain for "the many."

PLATO'S ELITISM

ONE OF THE recurring themes of a people's history of science
is the retardation that has often resulted from the domination of
knowledge by intellectual elites. The demise of Greek science
was primarily due to the structure of a slave-based society, not
to Plato's ideas. The Athens he lived in could afford the luxury
of "science" that made no contribution to productive activity. A
substantial class of slave owners living in the city while drawing
wealth from slave labor on their rural lands had ample time for
leisure activities, including abstract theorizing. Furthermore,
because "their slaves were their machines," the privileged classes had no economic motive for promoting technological
progress and in fact scorned all knowledge tainted by practical-
ity.[67] But even though Plato's ideology was a reflection of deeper social forces, it certainly played a significant role in a
two-thousand-year retardation of scientific thought—arguably
the greatest damage any scientific elite has ever inflicted on science in all of human history.

Plato was one of the most forthright elitists of all time. It is
ironic that he is often lauded as the greatest thinker produced by

Athenian democracy, because few have ever harbored a more passionate hatred of democracy. The eminent historian of science George Sarton described Plato, with good reason, as "a disgruntled old man, full of political rancor, fearing and hating the crowd." Sarton took Plato to task for praising the virtues of despotic Sparta and compared him to American right-wingers during World War II, "who carried their hatred of their own government so far that they were ready to admire the Fascists and the Nazis."[68]

Plato's elitism left its mark on science in at least two crucial ways. First of all, he developed a full-blown ideology of scientific elitism that ruled out usefulness as a goal of science and excluded from its practice people who work with their hands. Second, he gave his program a solid institutional base by founding a school, the Academy, which propagated the values of elitist science uninterruptedly for more than nine centuries. Although other important scientific institutions arose in the Greek world, the most influential of them—Aristotle's Lyceum and the Museum in Alexandria—adopted the Academy's elitist outlook and continued to follow its example in that regard.

Another key element of Plato's scientific elitism is the claim he makes in the *Republic* that "it is not the man who *makes* a thing, but the man who *uses* it, who has a true scientific knowledge about it." The political meaning and utility of this doctrine are not difficult to discern: "A slave who made things could not be allowed to be the possessor of a science superior to that of the master who used them." Plato set the history of science on an elitist path by fathering "the grotesquely unhistorical opinion later current in antiquity, that it was philosophers who invented the techniques and handed them over to slaves."[69]

The elitist arguments Plato leveled against the Sophists—teachers he perceived as rivals—deserve attention because they resulted in a historical injustice of long standing. His attacks on them were so effective that the word "sophist" has come to denote intellectuals who distort the truth by dishonest argumentation. Many of them—Hippias of Elis, Gorgias of Leontini, and Protagoras of Abdera, for examples—"made important contributions to knowledge" but they were "held up to scorn by Plato, who had an independent income, for teaching for money."[70] The aristocratic notion that anyone dependent on earning a living cannot be an impartial arbiter of knowledge unfortunately became part of the ideology of the *virtuosi* of the Scientific Revolution.[71]

THE "NOBLE LIE"

ALSO IRONIC IS Plato's public-relations image as a philosopher who personified the ideals of truthfulness and honesty as the highest virtues. In fact, Plato believed that government was only possible on the basis of a lie, and he "devoted his life to the elaboration" of that lie.[72] To uphold that falsehood, Plato urged that the books of the Ionian materialists be destroyed and "that his own fraudulent book [the *Laws*] should be imposed by the State as the one and only obligatory source of doctrine."[73] For dissenters who objected to his plans, he advocated the death penalty. This was Plato's idea of a political utopia, as he spelled it out in the *Republic*. "Who," Farrington asked, "with any sense of the human tragedy of the twenty-three centuries that separate us from Plato can read his proposals without a sense of horror?"[74]

What was the famous "noble lie" that Plato wanted to impose as the official doctrine of the state? Here is how he described it. Writing in dialogue form, he had one of his interlocutors ask him: "How then may we devise one of those needful falsehoods of which we lately spoke—just one royal lie?" He answered:

> I propose to communicate [the audacious fiction] gradually, first to the rulers, then to the soldiers, and lastly to the people. . . . Citizens, we shall say to them in our tale, you are brothers, yet God has framed you differently. Some of you have the power of command, and in the composition of these he has mingled gold, wherefore also they have the greatest honor; others he has made of silver, to be auxiliaries; others again who are to be husbandmen and craftsmen he has composed of brass and iron; and the species will generally be preserved in the children.[75]

Plato's "noble lie," then, was the ultimate ideological justification of elitism: that social hierarchies are immutable because they were created by God, and that the ruling class deserves to rule because God made its members out of superior material. The aristocrats are the Golden Men, whereas the farmers and artisans are composed of brass and iron. As part of this ideological program, Plato promoted two separate religions—a sophisticated, abstract one for the intelligentsia, and a cruder one, with the traditional anthropomorphic gods and goddesses, for the masses. To ensure the continued observance of the latter, Plato proposed sentencing disbelievers to five years in prison for a first offense and death for a second. Farrington commented, "Thus the advocacy of persecution for opinion made its first entry on the European scene."[76] Plato's successor, Aristotle, likewise understood the political usefulness of religious tradition; he called it "a myth" that was propagated "with a view to the persuasion of the multitude and to its legal and utilitarian expediency."[77]

Platonic elitism, unfortunately, is not merely a matter of ancient history; it is still drastically afflicting the human race even in the twenty-first century. The architects of American foreign policy who carried out the imperialist assaults on Afghanistan and Iraq are known to be zealous disciples of political philosopher Leo Strauss, an admirer of Plato. "The effect of Strauss's teaching is to convince his acolytes that they are the natural ruling elite," said Shadia Drury, who has written extensively on Strauss's ideas and their consequences.[78] "Leo Strauss," she continued, "was a great believer in the efficacy and usefulness of lies in politics" who "justifies his position by an appeal to Plato's concept of the noble lie." Straussian influence was all too apparent in the Bush administration's use of deception and blatant falsehoods to convince the American public of the need to go to war against Iraq. "The ancient philosophers whom Strauss most cherished believed that the unwashed masses were not fit for either truth or liberty, and that giving them these sublime treasures would be like throwing pearls before swine."[79]

ASSESSING PLATO'S CONTRIBUTION TO SCIENCE

IN NO SENSE was Plato a creative mathematician in his own right, but in his role as an intellectual impresario he made an undeniably valuable contribution to the mathematization of science. In spite of his disdain for practical work, however, subsequent Greek applications of mathematics to science were more often than not based on knowledge produced by artisans. In the third century B.C.E., for example, when Archimedes provided a mathematical basis for the science of hydrostatics, "a good deal of practical application had been going on for many years on such things as siphons, water-clocks and buoyancy devices."[80]

Mathematics aside, Plato's general influence on science has been condemned by some historians as thoroughly destructive. His and Socrates's "contempt for the physical world," Farrington charged, "was one of the main reasons for the death of science" in the Greek world. It represented a "complete revolt from physical enquiry" that "was one-sided and reactionary and had evil results." From that point forward, "mathematics, ethics, and theology become inextricably blended as *a priori* sciences independent of experience."[81]

As previously noted, Plato's idealism represented an aristocratic reaction against Ionian materialism. The Ionians sought knowledge by looking directly at nature and drawing conclusions from the evidence of their senses. Plato went in the exact opposite direction by basing the search for knowledge exclusively upon *a priori* truths, a method that paralyzes the scientific investigation of nature. As George Sarton explained,

> The Platonic point of view allured poets and metaphysicians, who fancied that it made divine knowledge possible; unfortunately, it made the more earthbound scientific knowledge impossible. The Platonic method of leading from the general to the particular, from the abstract to the concrete, is intuitive, swift, and sterile. . . . The opposite method . . . leading from known particulars to abstract notions of increasing generality, is slow but fruitful; it prepared very gradually the way for modern science.[82]

Plato's antiempirical approach to the knowledge of nature came to dominate science in the Greek world and was subsequently bequeathed to medieval Europe. Although today Plato is known as the author of a large corpus of writings including the *Republic* and the *Laws*, only one of his works, a peculiar dialogue titled the *Timaeus*, was known to European scholars before the middle of the twelfth century C.E. The *Timaeus*'s influence in early Europe "was enormous and essentially evil"; it "has remained to this day a source of obscurity and superstition" and "a monument of unwisdom and recklessness."[83]

The *Timaeus* offered a theory of the universe to justify the political ideas expressed in the *Republic*. In it Plato explicated his astral religion using quasimathematical reasoning that was mistaken by credulous scholars for valid astronomy:

> The success of Plato's astronomy, like that of his mathematics, was due to a series of misunderstandings: the philosophers believed that he had obtained his results by the aid of his mathematical genius. . . . He was speaking in riddles, and nobody dared to admit that he did not understand him for fear of being considered a poor mathematician. . . . Almost everybody was deceived, either by his own ignorance and conceit or by his subservience to fatuous authorities. The Platonic tradition is very largely a chain of prevarications.[84]

GREEK MEDICINE AND THE HIPPOCRATIC TRADITION

THE PLATONIC LEGACY has frequently led historians to assert that the Greek approach to investigating nature was purely theoretical, that empirical methods and experimentation were completely alien to Greek science. Among the many exceptions disproving that generalization, the medical sciences stand out. The famous medical school at Cos, the center of the Hippocratic tradition, is "the first scientific institution from which complete treatises have come down to us."[85] The sixty or so treatises that make up the so-called Hippocratic corpus represent an extensive body of knowledge derived from the observations, research, and experiments carefully recorded by many generations of medical practitioners. In those treatises "we find the science defended, as an observational and experimental one, against the encroachments of the philosophers who come with their ready-made views of the nature of man derived from cosmological speculation and attempt to base the practice of medicine upon them."[86]

The content of the Hippocratic corpus "is fully entitled to the name of science." Those writings reveal "a clear conception of medicine as based on observation of the behaviour of the human body in health and in disease, on experiment, and recording of results."[87] Its authors saw themselves as part of a collective endeavor that depended on the cooperative efforts of many generations of researchers and practitioners; they expressed that awareness in the famous aphorism *vita brevis est, ars longa*—"life is short, but art is long."[88]

The attribution of this scientific tradition to a single man named Hippocrates and his reputation as the "Father of Medicine" are the stuff of legend, not history. The school at Cos (a school of shared principles rather than a professional faculty of medicine) was in existence at the turn of the sixth century B.C.E., more than a century before Hippocrates was born, and the earliest parts of the collectively produced writings known as the Hippocratic corpus date from before his time. Even the famous Hippocratic Oath—the statement of ethical principles to which medical professionals continue to pledge allegiance—"is now known to have been written not by a Hippocratic doctor but by a medical follower of the ancient religious sect of Pythagoreans" in the fourth century B.C.E.[89] Furthermore, the Hippocratic corpus itself testifies to the prior antiquity of the medical sciences. The author of the treatise *On Ancient Medicine* declared that discoveries "made over a long period of time" showed that "medicine has for long possessed the qualities necessary to make a science."[90] In fact, the origins of Egyptian medicine were as far removed in time from Hippocrates as Hippocrates is from us.

The most significant rival of the school at Cos was located nearby at Cnidos. Both Cos and Cnidos were not far from Miletus; the medical practices that developed there in the sixth and fifth centuries B.C.E. must certainly be considered part of the Ionian enlightenment. As for what kind of people created the medical knowledge, the eminent historian of medicine Erwin Ackerknecht describes the "peculiar, unprotected social position of the Greek physician" as that of "a traveling craftsman" who "had to migrate from one city to another" to scrape together an adequate living. "The upper crust on which he lived was in general too thin to keep him alive permanently in one given locality." That explains why citizens of Cos appear so rarely as patients in the case studies produced by the Cos "school." Like other artisans, the Greek doctor "was trained not in school but through apprenticeship to an individual master."[91] Although these itinerant physicians were not a medical elite, they aspired to become one, as their orientation toward serving "the upper crust" suggests. Their writings therefore aimed at convincing the rich and powerful of the superiority of their kind of medicine over that of traditional healers.

In addition to the medical practitioners, another occupational group—gymnastics coaches, physical trainers, and directors of gymnasiums—made an especially noteworthy contribution to the Hippocratic tradition. It was they who

> learned to deal with fractures and dislocations, and it is no doubt largely as a result of their accumulated experience that the surgical treatises in the Hippocratic collection are on such a high level. . . . It would be impossible to stress too strongly the importance of these lines of experiment and research for the development of Hippocratic medicine.[92]

Greek medical science therefore accumulated knowledge in the course of practical, craftlike activities. That continued as its principal line of development until at least the third century B.C.E., when anatomy and physiology were brought into the realm of elite science by Herophilus and Erasistratus at the

Museum in Alexandria. Their research was based on the systematic dissection of human corpses. (The Alexandrian anatomists are also said to have performed vivisections on condemned criminals, but claims to that effect were first made several centuries after the fact.)

The medical tradition of later antiquity is dominated by the authority of Galen of Pergamon, who some time between 161 and 180 C.E. served as personal physician to the Roman emperor Marcus Aurelius. Although his individual contributions to medical knowledge were indisputably significant, the stifling weight of Galen's influence proved to be an obstacle to further progress. Ackerknecht underscored "the very paralyzing role played by his writings in the medicine of the Middle Ages and early modern times."[93]

Galen's early advances in human anatomy and physiology "were made at the cost of severe and disagreeable personal toil in dissecting dead and living animals."[94] But when he moved to Rome to take up his post at the imperial court, his elevation in social status apparently led to a transformation in the way he went about his work:

> Indicative of the beginning of the cleavage between surgery and medicine was the fact that Galen no longer practiced surgery to any great extent after coming to Rome. In that slave-holding society, manual labor was considered beneath the dignity of a gentleman, and surgery was regarded as a form of manual labor.[95]

The severance of surgery from medicine—so damaging to both—cannot be blamed on Galen alone; the social context dictated it. Medicine was treated as an elite profession, while surgery was "left to barbers, bath-keepers, hangmen, sow-gelders, and mountebanks and quacks of every description" for many hundreds of years. However, "surgery of a certain quality was kept alive until the sixteenth century, by which time the lowly barbers had themselves gathered enough strength and sophistication to enable them to contribute some of the greatest surgeons of history."[96] That it was "lowly barbers" who maintained this important field of knowledge from the second through the sixteenth centuries should not be overlooked.

As for how the paralysis of elite medicine began to be overcome in the sixteenth century, more is said in chapter 5.[97] Meanwhile, note that as heavily as the dead hand of Galen's authority weighed on medical science, it was not nearly as powerful a force of scientific retardation as the authority of the man known familiarly as "The Philosopher" to the scholars of the Middle Ages: Aristotle.

ARISTOTLE

PLATO'S LEGACY OF antiempiricism was somewhat moderated by his most important student, Aristotle, whose influence on subsequent European and Islamic scholarship far surpassed that of Plato himself. Aristotle joined Plato's Academy as a teenager and spent twenty years in his master's shadow. After Plato's death, Aristotle left the Academy and founded a rival school, the Lyceum, from which he criticized some of the more damaging aspects of Plato's scientific outlook. In his own extensive biological research, Aristotle utilized a down-to-earth approach—using the senses to investigate nature—that would have been anathema to Plato.

On the other hand, Aristotle was Plato's true successor in two significant ways. First, he took over the role of leading proponent of scientific elitism, which was reflected in the organization of the Lyceum. Farrington described how Aristotle's influence reinforced the idea of science as "the cultural preserve of an élite of citizens maintained by the labour of serfs and slaves," a science that consequently "became purely theoretical and lost its practical applications."[98]

In Aristotle's opinion, the manual workers who produced the necessities of life were not worthy of citizenship:

> For his exclusion of the producers from the citizen body he produces a striking argument. The producers, he says, are

necessary to the state but they do not form part of it, just as a field is necessary to maintain a cow but is not part of the cow. In a society where such arguments carried weight science was the preserve and privilege of the economically independent and was not felt to have a social function, but to be of value primarily as a discipline for the individual soul of one designed by nature to be a thinker.[99]

Second, Aristotle was the principal continuator of the Socratic revolution against the Ionians' materialist philosophy of nature. Although he broke with Plato's view that the ultimate reality consists of abstract mathematical relations, he championed a teleological view of nature that in the final analysis was no less idealistic in its philosophic consequences. The teleological conception derived from his biological observations. Noting that acorns invariably produce oak trees and chicken eggs invariably produce chickens, Aristotle reasoned that their development must be regulated by built-in plans. He then extended that supposition to argue that *all of nature* is guided by an inherent plan; everything that happens in nature, he believed, is aimed at reaching a predetermined goal. Does a cosmic plan imply a cosmic planner—a universal intelligence that controls everything? Aristotle insisted that his philosophy required no conscious supreme being, but his conception of nature unfolding according to an inner logic nonetheless made mental processes the central analogy for explaining natural phenomena. Mind, therefore, held priority over matter in Aristotle's philosophy of nature no less than in Plato's.

Aristotle's scientific legacy, although of mixed value, was potentially much more constructive than that of his teacher. On the negative side, his physics was based on the same kind of *a priori* method that rendered Plato's knowledge-seeking sterile. But unlike Plato, he was willing to admit the evidence of his eyes, hands, and other sense organs in the pursuit of biological and sociological knowledge. Ultimately, however, Aristotle's writings on biology and sociology "only explained to the learned in an orderly fashion what was known to every farmer, fisherman, or politician."[100]

Aristotle's fishermen sometimes misled him. His conclusion that warm water freezes more quickly than cold water was attributed to "the inhabitants of Pontus," who "when they encamp on the ice to fish (they cut a hole in the ice and then fish) pour warm water round their reeds that it may freeze the quicker, for they use the ice like lead to fix the reeds."[101]

Some of Aristotle's successors among the scientific elite of Alexandria put his empirical approach to good use, especially in the medical sciences. But as Aristotelian science traversed its historical path through Byzantium and the Islamic world and eventually back into medieval Europe, it hardened into a rigid orthodoxy that virtually paralyzed inquiry into the workings of nature. The intellectual elite of the European Middle Ages looked to Aristotle as the ultimate authority on all aspects of natural knowledge. Disputes over scientific questions were not resolved by appeal to observation or experiment, but by poring over the Philosopher's sacred texts.

Although the fault belonged not to Aristotle but to the oppressive conservatism of his successors, the scholastics of medieval Europe, Aristotelianism represented a very long detour in the history of science. It was thought to have solved all of science's general problems, but "the first task of modern science after the Renaissance was to show that for the most part these solutions were meaningless or wrong. As this process took the best part of 1,400 years it might be argued that Greek science was a hindrance rather than a help."[102]

CYNICS, STOICS, AND EPICUREANS

ARISTOTLE FOUNDED HIS Lyceum as a rival school to Plato's Academy, but it, too, was an elitist institution, open only to youth of the privileged classes. In opposition to both of them, three antielitist schools arose and sought support among the common people of Athens: the Cynics, the Stoics, and the Epicureans.

The Cynics in particular appealed to the downtrodden and oppressed. Their doctrine has been called "the philosophy of the Greek proletariate."[103]

The Cynics were spiritual forebears of the beatniks and hippies of the 1950s and '60s. Their social protest took the form of an ostentatious renunciation of prevailing social norms. Like the hippies, they were nonconformists who adopted provocatively antiestablishment lifestyles as a means of shocking the solid citizens. And also like the hippies, they attracted a lot of attention but were a passing phenomenon. Cynic teachers continued to profess their beliefs at least through the second century C.E., and the most famous Cynic, Diogenes of Sinope, remained a revered folk hero even longer. But the movement's thunder was stolen by the Stoics soon after Diogenes's death (c. 320 B.C.E.). The founder of Stoicism, Zeno of Citium, had been a pupil of Diogenes's chief disciple, Crates.

The Stoics' interest in nature was intimately connected with their social origins. The second most prominent leader of the school, Cleanthes of Assos, "was not only a proletarian but proud of it."[104] Most of the Stoic teachers were Asians— non–Greeks who were understandably repelled by the chauvinism of Plato's and Aristotle's schools, which was succinctly described by W. W. Tarn:

> Plato said that all barbarians were enemies *by nature;* it was proper to wage war upon them, even to the point of enslaving or extirpating them. Aristotle said that all barbarians were slaves *by nature*, especially those of Asia; they had not the qualities which entitled them to be free men, and it was proper to treat them as slaves.[105]

The Stoics' response appealed to an opposing view of nature, one that stressed the essential similarity of all human beings in support of the ideal of the solidarity of the human race.

But alas, the Stoics did not remain true to their precepts. They started out as militant opponents of the status quo, but as their movement grew and became established, they began to seek respectability and accommodation with the ruling powers. Ultimately, the "destiny of Stoicism, like that of Christianity later, was to become the mainstay of the type of society it had begun by attacking."[106]

The third school, the Epicureans, aspired to be a "science for the people" movement. The Epicureans sought to create "a movement to rally the courage and self-respect of the little people, of the average man,"[107] with science as its ideological weapon of choice.

The Epicureans represented late antiquity's main challenge to the Socratic revolution. The founder of the school, Epicurus, loathed Plato's use of religious superstition as a political tool. He derisively called Plato the "Golden Man," an allusion to the latter's noble lie.[108] "The immense difference between Plato and Epicuros," George Sarton pointed out, was "that the former was ready to exploit popular ignorance and credulity, while the latter did his best to eradicate them."[109]

Epicurus "is the first man known to history to have organized a movement for the liberation of mankind at large from superstition."[110] Later, in the second century B.C.E., the historian Polybius suggested that religion originated as a deliberate sham to tame the unruly masses:

> seeing that every multitude is fickle, and full of lawless desires, unreasoning anger, and violent passion, the only recourse is to keep them in check by mysterious terrors . . . . Wherefore, to my mind, the ancients were not acting without purpose or at random, when they brought in those vulgar opinions about the gods, and the belief in the punishments in Hades.[111]

This is a naïve idea; religion was not simply a ruling-class conspiracy. Nevertheless, the use of religion as a means of social control was a political reality at the end of the fourth century B.C.E. In the context of the civil strife and turmoil that accompanied the breakup of Alexander's empire following his death in 323 B.C.E., governing circles came to rely more than ever on religion as an instrument of state.

The Epicureans challenged that trend, but they were not political revolutionaries. They debunked superstition but avoided serious conflict with the authorities, allowing their school to survive for seven centuries. Epicureanism's political passivity and somewhat pessimistic outlook limited its appeal, preventing it from becoming a genuinely mass movement. Although unlike most other Athenian schools in admitting women and slaves to membership, its social base was neither plebeian nor patrician, but "people who were caught between the ruling powers and the lower orders in the collapsing city-states under the Alexandrian and Roman empires and seeking a moral sanctuary from the disorders and dangers of the times."[112]

In contrast to the Stoics, the Epicureans did not default on their principles; the consistency of their doctrine over the seven hundred years of their school's existence is remarkable. The kind of science they championed was that of the Ionian materialists Leucippus and Democritus, the originators of the atomic theory of matter, an abomination in the eyes of Plato and Aristotle.[113] The Epicureans were not practicing scientists and are not known to have made direct contributions to the knowledge of nature, but their challenge to the scientific authority of Plato and Aristotle was a significant one. Although that challenge had almost no impact in elite scientific circles for more than a millennium, the Epicurean legacy would eventually be recognized and restored to a place of honor by the "mechanical philosophers" of the seventeenth century.

ELITE SCIENCE AFTER ARISTOTLE

AFTER ARISTOTLE'S DEATH in 322 B.C.E., the Lyceum continued as the institutional focus of Greek science under the direction of his able disciple Theophrastus. But the school did not long survive Theophrastus, and the baton of elite scientific leadership then passed to the Museum at Alexandria. The Museum represented the introduction into the Greek world of state-sponsored science on a major scale. The Ptolemies, the Macedonian dynasty that ruled Egypt for almost three centuries between the breakup of

Alexander's empire and the rise of Rome (305–30 B.C.E.), spent lavishly to attract "the best brains" to the Museum, "and among the hundred or so regius professors that the establishment supported a truly remarkable number have left their names to posterity as benefactors of mankind":[114]

> The scientific work of the Museum . . . was far more specialized than any other had ever before been or was to be for another 2,000 years. It reflected the isolation of the Greek citizen to an even greater degree. The scientific world was now large enough to provide a small, appreciative, and understanding *élite* for works of astronomy and mathematics so specialized that even the average educated citizen could not read them, and at which the lower orders looked with awe mixed with suspicion.[115]

State financing was likewise provided for the Museum's sister institution, the renowned Library of Alexandria. Much of the knowledge it contained was stolen. The Ptolemies "confiscated any books found on ships unloading at Alexandria; the owners were given copies . . . and the originals went to the library." The social status of the people who staffed these twin bastions of elite science and kept them running should also be noted: "In the Greek world, white-collar work, like so many other forms of labor, was done by slaves."[116]

In contrast to the Platonic dictum that science should be useless, the Ptolemies expected practical results from the research they subsidized. Their interests, however, were focused on improving military and civil engineering practices rather than on discovering labor-saving techniques or finding ways to raise the general standard of living. For that reason, official Greek science in the era of the Museum, while not completely sterile, was confined to narrow channels by being primarily geared to serving the interests of an imperialistic ruling class. That remained generally the case into and throughout the Roman era. In spite of the noteworthy second-century C.E. contributions to astronomy and geography by Claudius Ptolemy (no relation to the previous Egyptian dynasty) and to anatomy and physiology by Galen, by then:

> science had ceased to be, or had failed to become, a real force in the life of society. Instead there had risen a conception of science as a cycle of liberal studies for a privileged minority. Science had become a relaxation, an adornment, a subject of contemplation. It had ceased to be a means of transforming the conditions of life.[117]

On the other hand, not all knowledge of nature in Alexandria was confined to the Museum and Library:

> It was in mechanics that the Hellenistic age furnished its greatest contribution to physical science. The first impetus probably came from the technical side. Greek workmanship, particularly in metals, had reached a high level before Alexander. . . . We know that a great crop of apparently new devices appeared around the third century [B.C.E.], but their origin is still obscure. They may well have come from the discovery by invaders of traditionally developed machinery of local craftsman, afterwards written up and further developed by literate Greek technicians.[118]

As for the materials sciences, the Alexandrian scholars' "unwillingness to concern themselves with anything that would dirty their hands prevented them from making any serious progress in chemistry."[119] Meanwhile, Hellenistic artisans were making significant contributions to that field; some of their chemical knowledge was recorded on papyri that were "evidently meant for the use of a skilled workman."[120] These documents were written in the third century C.E. but describe alchemical activities of as much as two centuries earlier.

Collectively, "the Alexandrian chemists showed an astonishing ingenuity in the invention of stills, furnaces, heating baths, beakers, filters, and other pieces of chemical equipment that find their counterparts in use today." It is noteworthy that "names of women appear among these alchemists." One in particular, Mary the Jewess, "is said to have invented much apparatus"; her name is immortalized in the *bain-marie*. "The Alexandrians who founded alchemy and influenced the thought of all who studied chemical changes for fifteen hundred years" certainly deserve a prominent place in the annals of science.[121] The line of continuity connecting Alexandrian alchemists to modern science passed through early modern Europe, where alchemy "was indistinguishable from what we would today call metallurgy, chemistry, or the science of matter."[122]

Before leaving this subject, a word about its "dark side"—the mystical aspect—is in order. Alchemy has, throughout the ages, been closely linked with mysticism, an association that was especially strong in Hellenistic Alexandria. The seemingly miraculous transformations that the artisans were able to produce in their fiery furnaces had such a powerful effect on the imagination of Neoplatonic philosophers that it induced the latter to spin fanciful metaphysical systems from alchemical metaphors. As a result, "to the already confused terminology of alchemy was added a still greater mass of philosophical speculation, using chemical terms, but with almost no chemical content." The mystical philosophers scorned the practical alchemists as mere "puffers" (an allusion to their use of bellows to stoke their fires). However, it was the practical alchemists "who preserved and advanced alchemy as a science until it became chemistry, while the others, lost in a cloud of obscure nomenclature and speculation, contributed nothing further to chemistry."[123]

ROMAN SCIENCE?

"ROMAN SCIENCE" HAS generally been considered an oxymoron by historians, but the Romans' impressive technological achievements demonstrate a great deal of knowledge of natural processes. Aside from a few famous architects or engineers, such as Vitruvius and Frontinus, most of the people who possessed that knowledge represented socially subordinate classes, and their contributions went unrecorded.

The major documented work of science in ancient Rome was Pliny the Elder's *Natural History*. Historians of science routinely consult Pliny's massive compendium to determine what was known in the first century C.E. about any given scientific subject. Pliny, Diodorus Siculus, and other Roman encyclopedists, however, depended upon information provided to them by working people. Their works "were hardly more than discursive catalogs," J. D. Bernal observed, "of the common observations of smiths, cooks, farmers, fishermen, and doctors."[124] For the most part, "these writers merely copied down what they read or were told, often without any real understanding of the work they were describing."[125] But the problem with the encyclopedic tradition was not merely its failure to *create* scientific knowledge; it "tended to *atrophy* scientific knowledge and pass it on as unchallengeable fact with no methodology at all."[126]

The Greco-Roman savants Ptolemy and Galen represented the high-water mark of elite science in late classical antiquity; from then on it declined. Science "had become an almost exclusively upper-class preserve and was consequently abstract and literary, for ingrained intellectual snobbery had barred the learned from access to the enormous wealth of practical knowledge that was locked in the traditions of almost illiterate craftsmen."[127]

When victims of Roman imperialism rose against Rome and destroyed it in the fifth century C.E., literate culture, including science, virtually disappeared in the Western world. Ancient technology, however, "in contrast to the sciences, lasted far better and lost less. Indeed, except where they depended on scale, like the making of roads and aqueducts, they were transmitted unchanged in essentials."[128] The oral traditions of illiterate artisans therefore proved more durable than the scholars' books, and the continuity of science owes more to the former than to the latter. As for the "barbarians" who conquered the western empire, it seems that their agricultural technology was superior to that of the Romans they displaced.[129]

Learned science did not cease to exist in the eastern, Greek-speaking portion of the Roman Empire, but it stagnated under the weight of Platonic and Aristotelian orthodoxy. The scholars of Byzantium preserved the traditions of Greek science until the explosive rise of Islam in the seventh century provided fertile fields for their renewal.

SCIENCE IN THE ISLAMIC WORLD

A COROLLARY OF the Greek Miracle doctrine is that nothing of importance happened in the history of science during the so-called "Dark" or "Middle" Ages—Eurocentric terms encompassing the sixth or seventh through the eleventh or twelfth centuries C.E. The Greek legacy, the traditional story goes, was preserved by non-Aryans in the Islamic world, who merely acted as its custodians until it could be transmitted back to Aryan Europe. There, once again, Greek science could take root and grow among people who had sufficient genius to appreciate it.

This is a thoroughly misleading picture. Arabic-Islamic scholarship was not simply a passive reflection of previous Greek triumphs; it also received major inputs from Persia, India, and China and was itself a source of original contributions to scientific culture. That can be seen most clearly (but not only) in the field of mathematics. Whereas Greek mathematicians kept almost exclusively to geometry and scorned arithmetic as a tool fit only for lowly practical pursuits, Muslim mathematicians adopted the base-ten, position-value number system from India and utilized it to further mathematical knowledge in ways unimaginable to their Greek predecessors. The evidence is embedded in our language: the words "algorithm" (which meant "arithmetic" in early modern Europe) and "algebra" are of Arabic origin. The Muslims' contribution to the development of trigonometry is preserved in the words "sine" and "cosine," which also derive from Arabic.

It is true that elite science in the Islamic world was built on Greek foundations; it began as an effort to translate the classic works of Greek science into Arabic. The extension of the Islamic empire to Egypt, Syria, and other Hellenized territories provided direct access to the principal texts of Greek learning. The translation movement initiated under the Umayyads in the seventh century accelerated when the Abbasid dynasty came to power in the middle of the eighth century. In 832 the Abbasid rulers founded a major research center in Baghdad, the Bayt al-Hikma, to which they recruited the leading scholars from throughout the Muslim world, and set them to work translating Greek medical, mathematical, and astronomical works.

The Abbasid dynasty came to an end in 1258 with the capture of Baghdad by non-Muslim Mongols, but the patronization of Arabic-Islamic science continued apace. The Mongol leader Hulagu established an observatory staffed by Muslim scholars at Maragha in Iran, "an event that marked the beginning of one of the longer-lasting and important episodes in the history of Arabic science."[130] The patrons' motive was the same as that of the Ptolemies in Alexandria many centuries earlier: they expected a payoff in practical applications that would enhance their statecraft.

But Muslim scholars were not simply translators and copyists; they added extensive critical commentaries, sometimes based on their own original scientific investigations, to the corpus of Greek science. When scholarship in Europe began to revive in the twelfth century, the works of Aristotle and Galen recovered from Arabic sources had been refracted through the interpretive lenses of Ibn Rushd (Averroës), Ibn Sina (Avicenna), al-Razi (Rhazes), and many others. The new learning, however, rapidly hardened into an orthodoxy that hindered further acquisition of knowledge of nature. Like Aristotle and Galen, some of the Muslim savants themselves were placed on pedestals and transformed into sacrosanct authorities by European scholars. Medical school professors and elite doctors of early modern Europe, for example, treated the works of Avicenna and Rhazes as virtually beyond challenge.

The transmission of Islamic science to the West is generally portrayed as the pacific work of scholarly translators, but it was also in part an act of violent expropriation, a corollary of the warfare that resulted in the destruction of Muslim power in Spain. "Although the lure of ancient philosophy was not the principal motive behind the West's Crusade in al-Anadalus—crusading hysteria and an appetite for booty were more effective inducements—the acquisition of Arab learning was one of the most important results of the *reconquista*."[131]

This brief outline of Islamic science serves to refute the Hellenophilic claim that it served a merely custodial function, but it is only of indirect relevance to a people's history of science. Most of the documented science of the Muslim world was the work of literate intellectual elites in the service of powerful ruling classes, and the record is dominated by a handful of Great Thinkers. But Muslim medical knowledge (for one example) did not begin with Avicenna and Rhazes or with the translation of Galen into Arabic:

> The pre-Islamic Near and Middle East possessed a popular medicine akin to that of the Mediterranean. . . . Cupping, cautery and leeches were employed for blood-letting; wounds were disinfected with alkali-rich saltwort, and ashes were applied to stanch bleeding. . . . Practical medicine was everyone's business, but those who, like bleeders and cuppers, possessed particular skills were paid for their services. . . . Only in the early ninth century did Arab-Islamic learned medicine take shape.[132]

Alchemy—another word of Arabic origin, and the root of "chemistry"[133]—was primarily a science of anonymous artisans. Among the Greeks it "had lived an underground existence because its practitioners—the fullers, the dyers, the glassmakers, the potters, the compounders of drugs—were outlawed from society."[134] In the Islamic world, however, much of their knowledge was recorded, beginning with eighth-century writings attributed to an alchemist named Jabir ibn Haiyan.[135]

Whether Jabir was a real person or the collective pseudonym of numerous practitioners cannot be known for sure.[136] In either case, the more than two thousand books attributed to this "father of Arabic chemistry" represented the collaboration of many alchemists over many centuries.[137] In them appeared the theory that the various metals differ according to the proportions of mercury and sulfur they contain—a hypothesis of great influence in the history of materials science.[138] They also document the knowledge of chemical reactions that, for example, allowed the alchemists to isolate arsenic and antimony from their sulfides and to prepare such substances as lead carbonate.

The knowledge embodied in Arabic-Islamic chemistry was further documented by the Persian physician Rhazes and other elite scholars. Rhazes's best-known book, the *Kitab Sirr al Asrar,* or "Book of Secret of Secrets," is a collection of chemical recipes that could have originated only in the workshops of a large number of artisans. The Islamic scholars' desire to tap the knowledge of the crafts foreshadows the Baconian program that was later to become a central element of the Scientific Revolution.

The rise of chemistry cannot be understood apart from its practical applications. The science developed in conjunction with "the first full-scale production in localized chemical industries in Islamic countries of such commodities as soda, alum, copperas (iron sulfate), nitre, and other salts which could be exported and used, particularly in the textile industry, all over the world."[139] The alchemists "discovered sal ammoniac, they prepared caustic alkalies." (The word *alkali* is a variant transliteration of *al-qili,* the Arabic term for sodium carbonate.) Furthermore,

> they recognized the properties of animal substances and their importance to chemistry, and they introduced on a broad scale the method of destructive distillation of these substances as a means of analyzing them into their "ultimate components." Their classification of minerals became the basis for most of the systems used later in the West. Chemistry owes the Arabic alchemists far more than has usually been recognized, and their contribution to the development of the science was a major one.[140]

"It was in chemistry," J. D. Bernal concluded, "that the Islamic doctors, perfumers, and metallurgists made their greatest contribution to the general advance of science. Their success in this field was largely due to their having escaped, to a considerable degree, from the class prejudices which kept the Greeks away from the manual arts."[141]

SCIENCE IN TRADITIONAL CHINA

MUSLIM ALCHEMISTS BUILT not only on the knowledge of their Hellenistic predecessors but also on inputs from China. More generally, if the creative role of technology in the history of science is properly understood, then it becomes obvious that "the river of Chinese science flowed into the sea of modern science."[142] The importance of that river cannot be overstated. China was the source of many very important technical innovations that eventually found their way to Europe and stimulated the Scientific Revolution:

> Not only the three which Lord Bacon listed (printing, gunpowder and the magnetic compass) but a hundred others— mechanical clockwork, the casting of iron, stirrups and efficient horse-harness, the Cardan suspension and the Pascal triangle, segmental-arch bridges and pound-locks on canals, the stern-post rudder, fore-and-aft sailing, quantitative cartography—all had their effects, sometimes earth-shaking effects, upon a Europe more socially unstable.[143]

More will be said about how the relative social instability of Europe allowed Chinese technological innovations to revolutionize science there but not in China (nor in the Islamic world, through which they passed). For now, let us simply note that the development of science in traditional China—from earliest times until the early twentieth century—was severely constrained by "a bureaucratic feudal class who monopolized learning and also largely sterilized it."[144] Mandarins—the intellectual elite that dominated the imperial bureaucracy—had little

A PEOPLE'S HISTORY OF SCIENCE

interest in promoting the trades or crafts and no interest at all in enhancing the social position of merchants: "The scholar-gentry systematically suppressed the occasional sprouts of mercantile capital."[145] In that context, technological advances were carefully controlled to *make sure* they had no revolutionary social consequences.

Accordingly, "certain sciences were orthodox from the point of view of the scholar-gentry, and others not." Among the orthodox sciences were astronomy, mathematics, and, "up to a point," physics. The attitude toward medicine was schizoid; prestige was conferred on Confucian physicians, but "on the other hand its necessary association with pharmacy connected it with the Taoists, alchemists, and herbalists." Alchemy "was distinctly unorthodox," as were other activities that required working with hands.[146]

But center stage in a people's history of science are the Chinese artisans whose knowledge of nature was implicit in the momentous technological advances mentioned earlier. As Joseph Needham, the pioneer Western historian of Chinese science, declared, "the Chinese civilisation had been much more effective than the European in finding out about Nature and using natural knowledge for the benefit of mankind for fourteen centuries or so before the scientific revolution."[147] It was stated above that the Hellenistic age's greatest contribution to physical science was in mechanics, but "the world owes far more to the relatively silent craftsmen of ancient and medieval China than to the Alexandrian mechanics, articulate theoreticians though they were."[148]

"We are," Needham explained,

not here dealing with philosophers, princes, astronomers or mathematicians, the educated part of the Chinese population, but with those concerned with the obscurer expanses of the trades and husbandries. . . . We can no longer leave out of account the mass of the workers and the conditions under which they laboured. They were the human material without which the planners of irrigation works or bridges, or vehicle workshops, or even the designers of astronomical apparatus, could have done nothing, and not seldom it was from them that ingenious inventors and capable engineers rose up to leave particular names in history.[149]

Chinese artisans were, to a greater degree than in other parts of the ancient world, under direct control of the state bureaucracy:

> partly because in nearly all dynasties there were elaborate imperial workshops and arsenals, and partly because during certain periods at least those trades which possessed the most advanced techniques were "nationalized," as in the Salt and Iron Authority under the Former Han. . . . At the same time there is no possible doubt that throughout the ages there was always a large realm of handicraft production independently undertaken by and for the common people.[150]

Needham particularly challenged the notion that Chinese achievements were *merely technical* and therefore somehow unworthy of being deemed scientific. "On the contrary," he stated, "there was a large body of naturalistic theory in ancient and medieval China, there was systematical recorded experimentation, and there was a great deal of measurement often quite surprising in its accuracy."[151]

The ninth-century discovery of gunpowder, he pointed out, "arose in the course of a systematic exploration of the chemical and pharmaceutical properties of a great variety of substances" by Taoist alchemists searching for longevity elixirs. The recognition that the magnetic compass needle does not align precisely with the earth's astronomical poles "would never have occurred unless the geomancers had been attending most carefully to the positions of their needles." The triumphs of Chinese ceramics "could never have been achieved without some fairly accurate form of temperature measurement and control, and the reproduction at will of oxidizing and reducing conditions inside the kilns."[152] Further examples abound:

The long succession of pharmaceutical experiments on animals carried out by the alchemists from Ko Hung to Chhen Chih-Hsü, or the many trials made by the acoustics experts on the resonance phenomena of bells and strings, or the systematic strength-of-material tests which internal evidence shows must have been undertaken before the long beam bridges across the Fukienese estuaries could have been constructed. Is it possible to believe that apparatus so complex as that of the water-wheel linkwork escapement clocks, or indeed much of the textile machinery, could ever have been devised without long periods of workshop experimentation?[153]

As in other parts of the world, the creation of these historically important innovations was not adequately documented because those responsible for them were most often illiterate and could record their activities only with the help of scribes:

> The fact that relatively few of these technical details have come down to us springs from social factors which prevented the publication of the records which the higher artisans most certainly kept, though we do get instances of such records from time to time: for example there was the *Mu Ching* (The Timberwork Manual) which was the basis of the *Ying Tsao Fa Shih*, the great classical work on architecture of 1102 [C.E.]. The *Mu Ching* was the work of a famous pagoda architect, Yü Hao, but it must have been dictated by him, because although he certainly could not read or write he was able to pass on his information. Another example is the famous *Fukien Shipbuilders' Manual*, a rather rare manuscript which shows that the artisans had friends who could write, and who could use the technical terms, and who wrote down in books what the artisans were able to tell them.[154]

The pioneering artisans of China, however, are not quite as anonymous as their counterparts elsewhere: "No classical literature in any civilization paid more attention to the recording and honouring of ancient inventors and innovators than that of the Chinese," to the point of virtual deification. Many of the attributions are purely mythical, but some of those so honored "were undoubtedly historical personages."[155] Their names are repeated here with the caveat that the crediting of great discoveries to individuals, no matter what their social status, almost always represents an injustice to many predecessors and collaborators of those so named.

The "historical personages" honored for their creative deeds by Chinese chroniclers are drawn from the entire social spectrum, from royalty to slaves, but "the greatest group of inventors is represented by commoners, master-craftsmen, artisans who were neither officials, even minor ones, nor of the semi-servile classes." The invention of movable-type printing is traditionally credited to a commoner ("a man in hempen cloth," which is to say, not dressed in silk) named Pi Shêng in about 1045 C.E. A minor military officer and swordsmith, Chhiwu Huai-Wên, "was one of the earliest protagonists, if not the inventor, of the co-fusion process of steelmaking." A third example is Yü Hao, the illiterate master architect mentioned earlier:

> Yü Hao was a man of the tenth century [C.E.]. but we may find his like in every dynasty. The second century saw the life of Ting Huan, renowned for his pioneer development of the Cardan suspension, the seventh century was the time of Li Chhun, the constructor of segmental arch bridges . . . and the twelfth century the time of the greatest naval architect in Chinese history, Kao Hsüan, who specialized in the making of warships with multiple paddle-wheels.[156]

Sometimes the chroniclers mention an inventor without recording his surname for posterity,

> an omission which makes one wonder whether such men were not living on the borders of one of the semi-servile groups

where surnames were not customary; for example there was the old craftsman (*lao kung*) who made astronomical apparatus in the first century [B.C.E.], or again that "artisan from Hai-chow" who presented to the Empress in 692 [C.E.] what was in all probability a complicated anaphoric clock.[157]

There are even a few cases of individuals "who came down in history as brilliant scientific or technical men and yet whose social standing in their own time was very low indeed." One such was Hsintu Fang, a sixth-century C.E. man of semiservile rank who "nevertheless left behind him a very high reputation in Chinese scientific history." Another was Kêng Hsün, also of the sixth century, who was a slave. "Kêng Hsün built an armillary sphere or celestial globe rotated continuously by water-power. The Emperor rewarded him for this by making him a government slave and attaching him to the Bureau of Astronomy and Calendar."[158]

The career of the engineer Ma Chün illustrates the inhibitory effect the scholarly elite had on science and technology. Ma Chün, Needham told us, "was a man of outstanding ingenuity." Among many accomplishments, he "improved the drawloom" and "invented the square-pallet chain-pump used so widely throughout the Chinese culture-area afterwards." In spite of his obviously keen intellect, he had not been educated in the minutiae of Confucian scholasticism. Therefore,

> Ma Chün was quite incapable of arguing with the sophisticated scholars nursed in the classical literary traditions, and in spite of all the efforts of his admirers, could never attain any position of importance in the service of the State, or even the means to prove by practical test the value of the inventions which he made.[159]

It was exceptional "to find an important engineer who attained high office in the Ministry of Works, at any rate before the Ming" (i.e., before 1368 C.E.). "This was probably because the real work was always done by illiterate or semi-illiterate artisans and master-craftsmen who could never rise across that sharp gap which separated them from the 'white-collar' literati in the offices of the Ministry above."[160]

Although artisans were the primary producers of natural knowledge in ancient and medieval China, its transmission was largely the work of merchants. The principles of Chinese alchemy, for example, found their way to the Islamic world through the medium of the *hu* merchants during the T'ang dynasty of the seventh though ninth centuries C.E. The *hu* merchants were Persian and Arabic traders who traversed the famous Silk Road between their native lands and China. "Since in T'ang times," Needham explained,

> foreign people and things were all the rage, there was hardly a city in China unfamiliar with the *hu* merchants. . . . It was said of Ch'ang-an that if one stayed there long enough one would meet representatives of every country in the known world. Not only Parthians, Medes, Elamites, and dwellers in Mesopotamia were there, but one could rub shoulders too with Koreans, Japanese, Vietnamese, Tibetans, Indians, Burmese, and Singhalese—and all had something to contribute on the nature of the world and the wonders thereof.[161]

To the T'ang Chinese, the "known world" apparently did not include the barbarous lands of Europe, but eventually, several centuries later, the alchemical knowledge carried home by the *hu* merchants made its way to the West by similar means.

But merchants were surely not the only carriers of Chinese science and technology to the West; exported slaves may also have played a significant role. Needham referred to the near-simultaneous appearance of numerous Chinese innovations in Europe as "transmission clusters," and suggests that fourteenth- and fifteenth-century clusters "had some connection with a slave trade which brought thousands of Tartar (Mongolian) domestic servants to Italy in medieval times and which reached its height in the first half of the fifteenth century. They may have brought all kinds of curious know-how with them."[162]

CHINESE ARTISANS' CONTRIBUTIONS
TO MODERN SCIENCE

AS PREVIOUSLY STATED, a large proportion of the technological advances that stimulated the Scientific Revolution in sixteenth- and seventeenth-century Europe were the creations of ancient and medieval Chinese artisans. Robert Temple, drawing on Needham's research, estimates that "more than half of the basic inventions and discoveries upon which the 'modern world' rests come from China." Many of those accomplishments—and the scientific knowledge they imply—are frequently attributed to Westerners, but

> Johann Gutenberg did *not* invent movable type. It was invented in China. William Harvey did *not* discover the circulation of the blood in the body. It was discovered—or rather, always assumed—in China. Isaac Newton was *not* the first to discover his First Law of Motion. It was discovered in China.[163]

The invention of movable type in the eleventh century by the "obscure commoner" Pi Shêng[164] certainly must be considered a watershed event in the history of science. In sixteenth-century Europe, as J. D. Bernal explained,

> printing was to be the medium for great technical and scientific changes by its setting out at large, for all to read and see, descriptions of the world of Nature, particularly of its newly discovered regions, and also, for the first time, of the processes of the arts and trades. Hitherto the techniques of the craftsmen had been traditional and never written down. They were passed on from master to apprentice by direct experience. Printed books made it first possible and then necessary for craftsmen to be literate. Their descriptions of technical processes, and even more their illustrations, helped to bring about for the first time close relations between the trades, the arts, and the learned professions.[165]

Although the early Chinese type fonts were made of wood rather than metal, it would be a mistake to think that they were inferior in their ability to generate large numbers of copies. In fact, even long before Pi Shêng's invention, Chinese printers had established "a woodblock printing industry which in the quantities produced rivaled the most modern efforts of our own times":

> The Confucian classics were printed in 953. Filling 130 volumes, they were the world's first official printed publications, sold to the public by the Chinese National Academy. By this time, printing had come of age. Vast quantities of certain works continued to be issued, ranging into many millions of copies. Of one Buddhist collection of the tenth century, over 400,000 copies still survive. So we can imagine what the initial print run must have been![166]

Exactly how block printing reached Europe in the fourteenth century is unclear, but "the circumstantial evidence is strong enough" to be confident of its Chinese origins.[167]

The few examples given next—grouped under the headings *Biology, Mechanics, Geology, Metallurgy, Chemistry, Agriculture,* and *Navigation*—are but a small sample of the knowledge of nature produced by Chinese artisans, but they provide a good indication of its breadth and its antiquity. Needham and Temple were not always able to specify the precise path by which any given scientific idea or technique traveled from China to Europe, but they presented cogent arguments in support of the assumption that such transmission did indeed occur.

BIOLOGY

The early Chinese printers and engravers used empirical methods to gain botanical knowledge relevant to their crafts:

> The blocks used by the Chinese for their printing tended to be made of fruit woods. Coniferous woods were found

unsuitable, because they were impregnated with resin which affected the evenness of the ink coating. For cutting delicate lines and illustrations, a favorite material was the extremely hard wood of the Chinese honey locust tree. For regular text, the soft and easily worked boxwood was often used. But the best all-round wood for block printing was pear: this has a smooth and even texture with a medium hardness.[168]

The subtlety of early Chinese observations of nature produced a useful correlation of mineralogical and botanical knowledge known as geobotanical prospecting. The presence of certain plants in a locality, it was discovered, could help to locate underlying deposits of zinc, selenium, nickel, or copper.[169]

The extent of biological knowledge in ancient China is also indicated by the rise of the silk industry some thirty-five hundred years ago, which could not have occurred in the absence of detailed knowledge of the biology of the silkworm. Further evidence is to be found in the practice of biological pest control. In the third century C.E., yellow citrus killer-ants were systematically used to protect mandarin orange trees from other insect predators.[170]

Mechanics

The silk industry also created a need for machinery to handle the long silk fibers, which stimulated mechanical thinking and prompted the development of the science of mechanics. In the latter part of the thirteenth century C.E., spinning wheels and other textile machines like those that had existed for two centuries in China suddenly appeared in Italian cities such as Lucca; Needham concluded that an anonymous Italian merchant must have "brought back the designs in his saddle-bags."[171]

The mechanical clock and the suspension bridge were inventions that also had major implications for the science of mechanics. Mechanical clocks date from the eighth century in China, but appeared in Europe only in the early fourteenth century. The stimulus then "seems to have been some garbled accounts of Chinese mechanical clocks which came to the West by way of traders."[172] The key invention was the escapement, the mechanism that precisely regulates the rate at which clockwork wheels turn. Because the first escapements were developed for application to waterwheels, "the mechanical clock may thus be said to owe its existence largely to the art of Chinese millwrights."[173]

As for the suspension bridge,

> In the case of this invention, we can trace its transmission fairly well. The suspension bridges of Kweichow ... came to the attention of the Jesuits and other Westerners who visited China in the seventeenth century. In 1655, Martin Martini described an iron-chain bridge over a river in Kweichow, which was incorporated in Blaeu's great *New Chinese Atlas* of that date.... Martini's remarks ... brought wide attention to suspension bridges in Europe.[174]

GEOLOGY

Geological knowledge was created and expanded by early Chinese pursuit of subterranean resources. Drilling for brine, an important source of salt, Chinese workmen discovered reservoirs of natural gas (methane) and petroleum. "It is probably a conservative estimate to say that the Chinese were burning natural gas for fuel and light by the fourth century [B.C.E.]," Temple said. "Bamboo pipelines did indeed carry both brine and natural gas for many miles, sometimes passing under roads and sometimes going overhead on trestles."[175] By the first century B.C.E., Chinese drillers had sunk boreholes as deep as 4,800 feet into the earth; depths of 3,000 feet were routine.

Dutch travelers apparently became aware of Chinese deep drilling in the seventeenth century, "but the first full description of the Chinese system sent to Europe was in the form of letters written in 1828 by a French missionary named Imbert." Soon thereafter a French engineer named Jobard put the Chinese method into practice, and within a decade and a half,

"Chinese drilling techniques had become properly established in Europe." It is possible that the path of transmission from China to America bypassed Europe:

> In 1859, a well exclusively for oil was drilled at Oil Creek, Pennsylvania, by Colonel E. L. Drake, using the Chinese cable method. . . . Drake and similar oil drillers in America may have obtained knowledge of the system not from France but from the hordes of Chinese indentured laborers used to build the railways in nineteenth-century America.

However the knowledge was transmitted, the oil drilling method in America "was exactly the same as the Chinese technique," so it is evident that "Western deep drilling was essentially an importation from China."[176]

Porcelain provides a particularly cogent example of how awareness of a Chinese technology stimulated European science—in this case, geological theory. By the third century C.E. at the latest, Chinese potters had invented a form of vitrified pottery that was far superior to ordinary earthenware. A thousand years later, Chinese porcelain became a prized commodity in Europe, but European merchants were unable to learn the closely kept secret of its manufacture. "The countless experiments" that European artisans

> carried on with various earths and solid substances in furnaces eventually had the most unpredictable result. Scientists and craftsmen began to notice that upon cooling down again, molten minerals could crystallize. Until this began to be observed, Western scientists had been convinced that crystals could only be formed from liquids. About the middle of the eighteenth century in Europe, the idea began to gain ground that perhaps the earth's rocks could have been formed from the cooling of molten masses of lava.[177]

In 1776 James Keir theorized that the "property in glass to crystallize" suggested "the great native crystals of *basaltes* . . .

have been produced by the crystallization of a vitreous *lava*, rendered fluid by the fire of volcanoes."[178] Further experimentation supported Keir's speculation. "And so," Temple concluded, "one of the great scientific advances in the Western world took place as a direct consequence of the attempts by Europeans to find the secret of porcelain manufacture."[179]

METALLURGY

Chinese methods for making steel from cast iron that date from the second century B.C.E.

> eventually led to the invention of the Bessemer steel process in the West in 1856. Henry Bessemer's work had been anticipated in 1852 by William Kelly, from a small town near Eddyville, Kentucky. Kelly had brought four Chinese steel experts to Kentucky in 1845, from whom he had learned the principles of steel production used in China for over two thousand years previously.[180]

CHEMISTRY

The most historically important Chinese chemical discovery was gunpowder. In about 850 C.E., the systematic investigations of Taoist alchemists led them to combine saltpeter (potassium nitrate), sulfur, and the carbon of charcoal. The volatile nature of the mixture prompted further experiments that resulted in the invention and then the mass production of weapons that transformed the art of war, first in China and later throughout the world.

Gunpowder's principal ingredient, saltpeter, "is not something which is just sitting there waiting to be used; it has to be recognized for what it is, differentiated from all other similar-looking chemical salts, and purified."

> How is it possible to know when one has real saltpeter, which looks more or less like any number of other chemicals? The

potassium flame test is crucial for detecting saltpeter, because it burns with a violet or purple flame. This was being used to test for saltpeter in China by at least the third century [C.E.]. . . . Later, an even more important test was developed, which is described by Sheng Hsüan Tzu in his book of 1150, *Illustrated Manual on the Subduing of Mercury:* "If you heat a piece of white quartz and then put a drop of the saltpetre on it, it will sink in."[181]

A chemical substance of biological origin, lacquer—"the most ancient industrial plastic known to man"—was discovered in China more than three thousand years ago.[182] "As early as the second century [B.C.E.], the Chinese had made important chemical discoveries about lacquer."[183] Artisans learned that they could prevent their lacquer from evaporating and hardening by throwing crabs into it. They had made the empirical discovery that something in crustacean tissue inhibits the solidification of lacquer.

Also by the second century B.C.E., the Chinese "were isolating sex and pituitary hormones from human urine and using them for medicinal purposes." This was not a small-scale laboratory process; it required evaporating hundreds of gallons of urine to obtain a few ounces of the desired precipitate.[184]

The invention of matches by "impregnating little sticks of pine wood with sulfur and storing them ready for use" is another impressive example of early Chinese chemical knowledge.[185] The inventors were "a group of anonymous Chinese women of the sixth century [C.E.]" whose creativity was necessitated by the hardships of a military siege.[186]

AGRICULTURE

The Industrial Revolution in Europe was predicated on a dramatic increase in agricultural productivity that historians have traditionally attributed to the scientific spirit of "improving landlords" who applied experimental methods to growing crops and livestock. That great leap forward, however, came about only because of the importation of Chinese ideas and inventions. The growing of crops in rows, intensive hoeing of weeds, the "modern" seed drill, the iron plow, the moldboard to turn the plowed soil, and efficient harnesses were all imported from China. Before the arrival from China of the trace harness and collar harness, Westerners choked their horses with straps round their throats.[187]

"There was no single more important element in the European agricultural revolution than Chinese plows with moldboards," originally introduced into Holland by Dutch sailors in the seventeenth century:

> And because the Dutch were hired by the English to drain the East Anglian fens and Somerset moors at that time, they brought with them their Chinese plows, which came to be called "Rotherham ploughs." Thus, the Dutch and the English were the first to enjoy efficient plows in Europe. . . . From England it spread to Scotland, and from Holland it spread to America and France.[188]

Seed drills were another crucial innovation. Prior to their adoption in the sixteenth century, standard procedure for European cultivators was to broadcast seed by hand. Temple estimates that "probably over half of Europe's seed was wasted every year before the Chinese idea of the seed drill came to the attention of Europeans."[189]

NAVIGATION

The navigational sciences and their place in the general history of science are the subject of chapter 4, but first the Chinese contributions should be identified: "Without the importation from China of nautical and navigational improvements such as ships' rudders, the compass and multiple masts, the great European Voyages of Discovery could never have been undertaken."[190]

The principle of the magnetic compass was discovered in China no later than the fourth century B.C.E. Documentary evidence of its use by Chinese sailors can be dated to 1117 C.E., whereas comparable evidence of European use does not appear until seventy years later.[191] Rudders were in use on Chinese ships by the first century C.E., but "Western ships had to make do with steering oars" until the rudder was adopted from China in the twelfth century.[192] The earliest use in China of leeboards, which help a tacking ship make headway by resisting leeward drift, was described in a book of 759 C.E.; they were not known in Europe until about 1570, when they were adopted by Dutch and Portuguese seamen engaged in the China trade.[193]

Another crucial innovation in ship design—dating from second-century C.E. China—was the separation of the hull into several watertight compartments, so that if a ship were to spring a leak, only a single compartment would fill with water, and the vessel would still retain enough buoyancy to stay afloat. Although the idea was transmitted from China by Marco Polo late in the thirteenth century, it was not put into practice by Europeans until almost five hundred years later. In 1787, Benjamin Franklin wrote a recommendation for the design of American mail-carrying boats wherein he stated that "their holds may without inconvenience be divided into separate apartments, after the Chinese manner, and each of these apartments caulked tight so as to keep out water."[194]

WHY EUROPE?

WHY DID TECHNOLOGICAL innovations that appeared in China, many of which were transmitted through the Islamic world, stimulate a Scientific Revolution in backward Europe rather than in the more culturally advanced Chinese and Islamic lands?[195] Paradoxically, it was their backwardness that gave the Europeans an advantage over the longer-established civilizations. For one thing, by the fifteenth century the cultural sophistication of the latter had long since ossified into the domination of tradition-bound intellectual elites that resisted new ideas about nature. But hidebound scholars were a symptom, not the cause, of the retardation of science. More to the point was the relative weakness of European political institutions.

European feudalism was a system of extreme political decentralization. The fragmentation of political power allowed the merchant class of Europe a certain degree of freedom that merchants did not enjoy in any other part of the world. The nominal imperial power in Europe was hardly formidable; Voltaire's famous one-liner derided the Holy Roman Empire as neither holy, nor Roman, nor an empire. Monarchs, landed aristocracies, and the Catholic Church all wielded a great deal of political power in late medieval Europe, but they were perpetually engaged in a three-way struggle that gave the merchants sufficient room to thrive in their relatively free cities.

The traditional ruling classes in Europe proved unable to stifle the growth of a social class that created immense wealth and fostered the development of new sailing and military technology, allowing Europeans to dominate world trade. Beginning in the fifteenth century, the new European economic system based on production for the market would eventually insinuate itself and force itself into every part of the globe, wiping out the traditional economic systems and creating a new, unified, worldwide system. Europeans and people of European descent would continue to hold most of the key positions in the new imperialistic world system.

At the same time, the rise of capitalism created the social preconditions for a scientific revolution. A vast increase in the markets for manufactured goods stimulated the inventiveness of artisans and promoted empirical methods of gaining knowledge of nature. The profit motive spurred competing merchant-manufacturers to seek labor-saving technical innovations, which required knowledge that would provide increased mastery over natural processes.

The creative role of merchants in the history of science has frequently been mentioned, but it was in early modern Europe that they collectively registered their most significant contribution.

By bringing to the fore a new economic system, they gave free reign to experimenting artisans. The success of the artisans in creating new and useful knowledge of nature inspired a few perceptive scholars to break with their innovation-stifling traditions and formulate new ways of perceiving the world. Thus were born the "mechanical philosophy," the "experimental philosophy," and the Scientific Revolution.

The merchants' development of long-distance seaborne trade was especially consequential for the history of science. Their transoceanic ventures gave rise to a European "Age of Discovery" that paved the way for the Scientific Revolution. Unfortunately, this story is often distorted by traditional histories that focus only on such figures as Prince Henry "the Navigator" and Christopher Columbus. These heroic narratives deserve closer examination.

NOTES

1. David Pingree, "Hellenophilia versus the History of Science," pp. 30-31.
2. Michael H. Shank, "Introduction" to *The Scientific Enterprise in Antiquity and the Middle Ages*, pp. 4–5.
3. See, e.g., Dick Teresi, *Lost Discoveries*. On the other hand, that the spirit of Greek-genius-worship is still alive and well is adequately demonstrated by Charles Murray, *Human Accomplishment*.
4. Martin Bernal, *Black Athena*, vol. I: *The Fabrication of Ancient Greece, 1785–1985*. Martin Bernal, by the way, is the son of J. D. Bernal, whose *Science in History* I frequently cite.
5. See esp. Mary Lefkowitz, *Not Out of Africa;* and Mary Lefkowitz and Guy MacLean Rogers, eds., *Black Athena Revisited*.
6. Mario Liverani, "The Bathwater and the Baby," p. 421.
7. The assessment of Neugebauer is from Robert Palter, "*Black Athena,* Afrocentrism, and the History of Science," p. 213. The quotation from Neugebauer is from Martin Bernal, "Animadversions on the Origins of Western Science," p. 77 (emphasis added).
8. Pingree, "Hellenophilia versus the History of Science," p. 38.
9. Stephen F. Mason, *A History of the Sciences*, p. 23.

10. J. D. Bernal, *Science in History*, vol. 1, p. 127.

11. Benjamin Farrington, *Science in Antiquity*, pp. 3–4.

12. Ibid., p. 8. For the text of the treatise, see James H. Breasted, *The Edwin Smith Surgical Papyrus*.

13. James H. Breasted, *The Conquest of Civilization*, p. 112.

14. Ibid., p. 113: "The Great White Race . . . always included . . . the dark-haired, long-headed 'Mediterranean Race'. . . . To this type belonged the Egyptians (notwithstanding their tanned skins)."

15. Herodotus, *The History*, book II, 50–51.

16. Isocrates, *Busiris*, section 28 (Loeb Classical Library, vol. 3, p. 119).

17. See the testimonials by ancient authors quoted in G. S. Kirk, J. E. Raven, and M. Schofield, eds., *The Presocratic Philosophers*, pp. 76–86.

18. Aristotle, *Metaphysics*, book I, chapter 1, 981b.

19. Plato, *Phaedrus*, 274.

20. Herodotus, *History*, book II, 109.

21. Strabo, *The Geography of Strabo*, vol. VII, pp. 269–271.

22. Proclus Diadochus, *Commentary on Euclid's Elements*, quoted in Morris R. Cohen and I. E. Drabkin, *A Source Book in Greek Science*, p. 34.

23. Quoted in M. Bernal, *Black Athena*, p. 106. It is interesting to note that Karl Marx made the same point: "Plato's *Republic*," he wrote, "is merely the Athenian idealisation of the Egyptian system of castes." Marx, *Capital*, vol. 1, p. 366.

24. M. Bernal, *Black Athena*, pp. 215–223.

25. Ibid., p. 215.

26. Johann Friedrich Blumenbach, *On the Natural Varieties of Mankind*.

27. M. Bernal, *Black Athena*, p. 219.

28. Ibid., pp. 218–219. Meiners's conception of "peoples' history" is quite distinct from "people's history." The placement of the apostrophe makes a great deal of difference.

29. Ibid., p. 33.

30 Quoted in Stephen Jay Gould, *The Mismeasure of Man*, p. 36.

31. Quoted in M. Bernal, p. 241.

32. Quoted in Stephen Jay Gould, *The Mismeasure of Man*, p. 36.

33. Quoted in ibid., p. 36.

34. Quoted in ibid., p. 45.

35. Quoted in ibid., p. 47.

36. Quoted in ibid., pp. 83–84.

37. Quoted in ibid., p. 84.

38. Quoted in ibid., p. 84.

39. Farrington, *Science in Antiquity*, p. 6. He is referring to the Rhind papyrus (mathematics) and the Edwin Smith papyrus (medicine), which were found together. See Breasted, *Edwin Smith Papyrus;* and T. Eric Peet, *The Rhind Mathematical Papyrus*.

40. M. Bernal, *Black Athena*, p. 15.

41. Herodotus, *History*, book II, 104.

42. See, e.g., Mary R. Lefkowitz, "Ancient History, Modern Myths," pp. 11–12.

43. Were the people of ancient Egypt "black"? What does that question mean? See the exchange between Frank M. Snowden, Jr., and Martin Bernal in the Fall 1989 issue of *Arethusa*.

44. M. I. Finley, *The Ancient Greeks*, p. 121.

45. Farrington, *Science in Antiquity*, p. 18.

46. "The original sources reflect a bias toward the scientific achievements of individuals (which has, moreover, been perpetuated by many modern commentators on the significance of the texts)." O. A. Dilke, "Cartography in the Ancient World: An Introduction," p. 140.

47. Simplicius, following Theophrastus, quoted in Kirk, Raven, and Schofield, *Presocratic Philosophers*, p. 86.

48. Finley, *Ancient Greeks*, p. 34.

49. Quoted in Benjamin Farrington, *Greek Science*, p. 80.

50. Aristotle, *Politics*, book I, chapter 11, 1259a. Aristotle's account suggests that Thales's scientific pursuits included astrology. It must be remembered that astrology was considered to be a fully legitimate science until relatively recent times.

51. Karl Kautsky, *Foundations of Christianity*, p. 205.

52. See Kirk, Raven, and Schofield, *Presocratic Philosophers*, pp. 144–145.

53. Ibid., pp. 197–198.

54. J. D. Bernal, *Science in History*, vol. 1, p. 167.

55. Otto Neugebauer, *The Exact Sciences in Antiquity*, p. 36.

56. Palter, "*Black Athena*, Afrocentrism, and the History of Science," p. 233.

57. Walter Burkert, *Lore and Science in Ancient Pythagoreanism*, pp. 216, 426.

58. Ibid., pp. 292–294.

59. Farrington, *Science in Antiquity*, p. 32.

60. J. D. Bernal, *Science in History*, vol. 1, p. 181, citing G. Thomson, *Studies in Ancient Greek Society* (London, 1949; emphasis in original).

61. Farrington, *Science in Antiquity*, p. 34.

62. J. D. Bernal, *Science in History*, vol. 1, pp. 181–183. "Parmenides' idealism," Bernal added, "is an extremely convenient [philosophical basis] for a minority ruling by 'divine' right."

63. Farrington, *Science in Antiquity*, p. 36.

64. Benjamin Farrington, *Science and Politics in the Ancient World*, p. 119.

65. Farrington, *Science in Antiquity*, p. 36.

66. Quoted in Farrington, *Science and Politics in the Ancient World*, pp. 29–30.

67. Farrington, *Science in Antiquity*, p. 114.

68. George Sarton, *A History of Science*, vol. I, p. 409. For an exposition of Plato's antidemocratic ideology, see Karl Popper, *The Open Society and Its Enemies*.

69. Farrington, *Greek Science*, p. 106.

70. Farrington, *Science in Antiquity*, p. 58.

71. The *virtuosi* are discussed in chapters 5 and 6.

72. Farrington, *Science and Politics in the Ancient World*, p. 126.

73. Ibid., p. 127.

74. Ibid., p. 94.

75. Plato, *The Republic*, book III, 414–415.

76. Farrington, *Science and Politics in the Ancient World*, p. 105.

77. Aristotle, *Metaphysics*, book XII, chapter 8, 1074b.

78. See Shadia Drury, *The Political Ideas of Leo Strauss;* and Drury, *Leo Strauss and the American Right.*

79. Shadia Drury, "Noble Lies and Perpetual War."

80. J. G. Landels, *Engineering in the Ancient World*, p. 189.

81. Farrington, *Science in Antiquity*, p. 7.

82. Sarton, *A History of Science*, vol. 1, pp. 403–404.

83. Ibid., pp. 420, 423, 430.

84. Ibid., p. 451.

85. Farrington, *Science in Antiquity*, p. 49.

86. Ibid., p. 52.

87. Ibid., p. 52.

88. "Aphorisms," in Hippocrates, *Hippocratic Writings*, p. 206.

89. John Henry, "Doctors and Healers," p. 193. That the Hippocratic Oath was "a Pythagorean manifesto" was first revealed by Ludwig Edelstein in *The Hippocratic Oath* (1943).

90. "Tradition in Medicine," in Hippocrates, *Writings*, p. 71.

91. Erwin H. Ackerknecht, *A Short History of Medicine*, p. 50.

92. Farrington, *Science in Antiquity*, p. 51.

93. Ackerknecht, *Short History of Medicine*, p. 72.

94. Farrington, *Science in Antiquity*, p. 134.

95. Ackerknecht, *Short History of Medicine*, p. 77. The word "surgeon" evolved from the French *chirurgien*, which derived from *cheir ourgos*, Greek for "hand worker."

96. Ackerknecht, *Short History of Medicine*, pp. 89–90.

97. See the section "Physicians, Surgeons, Apothecaries, and 'Quacks'" in chapter 5.

98. Farrington, *Science in Antiquity*, p. 142.

99. Ibid., p. 114. See Aristotle, *Politics*, book III, chapter 5, 1278a, and book VI, chapter 4, 1319b, for his attitude regarding citizenship for "mechanics or traders or labourers."

100. J. D. Bernal, *Science in History*, vol. 1, p. 44.

101. Aristotle, *Meteorology*, book I, chapter 12, 348b–349a.

102. J. D. Bernal, *Science in History*, vol. 1, pp. 167, 169.

103. T. Gomperz, *Greek Thinkers*, vol. 2, p. 148. The meaning of the word "proletariat" in the ancient context is closer to "urban poor" than to "working class."

104. Farrington, *Science and Politics in the Ancient World*, p. 122.

105. W. W. Tarn, *Alexander the Great and the Unity of Mankind*, p. 4 (emphasis added). See Aristotle, *Politics*, book I, chapter 5, 1254b.

106. Farrington, *Science and Politics in the Ancient World*, p. 113.

107. Ibid., p. 155.

108. Ibid., p. 98.

109. Sarton, *A History of Science*, vol. I, p. 592.

110. Farrington, *Science and Politics in the Ancient World*, pp. 125–126.

111. Polybius, *The Histories* (book 6, section 56), vol. 1, p. 506.

112. George Novack, *The Origins of Materialism*, pp. 258–259.

113. Leucippus and Democritus (fifth century B.C.E.) proposed the atomic theory of matter as an answer to the idealist notion that because the material world lacks permanence (rocks erode, water evaporates, people die), it cannot be real. The atomists theorized that all material objects are composed of imperceptibly tiny particles ("atoms") that are permanent and eternal and therefore real. What our senses perceive as lack of permanence in the material world, they said, is simply a rearrangement of atoms.

114. Farrington, *Science in Antiquity*, p. 102.

115. J. D. Bernal, *Science in History*, vol. 1, p. 212.

116. Lionel Casson, *Libraries in the Ancient World*, pp. 35, 70.

117. Farrington, *Greek Science*, p. 302.

118. J. D. Bernal, *Science in History*, vol. 1, p. 218.

119. Ibid., p. 222.

120. Henry M. Leicester, *The Historical Background of Chemistry*, p. 38.

121. Ibid., pp. 41, 44, 46.

122. Pamela H. Smith, *The Body of the Artisan*, p. 16.

123. Leicester, *Historical Background of Chemistry*, p. 47.

124. J. D. Bernal, *Science in History*, vol. 1, p. 235.

125. Leicester, *Historical Background of Chemistry*, p. 38.

126. Edward Peters, "Science and the Culture of Early Europe," p. 9 (emphasis added).

127. J. D. Bernal, *Science in History*, vol. 1, p. 267.

128. Ibid., p. 235.

129. Ibid., p. 234.

130. A. I. Sabra, "Situating Arabic Science," p. 226.

131. William Eamon, *Science and the Secrets of Nature*, p. 39.

132. Roy Porter, *The Greatest Benefit to Mankind*, pp. 93–94.

133. But "chem" seems to have come into the Arabic language from the Egyptian word *khem*, reflecting the much earlier origins of materials science. Chinese etymologies have also been suggested; see Needham, *Science in Traditional China*, p. 59.

134. Farrington, *Greek Science*, p. 309.

135. Islamic tradition has its own "Great Man" history of science, which attributes the origin of alchemy to an Umayyad prince, Khalid ibn Yazid, and to

the sixth Imam, Ja'far al-Sadiq, who was supposedly Jabir's teacher. However, "there is no evidence that the actual Khalid or the actual Ja'far ever concerned themselves with alchemy"; Leicester, *Historical Background of Chemistry,* pp. 62–63.

136. The shadowy nature of Jabir's existence is such that he has been placed by various authors in the eighth, ninth, and tenth centuries, and has frequently been confused with a thirteenth-century Latin author who borrowed his name (the Latin variant is "Geber").

137. P. Kraus, *Jabir ibn Hayyan,* vol. 1.

138. The "mercury" and "sulfur" of this theory were not specific elements but abstract qualities. Sulfur, for instance was the "inflammable principle."

139. J. D. Bernal, *Science in History,* vol. 1, p. 280.

140. Leicester, *Historical Background of Chemistry,* p. 72.

141. J. D. Bernal, *Science in History,* vol. 1, p. 279.

142. Needham, *Science in Traditional China,* p. 9. Needham's magisterial multi-volume work *Science and Civilisation in China* will undoubtedly stand for a long time to come as the foremost authority in the Western world on the history of Chinese science. According to Susan Bennett of the Needham Institute, "So far 21 volumes have been published in the series, and work is continuing on volumes in the project that Dr Needham conceived during his lifetime. . . . It is expected that there will be a total of 28 volumes in the series." Personal correspondence, 2003. There is also an abridged version, *The Shorter Science and Civilisation in China.* I cite three shorter books in which Needham summarized the results of his lifetime of study: *The Grand Titration, Science in Traditional China,* and *Clerks and Craftsmen in China and the West.* Needham never failed to acknowledge the collective nature of his research and gave credit to his many Chinese collaborators, especially Lu Gwei-Djen, Wang Ching-Ning, Tshao Thien-Chhin, and Ho Ping-Yü. He calls Lu Gwei-Djen, a young student of his when he first met her, "my chief collaborator"; it was she, he says, who first introduced him to the history of Chinese science.

143. Joseph Needham, *The Grand Titration,* p. 11. The Cardan suspension is "the system of linked and pivoted rings" named after the Italian Girolamo Cardano (Jerome Cardan) that was "commonly used in China a thousand years before Cardan's time." Also, "the triangle called by Pascal's name was already old in China in 1300 [C.E.]" (p. 17).

143. J. D. Bernal, *Science in History,* vol. 1, p. 157. The mandarinate, by the way, "was not as classless as has sometimes been made out, for, even in the best and most open periods, boys from learned homes which had good private libraries had a great advantage"; Needham, *Science in Traditional China,* p. 24.

145. Needham, *Grand Titration,* p. 117.

146. Needham, *Science in Traditional China,* pp. 24–25.

147. Ibid., p. 3.

148. Needham, *Grand Titration,* p. 58.

149. Needham, *The Shorter Science and Civilisation in China*, vol. 4, pp. 3–4.

150. Needham, *Grand Titration*, p. 24.

151. Ibid., p. 62.

152. Ibid., p. 70.

153. Ibid., p. 50. Needham's transliteration of Chinese names, which I have followed when citing his works, utilizes a second "h" instead of the more familiar aspirate apostrophe; i.e., "Chhen" rather than "Ch'en."

154. Needham, *Science in Traditional China*, pp. 22–23.

155. Needham, *Grand Titration*, p. 267.

156. Ibid., p. 28. For a full discussion of the social status of the early Chinese artisans and engineers, see vol. 7, part 1 of Needham's *Science and Civilisation in China*.

157. Needham, *Grand Titration*, p. 28.

158. Ibid., pp. 29–30.

159. Ibid., p. 31.

160. Ibid., p. 27.

161. Needham, *Science in Traditional China*, pp. 73–74. The cosmopolitanism of T'ang China makes an interesting contrast with the cultural insularity of Paris a thousand years later, as described by Jonathan D. Spence in *The Question of Hu*. Eighteenth-century Parisians found a Chinese man named Hu as exotic as we might find a visitor from another planet. They attributed his different way of perceiving the world to insanity.

162. Needham, *Clerks and Craftsmen in China and the West*, pp. 61–62.

163. Robert Temple, *The Genius of China* (with an introduction by Joseph Needham in which he describes Temple's book as "a brilliant distillation of my *Science and Civilisation in China*"), p. 9.

164. Temple, *Genius of China*, p. 114. Temple's source is Shen Kua, *Dream Pool Essays* (1086 C.E.).

165. J. D. Bernal, *Science in History*, vol. 1, p. 327. The impact of printing on the history of science is discussed further in chapter 5.

166. Temple, *Genius of China*, p. 112.

167. Ibid., p. 115.

168. Ibid., pp. 112–113.

169. Ibid., p. 59.

170. Ibid., pp. 94–95. Temple's source is Hsi Han, *Records of the Plants and Trees of the Southern Regions* (304 C.E.).

171. Quoted in Temple, *Genius of China*, p. 120.

172. Temple, *Genius of China*, p. 103.

173. Needham, *Clerks and Craftsmen in China and the West*, p. 204. For a critique of Needham's assertion of the continuity of Chinese and European mechanical time-keeping devices, see David Landes, *Revolution in Time*, pp. 17–25. Landes acknowledged the thoroughness and veracity of Needham's research, but argued that he has jumped to an unwarranted conclusion regarding the transmission of mechanical clockwork.

174. Temple, *Genius of China*, pp. 61–62.

175. Ibid., p. 78.

176. Ibid., p. 54.

177. Ibid., p. 93.

178. Quoted in Temple, *Genius of China*, p. 93.

179. Temple, *Genius of China*, p. 94.

180. Ibid., p. 49.

181. Ibid., pp. 224–225.

182. Ibid., p. 75.

183. Ibid., p. 76.

184. Ibid., p. 127.

185. T'ao Ku, *Records of the Unworldly and the Strange* (c. 950), quoted in Temple, *Genius of China*, p. 98.

186. Temple, *Genius of China*, p. 99. The women were "impoverished court ladies . . . in the short-lived Chinese kingdom of the northern Ch'i."

187. Ibid., p. 11.

188. Ibid., p. 20.

189. Ibid., pp. 11, 25.

190. Ibid., p. 9.

191. Ibid., pp. 149–150.

192. Ibid., p. 185.

193. Ibid., p. 188.

194. Quoted in Temple, *Genius of China*, p. 190.

195. This has been called "the Needham problem" because it is the central question that Joseph Needham set out to answer. See Kenneth Boulding, "Great Laws of Change," p. 9.

4

Blue-Water Sailors and the Navigational Sciences

THE STORY OF Prince Henry . . . has the grandeur and the dramatic unity of an epic poem. We see the princely hero patiently acquiring the needful qualifications for his task, and then devoting his life to the discovery of a way to India round the southern point of Africa. . . . We see him cultivating science, collecting information, training cadets, and inspiring every one who came in contact with him.

—CLEMENTS R. MARKHAM, *The Sea Fathers*

FROM THE BEGINNING, Prince Henry the Navigator called upon mathematicians and astronomers to answer the problems, and during the following three centuries it was the same group of men, never the practical tarpaulins who sailed before the mast, but always the astronomers and mathematicians, who taught the navigators.

—RICHARD WESTFALL, "Science and Technology during the Scientific Revolution"

O NE OF THE most persistent of the heroic history-of-science legends is that we are indebted to Prince Henry of Portugal (1394–1460), a.k.a. "Henry the Navigator," for the knowledge that made oceanic navigation possible—that is, for creating blue-water sailing by teaching seamen how to sail out of the sight of land. That tradition originated with Henry himself. It grew from the seed of panegyric narratives written by Gomes Eanes de Zurara, who was a publicist in Prince Henry's employ.[1] By producing paeans to the prince, Zurara was simply doing his job, but subsequent biographers and historians— from Richard Hakluyt and Samuel Purchas to R. H. Major and

C. R. Beasley—should have known better than to accept his claims at face value. Nevertheless, Zurara's chronicles served as the basis for almost everything written about Henry's exploits for five hundred years. Not until well into the twentieth century were his activities and putative accomplishments subjected to critical analysis.

One thing that cannot be blamed on Zurara is Henry's familiar sobriquet, "the Navigator," which was the embellishment of a nineteenth-century German geographer.[2] It could hardly be less appropriate. Not only was Henry not a navigator, but he rarely set foot on a ship at all. When he did, it was not as one responsible for guiding a craft through unfamiliar waters but as a royal passenger on routine voyages.

Henry's English-language admirers have until recently propagated the myth of his scientific prowess. Fortunately, a twenty-first-century biography by Peter Russell provides a more credible account of his life and describes how his legend was crafted. "The cult surrounding the ever more mythical figure of Prince Henry," Russell explained, "has proved capable of continuous remodelling over the centuries to serve the purposes of succeeding ruling élites in Portugal."

It began with Zurara's contemporary glorification of Henry's imperialistic exploits as the source of the country's prosperity. Then, in the sixteenth century, João de Barros initiated the notion of Henry "as a scholarly student of geography and the science of navigation." The same author was also responsible for "the romantic canard" that Henry "took up permanent residence at Sagres, near Cape St Vincent . . . so that he could pursue his studies of cosmology and stellar navigation far away from the world's affairs." Later authors reworked the fable into its final form, which maintains "that this royal prince had set up on one of the Sagres capes a formal school where, acting as a domine himself, he personally taught the science of oceanic navigation to sea-captains and pilots."[3] Some also asserted, on no evidence, that Henry founded a chair of mathematics at the University of Lisbon to further the maritime sciences.[4]

Henry's will contained many bequests to religious institutions and did establish a university chair of theology, but it made no mention whatsoever of any scientific endeavors.[5] If there is a small kernel of truth in the story that Henry created a school of navigation science at Sagres, it is that he may have provided funding and a place to attract vanguard seamen to congregate, exchange information, and accumulate a body of navigational knowledge. But if there was any teaching going on, it was not Henry who taught the sailors; it was they who taught him—or brought him the information he sought. Henry's contribution to oceanic science was confined to patronage of navigators, cartographers, mathematicians, and astronomers.

But just as the Medicis' financial support to Michelangelo did not make them artists, neither did Prince Henry's patronizing make him a scientist. Furthermore, whereas those who gave financial support to artists may in some cases have genuinely wanted to underwrite the creation of aesthetic beauty, Henry's patronage did not spring from a passion for abstract scientific truth. He had an ulterior motive; his purpose, though refracted through the crusader's ideology of holy war against the Muslim world, was colonial conquest and imperial glory. The expeditions he sponsored on the high seas were designed to promote Portuguese imperialism, and the science he sought was simply a means to an end: domination of the Atlantic islands and Africa. His successors extended Portuguese ambition to include "the Indies."

Henry did not *create* the important scientific knowledge for which he is often praised; he *bought* it. And even that gives him too much credit, because he did not pay for it all. Some of it he stole, and in the most brutal manner: those whose knowledge he coveted were kidnapped at his command and interrogated to extract the valuable information they possessed.

The most celebrated of Henry's alleged achievements was the passing of Cape Bojador on the African coast. According to the Henrican legend, sailors had always refused to sail south of that point until he overcame their ignorance and fear, urged them onward, and thereby proved to them that they could do it.

But in fact, a charter he issued in 1443 *prohibited* any sailors from passing Cape Bojador *without his permission*. That "indicated a desire on the part of navigators to do a little exploring on their own without reporting back to Prince Henry; in other words, they had to be *restrained* in the interests of the kingdom."[6] The charter illustrates Henry's intent to monopolize the knowledge his sailors gained.

None of this will be new revelation to anyone who has carefully studied Henry's life. At the end of the nineteenth century, some scholars began to publish massive collections of documents from the era of Portuguese exploration that would eventually allow a realistic assessment of what Henry had and had not done.[7] Nonetheless the "traditional encomiastic presentation" of Henry persisted deep into the twentieth century as most authors "went on presenting him without serious reservation as a paragon of . . . intellectual achievement."[8]

Prince Henry is one of a handful of Great Men whom historians have traditionally credited with creating the navigational sciences. The most famous name on the list, of course, is Christopher Columbus. Columbus, unlike Henry, was a competent navigator, and his pioneering transatlantic voyage was a significant feat worthy of recognition. Portraying him as a transcendent genius who stood far above his contemporaries in knowledge and ability, however, is simply untenable. His achievement cannot be isolated from the general context of the navigational practice created by many thousands of his seafaring peers and predecessors. They were, with few exceptions, drawn from the less privileged layers of society and are therefore appropriate subjects for a people's history. Although the captains of larger ships were sometimes of the class of gentlemen—the ship's owner, a wealthy merchant keeping a close eye on his investments, or an important government official—they rarely took a direct part in navigational operations. Those who did were the ships' pilots and officers, the maritime analog of skilled artisans. The most numerous category among the working mariners were the ordinary seamen who were the equivalent of semiskilled laborers.

SHORE-HUGGING

An essential part of the Henrican legend held that before the prince enlightened them, sailors "always coasted along in sight of land, and the entrance of an unknown sea was contemplated with terror."[9] This notion represents seamen as generally ignorant, timid, and conservative by nature, but it is as thoroughly false as the similar absurdity that Columbus's crew feared sailing over the edge of the earth.[10] The customary denigration of sailors as stubbornly conservative for resisting the imaginative schemes of armchair scholars is unwarranted. The sailors themselves had to evaluate novel proposals in terms of risks that could potentially cost them their lives. Their line of work was already dangerous enough; their tendency to proceed cautiously in departing from tradition was completely rational.

Commodore Collins of the British Admiralty, commenting on "that persistent myth that the first sailors navigated by 'hugging the shore,'" indicated where ignorance of the sea really lay:

> Those words could never have been written by a sailor. Nothing is more fraught with peril, and therefore the more assiduously avoided on a little-known coast, than hugging the shore. The myth is based on the assumption that the mariner had neither the means nor the ability to find his way out of sight of land. Now this assumption is shown to be groundless. . . . Nothing is more sure, by whatever means they achieved it, than that the sailors of all ages have navigated in deep waters.[11]

"When Prince Henry died," one of his adulators wrote, "he had only just persuaded his people to venture out of sight of land for a couple of days, from the Canaries to the African coast."[12] But in fact, long before Henry was born, voyages to and from the Canary Islands were commonplace. Phoenician sailors had been there two millennia earlier, and by the second century C.E., Claudius Ptolemy could ascribe knowledge of their position to "the continual measuring of this distance by voyagers."[13] They were subsequently lost to European memory but were rediscovered in about 1270 by Genoese seamen who also traveled to the Azores and Madeiras. At least nine round-trips were made to the Canaries between 1336 and 1393.[14] Voyages like these were undoubtedly where Iberian navigators developed the crucial knowledge and skills that allowed them eventually to cross the Atlantic. The north Atlantic, meanwhile, had been repeatedly traversed by Viking ships much earlier, before 1000 C.E. It is remarkable "that these great ocean passages were made and remade by the Norse seamen without quadrant or astrolabe, without magnetic compass, and without sea-charts."[15]

Archaeological evidence indicates that seamen were sailing out of sight of land in the Red Sea some four thousand years ago. During the second millennium B.C.E., Minoan sailors from Crete regularly crisscrossed the Mediterranean. *The Odyssey* provides literary corroboration; although Odysseus is a mythological character, Homer's detailed descriptions of how he steered his ship by the stars across the sea for seventeen days testifies to an established maritime tradition. In about 600 B.C.E., Hamilcar, a Carthaginian captain, reported that natives of Brittany in boats much inferior to his own were making frequent crossings to Ireland, a three-hundred-mile open-sea journey.[16] Phoenician sailors plied regular trade routes across the Mediterranean and the Black Sea, ventured into the Indian and Atlantic Oceans, and reportedly even circumnavigated Africa.[17] And on the other side of the globe, as was mentioned in chapter 2, the prehistoric peopling of the Pacific Islands testifies to the existence of sophisticated navigational techniques even in the absence of written language and metal. Long before Captain Cook "discovered" the Hawaiian isles, Polynesians had been sailing there from Tahiti and from the Marquesas, voyages of some two thousand nautical miles.

The shore-hugging myth arose from the fact that the Portuguese sailors Henry ordered to find a sea route to Guinea—sub-Saharan West Africa—proceeded systematically southward along the contours of the continent, but they were by no means always within sight of land. The practice of "coasting" entailed periodically sighting land as a means of reaffirming the position of the ship, not remaining continuously close to the shore.[18] Pilots used the magnetic compass and took frequent soundings with the lead and line to find their way from one familiar cape or prominent landmark to another.

The contention that mariners before Henry's time were incapable of sailing out of sight of land would have seemed uproariously funny to the deep-sea fishermen whose industry then constituted an essential part of the Portuguese economy. Their livelihood depended on their being able to pursue their quarry far out into the ocean—"fishermen follow the fish"—and as a consequence, they had come to know the waters of the eastern north Atlantic very well.[19]

WINDS AND CURRENTS

THE KNOWLEDGE THAT was of most value in facilitating blue-water navigation involved understanding the regularities of the oceans' winds and currents. The recognition that wind and current patterns are predictable to a significant degree was the initial insight that made it possible for mariners to record them on charts and use them to go where they wanted to go. The reliability of the winds of the Red Sea "both in direction and character" helps to explain the early appearance of open-sea sailing there.[20]

It is patently evident that information of that sort could not have originated with land-based scholars but must have first been gathered and interpreted by those who spent their working lives at sea:

> This exploration of the winds, which set down the world's trade routes until the coming of steam, was hardly less important than the exploration of new lands. And this could only be done at sea, by mariners constantly learning, recording their observations and passing them on to others. . . . [W]ith methods so rudimentary and achievements so colossal, we must regret the more how little we know of the men who made them.[21]

Until recent centuries this knowledge was preserved and transmitted orally, "for the sailor was a craftsman, learning as a youth how to pilot his ship by working beside his master. Nothing was written down."[22] The difficulty this creates for historians searching for documentary evidence is apparent. A historian of cartography has observed that

> next to professional lawbreakers, no group of people in the history of mankind has been more reluctant to keep records than professional sailors. They were philosophers without benefit of school tie, mathematicians of necessity rather than conviction, astronomers without an observatory other than the deck of a ship. They kept their knowledge to themselves.[23]

Who but those with practical seafaring experience could have figured out that the best and fastest way to return from the Canaries to the Iberian coast would be to begin by sailing farther out to sea on a northwesterly course rather than to head directly homeward via a northeasterly course? The direct course required battling the northeast trade winds and the current all the way, which could be accomplished only by laborious tacking, while sailing northwest to pick up the prevailing westerlies —the method the Portuguese sailors called the *volta do mar*— produced a relatively quick and easy way home. Those who first discovered this counterintuitive route are unknown, but they were certainly working mariners rather than mathematicians or astronomers.

It was knowledge of the *volta do mar* and by extension the larger clockwise circular pattern of the north Atlantic's winds— what oceanographers call a gyre—that made possible the voyages to and from the Western hemisphere by Columbus and his successors. Furthermore, "prevailing patterns of thought grew up to match the patterns of prevailing winds, and Iberian sailors used the *volta* as a template with which to plot their courses to Asia, to the Americas, and around the world."[24] Below the equator, in the south Atlantic, they discovered a counterclockwise gyre mirroring the one in the north, and in the next generation found gyres in the north and south Pacific analogous to those in the Atlantic.

Imagine sailors attempting to go from Portugal to India by rounding the southern tip of Africa. Now imagine them unexpectedly bumping into Brazil! If you were to assume (as numerous historians have) that they had blundered way off course, you would be wrong. In 1500 Pedro Alvares Cabral and his crew—following directions that three years earlier had proven successful—intentionally sailed hundreds of miles to the west in order to ride the south Atlantic gyre around Africa and into the Indian Ocean. In the middle of their journey they were surprised to see land birds flying overhead, and followed them. Thus was Brazil "discovered" by Europeans, an unanticipated boon for Portuguese imperialism. But if Cabral's crew had not stumbled on Brazil at that time, other sailors would have soon enough.[25]

The first Iberian sailors to utilize the south Atlantic gyre to reach India were those under the command of Vasco da Gama in 1497. Although da Gama is traditionally portrayed as a navigational genius for pioneering the bold maneuver, there is no evidence suggesting that he himself supplied the knowledge necessary to attempt it. Da Gama, a gentleman of the king's household with limited navigational experience, was a royal appointee in charge of a fleet of ships for the first time. A Portuguese historian notes that although the chronicler Castanheda claims da Gama was chosen for the mission "because 'he was experienced in the things of the sea in which he had rendered much service to D. João II,' we cannot find in chronicle or document a single reference to any such experience or 'service' worth mentioning."[26] It was not his aptitude for seamanship that recommended da Gama to the Portuguese king, but his dependability as a vassal and his ruthlessness in pursuit of plunder and conquest.

After his initial reconnoitering voyage, da Gama was sent back to India in 1502 on a colonizing mission. One of his crew members reported an encounter with a ship full of Muslims returning from a religious pilgrimage to Mecca: "We took a Meccha ship on board of which were 380 men and many women and children, and we took from it at least 12,000 ducats, and at least 10,000 more worth of goods, and we burnt the ship and all the people on board with gun powder." To intimidate local rulers, da Gama captured unarmed fisherman, hacked them to pieces, and delivered the severed heads, hands, and feet with a written message declaring that any resistance would be futile.[27] Portugal's dominance in the Indian Ocean was accomplished by what today would be called "state-sponsored terrorism."

Sailors under the command of Ferdinand Magellan were the first Europeans to make systematic use of the gyres of the Pacific Ocean. Magellan is conventionally credited with being the original circumnavigator of the globe, but that feat was in fact accomplished by his Basque pilot, Juan Sebastián Del Cano, "an ordinary man, without wealth or genius to help him."[28] Magellan was killed in the Philippines well before the end of the journey. Del Cano "drew on all the lessons of the winds learned by the anonymous sailors of the Mediterranean Atlantic . . . and by the unknown ancients who first sailed the Asian seas."[29] As we saw in an earlier chapter, one of the primary methods by which the Europeans "drew on" the knowledge of the Asians was kidnapping.

Del Cano was the first navigator to complete a voyage around the globe, but the first human being actually to circle the earth and return to his point of origin was probably Magellan's slave, Enrique. Enrique's journey did not begin and end in Europe. Ten years before the famous expedition, Magellan purchased Enrique in a slave market in Malaysia and took him through Africa to Portugal and eventually to Spain, from whence they set out westward on the trip around the world. When they reached the Philippines, Enrique surprised his owner by being proficient in the local language. Enrique, it seems, had returned home after an extended involuntary absence.[30]

Magellan's death resulted from his own imperialistic belligerence. Attempting to conquer Mactan, one of the Philippine islands, he "anticipated a ragged band of nearly naked warriors who would flee the moment he fired his artillery, and whose flimsy bamboo spears would be useless against impenetrable Spanish armor." The islanders, however, surprised him; their leader, Lapu Lapu, sent fifteen hundred warriors against the Europeans' force of about fifty. Magellan himself was hacked to pieces in the battle:

> Today, in the Philippines . . . Magellan is not regarded as a courageous explorer; instead he is portrayed as an invader and a murderer. And Lapu Lapu has been romanticized beyond recognition. By far the most impressive sight in Mactan harbor today is a giant statue of Lapu Lapu, his bamboo spear at the ready, as he gazes protectively over the Pacific.[31]

As for Enrique, after his master's death he proclaimed himself a free man as explicitly provided by Magellan's will. But when Magellan's immediate successors insisted that he was still a slave in their service, Enrique angrily deserted the expedition and shortly thereafter disappeared from the historical record.

The scientific consequences of the first circumnavigation went far beyond increased knowledge of the oceans' wind patterns. Although Magellan and his sailors

> failed to understand much of what they had experienced, they had made records for others to study, enlarging the Europeans' knowledge of the world. They had circled the globe, only to demonstrate that the world was now a larger place than previously imagined, not smaller. Seven thousand miles had been added to the globe's circumference, as well as an immense body of water, the Pacific Ocean. They had learned that beyond Europe, people existed in astonishing profusion and variety, as tall as the giants of Patagonia and as short as the pygmies of the Philippines.[32]

The expedition improved knowledge of the natural world not only by addition but also by subtraction. Its reports served to correct innumerable mistaken ideas that had previously been widely held. "Banished were phenomena such as mermaids, boiling water at the equator, and a magnetic island capable of pulling the nails from passing ships." Most important of all, its exposure of many glaring errors and omissions of Aristotelian natural philosophy prepared the way for the demise of the elite science of the day. "All these discoveries," one author reminds us, "came at the cost of over two hundred lives."[33] And that only takes the lives of Magellan and his crew into account; who knows how many of the non-Europeans they encountered perished?

HIPPALOS AND THE INDIAN OCEAN MONSOON

MOST INSTANCES OF discovery of the oceans' wind patterns are unrecorded, but one of the earliest and most important has been attributed to a Greek sailor from Alexandria named Hippalos. According to a first-century C.E. document, the *Periplus Maris Erythraei*,[34] Hippalos discovered how to use the monsoon winds of the Indian Ocean to make the journey from the east coast of Africa to India and back in under a year, a round-trip that had previously taken Greek merchants three years to accomplish. What Hippalos found was that rather than tracing a path along the contours of the continents, it was possible to directly traverse the Indian Ocean by taking timely advantage of the monsoon winds. In the summertime, the monsoon blows from southwest to northeast, and a mariner can sail before the wind from Africa to India; in the winter, it blows in the opposite direction, facilitating the return voyage. As the anonymous author of the *Periplus Maris Erythraei* explained,

> men formerly used to sail over in smaller vessels, following the curves of the bays. The ship captain Hippalos, by plotting the location of the ports of trade and the configuration of the sea, was the first to discover the route over open water. . . . In this locale the winds we call "etesian" blow seasonally from the direction of the ocean, and so a southwesterly makes its appearance in the Indian Sea.[35]

Hippalos did for the Indian Ocean what Columbus's first crossing was to do for the Atlantic many centuries later: he opened it up as a trade route for Europeans. As soon as Hippalos demonstrated its feasibility, other Greek and Roman seamen quickly followed in his path. Precisely when he accomplished this revolutionary breakthrough is not known, but it must have been about a century and a half before the *Periplus Maris Erythraei* was written. According to the Greek geographer Strabo, in about 116 B.C.E. Eudoxus of Cyzicus made use of the monsoons to get to India and back.[36] Perhaps Hippalos was the ship's captain or navigator on that voyage—Strabo does not name him—but the beginning of regular round-trips on that route suggests that if Hippalos was indeed the first Greek to navigate across the Indian Ocean, he must have done so in Eudoxus's time.[37] Soon other Greek seamen began to extrapolate from their newfound knowledge of the monsoon winds to sail much further down the coast of East Africa than they had previously ventured. By the middle of the first century C.E., "they had extended their range of trade as far south as Rhapta, somewhere in the vicinity of Dar es Salaam."[38]

Whether Hippalos was or was not the first Greek to comprehend the weather patterns of the Indian Ocean, he was certainly not the creator of that knowledge; he received it from anonymous Arab and Indian navigators whose use of the monsoon drew on a tradition that predated him by as much as two thousand years. When Greek ships first began trading in the Indian Ocean, they were perceived as rivals by those who were already there. The Arab and Indian seamen were able to keep their knowledge of the winds from their Greek counterparts until Hippalos or Eudoxus learned their trade secret. According to Poseidonius and Strabo, that occurred when

> an Indian was brought before the King by the guards of the Arabian Gulf (i.e., the Red Sea), who said they had found him half-dead and alone on a ship that had run aground. . . . In gratitude for the hospitality accorded him, he declared himself willing to show the sea-route to India to men appointed

for this purpose by the King and Eudoxus became also one of these.[39]

With the disappearance of the western Roman Empire, the knowledge of the monsoon winds was lost to Europeans for many centuries.[40] When Vasco da Gama entered the unfamiliar waters of the Indian Ocean in 1497, he encountered there, as his Greek predecessors had, a sophisticated tradition of navigation that did not willingly share its secrets. But the superiority of da Gama's weaponry allowed him to seize the knowledge he needed, once again utilizing the kidnapping method. He attacked Arab ships, captured a pilot, and forced his hostile prisoner to guide the Portuguese northward. At Melindi (in present-day Kenya), he was able to hire or coerce another navigator who knew the monsoon well.[41] This expert seafarer provided the Portuguese with "the means to surmount their ignorance of the winds and currents that were confounding their efforts to reach the riches of India."[42] But reach them they did: da Gama reportedly made a profit of 10,000 percent on the cargo he carried from India to Portugal.[43] As for the scientific significance of the enterprise, "The Portuguese entry into the Indian Ocean was more in the nature of an occupation than an exploration. They *took over existing knowledge* rather than extended it."[44]

CHARTING THE GULF STREAM

MARINERS WERE RESPONSIBLE for discovering the regularities not only of the oceans' winds but also of their currents. Winds are the primary but not the only cause of currents on the ocean surface, and the relationship of a wind to its associated current is not always simple and direct. Although it is true that currents are driven by winds, other factors—especially intervening landmasses—can cause currents and winds to thwart one another.

The way knowledge of the currents came into being is well illustrated by Benjamin Franklin's publication of the first chart of the Gulf Stream, the powerful current that explodes out of the narrow channel between Florida and the Bahamas, proceeds northward parallel to the coastline of the United States, and in the north Atlantic deposits waters whose warmth is carried by winds all the way to the shores of Ireland and England. Sailors had discovered no later than the early sixteenth century how to utilize the Gulf Stream to their advantage, but it remained their craft secret until Franklin published his chart more than two and a half centuries later. Franklin, one of the familiar Great Men of Science, had the good grace to report honestly how he came into possession of the knowledge embodied in his chart.

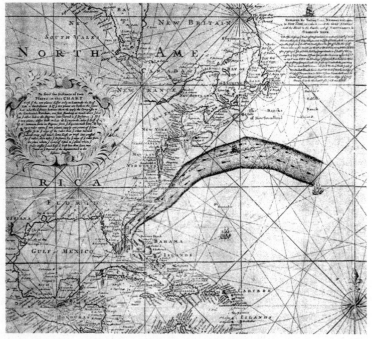

Benjamin Franklin's chart of the Gulf Stream

In 1769 a complaint was brought to Franklin's attention, in his capacity as head of the postal service for the American colonies, that it took mail boats ("packets") two weeks longer to make the trip from Falmouth to New York than it took merchant ships to travel from London to Rhode Island. Franklin was puzzled

> that there should be such a difference between two places
> scarce a day's run asunder, especially when the merchant
> ships are generally deeper laden, and more weakly manned
> than the packets, and had from London the whole length of
> the river and channel to run before they left the land of
> England, while the packets had only to go from Falmouth.[45]

Franklin said that he "could not but think the fact misunderstood or misrepresented," but he investigated and to his surprise found it to be true. It was Franklin's good fortune to have a cousin, Timothy Folger, who happened to be a Nantucket whaleboat captain. Folger explained the nature of the Gulf Stream to him:

> "We are well acquainted with that stream," says he, "because
> in our pursuit of whales, which keep near the side of it, but are
> not to be met with in it, we run down along the sides, and fre
> quently cross it to change our side; and in crossing it have
> sometimes met and spoke with those packets, who were in the
> middle of it, and stemming it. We have informed them that they
> were stemming a current, that was against them to the value of
> three miles an hour; and advised them to cross it and get out of
> it; *but they were too wise to be counselled by simple American fisher
> men.* When the winds are but light," he added, "they are car
> ried back by the current more than they are forwarded by the
> wind; and, if the wind be good, the subtraction of seventy
> miles a day from their course is of some importance."[46]

Franklin then remarked to Folger that it was too bad this current was not on any charts and "requested him to mark it out for me, which he readily complied with."[47] Folger drew for Franklin "the dimensions, course and swiftness of the Stream from its first coming out of the Gulph when it is narrowest and strongest, until it turns away to go to the southward of the Western Islands, where it is broader and weaker," and added "written directions whereby ships bound from the Banks of Newfoundland to New York may avoid the said Stream."[48] Franklin published this information "for the benefit of navigators," fully acknowledging his debt to "the Nantucket whalemen" who were "extremely well acquainted with the Gulf Stream, its course, strength and extent, by their constant practice of whaling on the edges of it, from their island quite down to the Bahamas."[49]

The economic value of this natural knowledge was evident; it meant that "a vessel from Europe to North America may shorten her passage by avoiding to stem the stream . . . and a vessel from America to Europe may do the same by . . . keeping in it."[50] This relatively recent (late-eighteenth-century) example of how the charting of an important ocean current depended upon the input of "simple fishermen" should not be minimized as anecdotal evidence but rather should be recognized as a rare instance in which the source of the essential knowledge in question can be documented.

THE TIDES

THE KNOWLEDGE OF winds and currents that allowed bluewater sailors to venture forth upon the high seas would not have been worth much if they had been unable to return to port without running aground or smashing into the rocks. For that, it was necessary to be fully aware of exactly when, where, how fast, and how high the tides rush in and out. Certain regularities, such as their twice-daily cyclical nature, must have been obvious to mariners and port dwellers very early on. Prehistoric Maoris in New Zealand recognized the moon's influence; they called Rona, the woman in the moon, the "tide-controller."[51] Similarly, more than four thousand years ago, Babylonians ascribed power over the tides to the moon goddess Ishtar.

Useable predictive powers, however, depended on an exact understanding of the correlation between the tides and the moon's position. Sailors in many parts of the ancient world no doubt independently worked out that relation, but the first recorded allusion to it, in the fourth century B.C.E., was by Aristotle's disciple Dicaiarchus, who got it from a navigator named Pytheas of Massalia.[52] Two centuries later, it was given fuller expression by a Stoic philosopher, Poseidonius, and by a Babylonian astronomer, Seleucus.[53] Seleucus associated tidal phenomena in the Indian Ocean with the phases of the moon, an inference that could not have been made without the extensive seafaring or seaport experience of mariner-informants. Poseidonius made his own systematic observations of the moon's effect on the tides at Gades (modern Cadiz, Spain), but not before learning about it from the local inhabitants. Strabo reported his conclusions:

> Poseidonius says that the movement of the ocean is subject to periods like those of the heavenly bodies, since, behaving in accord with the moon, the movement exhibits first the diurnal, secondly the monthly, and thirdly the yearly period; for when the moon rises above the horizon to the extent of a zodiacal sign [30 degrees], the sea begins to swell, and perceptibly invades the land until the moon is in the meridian; but when the heavenly body has begun to decline, the sea retreats again, little by little.[54]

That the sailor Pytheas was among the first Greeks to give serious thought to the lunar influence on the tides may be attributable to his pioneering navigating experience in the Atlantic. In the Mediterranean, "the tides were too small to attract attention," but on the Atlantic shores "the tides were high and as ancient people (not only the educated ones but also farmers and shepherds) observed the moon carefully, they could not have failed to notice any relation that might exist between the lunar cycle and the tidal one."[55]

The magnitude of the tides on the Atlantic coasts of northern Europe made understanding them especially crucial for sailors leaving or entering ports there. Atlantic seamen in the early modern era were able to correlate high-water times with the moon's position to calculate precise tidal patterns for each day of the lunar calendar. These sailors were for the most part illiterate and had no clocks to tell time by, but despite those handicaps, they developed techniques that served their purposes well. Lacking the means to specify clock times for when tides reached their high point, they instead designated those moments according to the moon's age and compass bearing. Historian of navigation J. H. Parry described their method:

> The moon's reflection on the surface of the sea was a dramatic and obvious track, whose bearing was simple to observe; and since the highest high waters, or spring tides, were seen to occur on days of full and new moon—at "full and change"— the bearing of the moon at these times became the tidal establishment of the port. The establishment of Dieppe, for example, was expressed: "Dieppe, moon north-north-west and south-south-east, full sea," which meant "high water occurs at Dieppe on days of full and new moon when the moon bears north-north-west or south-south-east."

Armed with that knowledge, a sailor could determine for any given day and any given port exactly when high tide would occur by adding a "retardation" factor for each day that had passed since full or new moon. Fortunately, the retardation factor in minutes appropriate for the north Atlantic "was conveniently equivalent to one point of the magnetic compass. . . . To the establishment of the port, expressed as a compass bearing, [the seaman] added one point for every day of the moon's age."[56] As this method of tidal computation illustrates, the magnetic compass served sailors as far more than simply a direction-finding instrument.

The documentary record contains no tide tables earlier than the twelfth century C.E., although their origins certainly lie much farther in the past than that. The *Catalan Atlas* (c. 1375) contains a sophisticated tide table that was probably the work of a skilled chart- and instrument-maker named Abraham Cresques, "who had every opportunity for gathering from seamen correct information about the tides at the various ports they visited." This tide table based on extensive sailors' observations was superior to contemporary tables that were "mechanically built up from a single observation" by mathematically-minded scholars.[57]

The precision with which many centuries of mariners' observations linked the tides to the moon was problematical for Galileo, who adamantly refused to admit a causal connection. His erroneous theory that tides are caused only by the earth's motions was central to his defense of Copernican astronomy—so much so that his major work on the subject, *Dialogue Concerning the Two Chief World Systems*, had originally been entitled *Dialogue on the Tides*.[58] He ridiculed the "many who refer the tides to the moon" by putting that opinion in the mouth of a character in his dialogue whom he named Simplicio (i.e., Simpleton), and had Salviati, the character representing his own views, declare, "That concept is completely repugnant to my mind."[59]

In his eagerness to attribute the "flux and reflux of the sea" solely to "the annual and the diurnal motions which belong to the entire terrestrial globe," Galileo went to great lengths to explain away the lunar correlation implicit in the tide tables. After positing an "irregularity" in the combined orbital motion of the earth and moon as a possible explanation of why the moon and tides might *seem* to be associated, he had another character in his dialogue object that "such an irregularity ought to have been observed and noticed by astronomers, but I do not know that this has occurred." His rather feeble response was that "although astronomy has made great progress over the course of the centuries" there are "many things still remaining undecided." At the same time, he arrogantly downplayed the value of the sailors' contributions. "I am content to have noticed that incidental causes do exist in nature," he sniffed; "I shall leave their minute observation to those who frequent the various oceans."[60]

GEOGRAPHY AND CARTOGRAPHY: FROM
PERIPLOI TO *PORTOLANI*

A HISTORIAN UNDER the influence of the Greek Miracle tradition asserts that Anaximander and Hecataeus, two natives of Miletus in the sixth century B.C.E., "launched the twin sciences of cartography and geography." They did so, we are told, by "collating the material" they gathered after they "haunted the bustling quays" of the seaport and "listened to the reports of returning sailors."[61] The acknowledgment of the seamen's contributions is worth noting, but as with most origins-of-sciences stories, the matter is not that simple.

"Cartography was not born full-fledged as a science," wrote an eminent historian of that field of knowledge; "it evolved slowly and painfully from obscure origins."[62] Fishermen and other mariners furnished the information not only for tabulating tides and mapping currents but for drawing up sea charts, an essential element in the development of cartography. This is evident in cases involving the most recently mapped areas of the world, namely polar regions, the history of which is less obscure. As in the Gulf Stream example, whaling captains were the primary source of data for Gerard van Keulen's 1714 map of the uninhabited Arctic islands of Spitzbergen. Swedish cartographers who later mapped the same area drew heavily on the local knowledge of Russian trappers and walrus hunters and Norwegian sealers. Even in the late nineteenth century, the expertise of sea captains guided the work of professional Arctic geographers who made the geodetic and astronomical measurements from which precise maps were constructed.[63]

But the origins of cartography precede the professionalization of science by many millennia, and the farther back one investigates, the more crucial the contributions of seamen and seagoing merchants are found to be. The previously mentioned *Periplus Maris Erythraei* illuminates one way early traveling salesmen contributed to the history of science; it was a merchant's handbook by an anonymous author who was engaged in seaborne trade between Egypt, eastern Africa, India, and

Arabia, and his intention was to pass along valuable geographic and ethnographic information to others involved in that trade. The *Periplus Maris Erythraei* was an early example of travelers' narratives of the sort most famously exemplified by the writings of the Venetian merchant Marco Polo.

An even earlier example—and this one was not anonymous—was a famous book, *On the Ocean*, written in the fourth century B.C.E. by a Greek merchant-mariner named Pytheas of Massalia.[64] Pytheas was mentioned in conjunction with knowledge of the moon's influence on the tides, but his most significant contribution was to the spread of geographic knowledge. Pytheas wrote about a voyage he made out into the north Atlantic to what would later be known as the British Isles and to "Ultima Thule," which may have been Iceland. His account was based on his own firsthand experience as a sailor, but his observations necessarily reflected the collective experience of local seamen, who knew the waters of the Atlantic far better than he did. His book gave the people of the Mediterranean world their first glimpse of what lay beyond the shores of continental Europe and shed light on the mystery of where such important commodities as tin and amber came from. Unfortunately, the text of *On the Ocean* has not survived, but it was cited and quoted extensively by numerous authors of antiquity.[65]

Not all of Pytheas's readers accepted his testimony. After all, he was reporting that the far-north lands that all scholars had always considered uninhabitable were in fact populated. "Armchair geographers," a modern commentator observed, "knowing nothing of the effect of the Gulf Stream and the moist Atlantic winds on climate, found the facts he reported hard to believe."[66]

The first major authors to cite Pytheas's work, Hipparchus and Eratosthenes, did so approvingly, but his reputation suffered at the hands of the historian Polybius, who bitterly challenged Pytheas's veracity on the grounds that Pytheas was merely "a private individual—and a poor man too."[67] The implication was that only elite authors like Polybius himself, who was favored by the patronage of important people, should be considered trustworthy. Even worse from his point of view was that Pytheas was of the merchant class, which he considered to be intrinsically dishonest.

Strabo, whose *Geography* was "compiled largely from the works of others," and whose main purpose was to assemble knowledge that the Roman emperors could use to extend and maintain their dominion, simply adopted Polybius's negative assessment of Pytheas and further maligned him. Nonetheless, Pytheas's observations and descriptions have for the most part been verified by archeological evidence, and "modern scholars are unanimous in their acceptance of his truthfulness and accuracy."[68]

Pytheas, Hippalos, and Marco Polo are but individual representatives of thousands of mariners and merchants who gathered and reported knowledge of faraway places, thereby providing the raw material for geography and cartography as well as for other scientific disciplines. In his *Geography*, Ptolemy declared that "the first essential" of that science is "reference to the history of travel, and to the great store of knowledge obtained from the reports of those who have diligently explored certain regions."[69] He cited merchants and navigators, sometimes by name: "a certain Macedonian named Maen, who was also called Titian, son of a merchant father, and a merchant himself, noted the length of this journey"; "the merchant Philemon whose reckonings make the length of the island of Hibernia from east to west a twenty days' journey"; "all who navigate these parts unanimously agree. . . ."[70]

The *Periplus Maris Erythraei* was not a typical *periplus*. Most *periploi* were written not by merchants but by sailors. They were port-books or pilot's books—in other words, practical guides that provided detailed sailing instructions for working mariners. Very few *periploi* survive from ancient times, but those that do constitute a rare resource for a people's history of science: documentary evidence. They reveal the cumulative navigational and geographic knowledge of a particular time and place and testify to the mariners' role in compiling that knowledge. The few surviving *periploi* are obviously representative of a large class of documents that served as primary source material for academic geographers of antiquity such as Hipparchus, who "trusted the sailors" for much of his data.[71] Strabo, Marinus of Tyre, and Ptolemy all drew on *periploi* attributed to Timosthenes, a third-century B.C.E. mariner.[72]

The earliest known example of such a work is one ostensibly authored by Scylax of Caryanda that dates from the fourth century B.C.E.[73] In spite of its attribution to Scylax, a sea captain famous for voyages more than a century earlier, this account of his circumnavigation of the Mediterranean was a collaborative production of many anonymous seamen. It presented data about the waters and ports along the north African coast from the mouth of the Nile to the Pillars of Hercules (now Gibraltar) at the entrance to the Atlantic. "Sailing along from Hermaea," for example, "it is a day and a half to Carthage. There are islands off the Hermaean cape, Pontia island and Cosyrus. From Hermaea to Cosyrus is a day's sail."[74]

At least five centuries and probably more separate the *periplus* attributed to Scylax from the *Stadiasmus of the Great Sea*, another of the very few surviving examples of the genre. The *Stadiasmus* was somewhat more detailed and (as its name suggests) gave distances between ports in *stadia:*

> From Hermaea to Leuce Acte, 20 stadia; hereby lies a low islet at a distance of two stadia from the land; there is anchorage for cargo-boats, to be put into with west wind; but by the shore below the promontory is a wide anchoring-road for all kinds of vessels.[75]

The *Mu'allim* or "Pilot in the Arabian Sea," a Sanskrit document of 434 C.E., provides an indication of the extensive natural knowledge accumulated by early mariners of a culture other than Greco-Roman. The navigator, it says,

> knows the course of the stars and can always orient himself; he knows the value of signs, both regular, accidental and

abnormal, of good and bad weather; he distinguishes the regions of the ocean by the fish, the colour of the water, the nature of the bottom, the birds, the mountains, and other indications.[76]

Although no maps or charts accompany the surviving *periploi*, there can be no doubt that the pilot's books were used in conjunction with graphic representations of some sort. Ancient documents, it must be remembered, have not come down to the present intact in their original form but only as copies of copies of copies made over the course of many centuries. Because few scribes were skilled illustrators, most of the graphic images embedded in manuscripts of antiquity, including maps, have disappeared; only text survives.

The rarity of surviving *periploi* and the complete absence of their accompanying charts is not difficult to account for. The information they contained was "money in the pocket in ancient times," as a historian of cartography has explained:

> In the beginning, mariners made their own charts, by dint of hard labor and at the peril of their own lives. Such trade secrets as they contained were guarded with care, and early sea charts were either worn out from constant use or were destroyed— willfully and with malice aforethought. And why not?[77]

These were trade secrets that Prince Henry of Portugal devoted his wealth and royal power to confiscating. And once sea charts came under the sway of governments, they gained the aura of state secrets:

> They were much more than an aid to navigation; they were, in effect, the key to empire, the way to wealth. As such, their development in the early stages was shrouded in mystery, for the way to wealth is seldom shared. There is no doubt that the complete disappearance of all charts of the earliest period is due to their secret nature and to their importance as political and economic weapons of the highest order.[78]

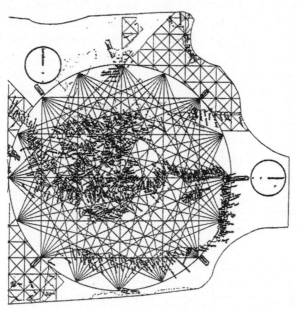

An enhanced drawing of the Carta Pisana, oldest surviving maritime chart, circa 1275.

The international competition for cartographical knowledge was intense in the sixteenth century, "and *piracy* is the term which most adequately describes the technique employed in acquiring it." To the privateers, "genuine Spanish charts of any part of the Americas were real maritime prizes, rated as highly by the French and English as the gold bullion which might be in these ships' strong rooms."[79]

In 1503, the Spanish monarchy took a major step toward institutionalizing the robbery of mariners' knowledge by forming a special agency called the Casa de la Contratación de las Indias. Its mission was to regulate commercial relations with Spanish colonies in the Americas by imposing strict controls over all navigational activities. One of its primary concerns was to exercise a monopoly over the valuable information embodied in sea charts. By law, navigators could not sail from Spain without a license issued by the Casa, and they would be fined if they were caught using any charts other than officially provided ones. On return from overseas, they were required to turn in their charts, upon which they were obliged to have recorded "every land, island, bay, harbour, and other thing, new, and worthy of being noted."[80] This information was compiled and kept under lock and key, with the Casa's pilot-major and cosmographer-major having the only keys. The regulations were counterproductive, however, because the mariners evaded them:

> Navigators were reluctant to put down on paper what they had discovered, with the result that printed maps and charts were always scarce and there was often a lag of from two to twenty years between the date of a new discovery and the time it became incorporated on the map.[81]

The connection between cartography and imperialism did not begin with the Portuguese or Spaniards. The earliest surviving maps, inscribed on clay tablets, were associated with the conquest of independent city-states in Mesopotamia about 2300 B.C.E. by the "first imperialist," the Semitic chieftain Sargon, founder of the Akkadian empire. When Alexander the Great embarked on his campaign of world domination, "the whole expedition was planned with the deliberate aim of expanding existing geographical knowledge."[82] The same kind of knowledge was essential to the Roman empire as well, but Strabo had fortunate connections that allowed him to circumvent the usual military secrecy:

> Since the Romans have recently invaded Arabia Felix with an army, of which Aelius Gallus, my friend and companion, was the commander, and since the merchants of Alexandria are already sailing with fleets by way of the Nile and of the Arabian Gulf as far as India, these regions also have become far better known to us of to-day than to our predecessors.[83]

Although Phoenician sailors most likely used maps of some sort to guide their oceanic voyages three thousand years ago, the earliest existing examples of sea charts are only a little more than seven hundred years old. The oldest, the *Carta Pisana*, dates from 1275 c.e. Not until about 1300 do several copies of a chart of the Mediterranean and the Black Sea suddenly appear in the documentary record, but its level of sophistication clearly reveals it to be a collaboration of many people over many years: "it was much too accurate and detailed to be the work of any one man or any one group of navigators, nor could it represent the surveys of any one generation; the area is much too large and the details too complex." The evidence suggests that it "originated in Genoa as a compilation of smaller charts of limited stretches of coast line sailed and charted by local fishermen and the skippers of small, coastwise traders."[84]

GENOA, VENICE, AND THE PORTOLAN CHARTS

AT THE END of the fourteenth century, the most advanced sailors in the world were Chinese. China's fleets of that era were not equaled in size anywhere until the twentieth century, and their range of action extended from Taiwan through the China Seas, into the Indian Ocean, over to the Red Sea, and down the east coast of Africa. Some of their ships were more than four hundred feet long—so big that Columbus's *Niña*, *Pinta*, and *Santa María* could all have fit on their decks with room left over. We have already noted in chapter 3 the contributions of Chinese sailors and shipwrights to European nautical knowledge. But in 1433 Ming dynasty Emperor Gaozong decided China had nothing to gain from trading with "barbarians" (i.e., the rest of the world) and ordered that the fleets be scrapped. The Chinese thus for the most part withdrew from the oceans, which meant that from then on, development of the navigational sciences would occur elsewhere.[85]

In the Mediterranean, meanwhile, the leading fourteenth-century seamen hailed from Genoa and Venice. In 1317, long before Prince Henry's birth, the king of Portugal created his country's navy by hiring Genoese navigators and sailors, thereby purchasing their knowledge of the seas. And soon thereafter,

"Portuguese nautical science was . . . also in the possession of Spain, if only because so many Portuguese pilots and shipmasters offered their services to the rival Crown."[86] It was a science transmitted not by books or university courses but through the daily work of mariners.

The academic geographers of early modern Europe—they called themselves "cosmographers" or "cosmologists"—were less inclined than their classical counterparts to "trust the sailors." As a result, "the cosmographers' maps of the world, filled with beguiling speculation, were often useless for actual navigation."

> Although it might be expected that pilots worked closely with cosmologists, that was far from the case. Pilots were hired hands who occupied a lower social stratum. Many of them were illiterate and relied on simple charts that delineated familiar coastlines and harbors, as well as their own instincts regarding wind and water. The cosmologists looked down on pilots as "coarse men" who possessed "little understanding." The pilots, who risked their lives at sea, were inclined to regard cosmologists as impractical dreamers.[87]

The pilot's books of that early era, *portolani*, demonstrate the superiority of the seamen's knowledge to that of the cosmologists. Derived from "the richness of detail and refinement of bearings that were now accumulating in the notebooks of literate sea-masters, or were stored in the memories of local pilots,"[88] the *portolani* were designed for use at sea to find targeted harbors and to avoid reefs and shoals while entering those harbors.

The intense mercantilism of the Italian ports created a social climate favoring the likelihood that

> some group of men would be found to set navigation on a firm mathematical basis. There is no clue to the names of any such men, but a threefold advance took place which could hardly have occurred at random. In the first place, the scattered sailing directions to be found all over the Mediterranean and Black Seas were collected into a single coherent whole. In the

second place a scale chart of the corresponding area was drawn for use with the existing advanced type of magnetic compass. And in the third place a method was devised by which a ship-master, sailing with such directions, chart and compass, could work out arithmetically his course made good. All three appear to be Italian in origin.[89]

The charts accompanying the *portolani* were highly accurate representations of stretches of coastline, with scales of distances and directional lines ("rhumb lines") by which a pilot could determine the bearings to set a course for a desired destination. Use of these charts required of the mariners a certain degree of mathematical aptitude; to plot their courses, they had to be proficient with "the two instruments that always lay to the hand of the practical geometer—hitherto only the architect or master-mason and the surveyor—namely the ruler and pair of dividers or compasses."[90]

The mathematical knowledge of the navigators was not, as is often assumed, simply a trickled-down, low-culture derivative of high-culture university mathematics. To the contrary, the early development of practical mathematics was relatively untainted by scholarly input.[91] With regard to the use of mathematics in the "great Affairs of *Navigation*, the Art *Military*, &c." one late-seventeenth-century commentator observed that "we see those Affairs are carried on and managed by such, as are not great Mathematicians; as Seamen, Engineers, Surveyors, Gaugers, Clock-makers, Glass-grinders, &c." By contrast, "the Mathematicians are commonly speculative retired, studious Men, that are not for an active Life and Business, but content themselves to sit in their Studies, pore over a *Scheme*, or a *Calculation*."[92]

Portolan charts were "conceived by seafaring men and based strictly on experience with the local scene, that is, with the coasts and harbors actually used by navigators to get from one place to another."[93] They "represented an aggregation of decades, in some cases centuries, of empirical observations by seafarers."[94] Eventually, by the sixteenth century, sea charts

Portolan chart depicting the coasts of Portugal and West Africa, and the Atlantic islands. Grazioso Benincasa, 1462.

would be commercially produced by skilled draftsmen, but "it was upon the trading ships that they relied for the precise data they worked upon."[95] Italian seamen were the first to chart carefully the coasts of the Mediterranean and Black Sea. Later, "Dutch skippers learned the coast and harbors, the prevailing winds and currents, the reefs and shoals of Western Europe as they were never known before," making it "logical that the Dutch should produce the first systematic collection of navigational charts bound together in book form."[96]

The portolan charts were the building blocks that combined

to reveal the true shapes of the continental landmasses, an obvious prerequisite of accurate world maps. But the mariners' contributions to cartography encountered resistance on the part of the elite cosmologists, who—even after the discovery of new continents across the Atlantic—were still reluctant to abandon their allegiance to ancient authority: "Geographers of the fifteenth and sixteenth centuries leaned on [Ptolemy's *Geography*] so heavily, while ignoring the discoveries of maritime explorers, that it exerted a powerful retarding influence on the progress of cartography."[97] That is undoubtedly why the path of development of scientific mapmaking went not through the universities but through the workshops of skilled craftsmen who could assimilate the new information with less difficulty. Some of these "superior manual laborers," such as Gerard Mercator and Abraham Ortelius, gained patronage, wealth, and renown because the geographical knowledge they had assembled from seafarers was of immense economic and military value to the ruling classes.

But as cartographers, Mercator and Ortelius were sixteenth-century beneficiaries of the labors of many predecessors. Who were the many not-so-famous chart-makers of the fourteenth and fifteenth centuries? Of the forty-six whose names are known to us, at least one, Grazioso Benincasa, was a nobleman, and another, Pietro Vesconte, came from a Genoese ruling family. "It would be quite wrong," however, to assume

a similar profile of high birth and exalted social status for all the other named and nameless chartmakers of the period. It is precisely patricians like Benincasa whom history remembers; his humbler colleagues have no memorials but their charts. A fairer picture of a chartmaker's true social position is probably the one that emerges from Agostino Noli's petition of 1438. Describing himself as "very poor," Noli managed to persuade the Genoese authorities to grant him ten years' tax exemption—among other reasons because they accepted that his work, though time-consuming, was not very lucrative.[98]

Mariners would continue to hold the key to the improvement of cartography for at least two centuries following the era of Mercator and Ortelius. "Cartographic inaccuracies," David Landes explained, "persisted into the nineteenth century" in spite of the mapmakers' best efforts. "Only navigators could correct this plague of errors; for competent astronomers were few, and fewer still were those among them prepared to sacrifice their comfort and risk their lives on long voyages to distant and unknown places."[99]

It is not necessary simply to assume that improvements to the charts were based on sailors' input; in 1403 a chart-maker named Francesco Beccari explicitly attributed "the marrow of the truth having been discovered" to

> the efficacious experience and most sure report of many, i.e. masters, ship-owners, skippers and pilots of the seas of Spain and those parts and also many of those who are experienced in sea duty, who frequently and over a long period of time sailed those regions and seas.[100]

Before leaving the subject of cartography, let us also acknowledge the contributions of a nonnautical plebeian occupational group. The sailors provided the data that allowed the contours of the land to be accurately drawn, but how did the interior lands come to be mapped? Who filled in the blanks? The detailed knowledge that made comprehensive mapping of the terrain possible was produced by multitudes of anonymous surveyors. The "vast program of colonization" that accompanied the foundation of the Roman Empire stimulated the creation of a land-surveying profession, and the expansion of the empire led to an "upgrading of the role of the land surveyors (*agrimensores*) working for the Roman state as well as of that of surveyors on other applications of large-scale mapping in the towns and in engineering projects." Wealthy patricians also mobilized teams of *agrimensores* to draw up maps of their country estates. "From such detailed plans of individual sites, it was a logical step to large-scale representations of entire towns."[101]

Another well-documented example, from a much later period, has to do with the mapping of the North American continent: "One has only to observe the unusual occurrence of straight lines all over the map of the United States, in roads as well as in state boundaries, to realize that here is a country more heavily indebted than any other to the work of the surveyor."[102] As noted in chapter 2, much of the initial information was provided by Amerindians, but the job was taken up and completed by "self-taught men engaged in a variety of scientific endeavors." The colonial American mapmakers and surveyors were underappreciated in their own time because "the practical sciences were regarded as mundane endeavors unworthy of academic attention":

> Nevertheless, it was they who charted the wilderness, navigated the unknown waterways inland and along the coasts, established the boundaries of personal properties and of provincial territories, made the instruments with which these tasks were accomplished, and instructed others in these skills. . . . Inasmuch as the combined efforts of all these "little men of science" made a significant impact on the formation of the early settlements and their unification first into colonies and then into a nation, and since in a true sense they were indeed participating members of a diverse scientific community, they deserve more than a footnote in history.[103]

SAILORS AND ASTRONOMY

ASTRONOMY IS AMONG the purest of people's sciences in its origins, because its earliest data acquisitions occurred long before the rise of distinctions between elites and commoners.[104] But it is often thought of as the most elite of sciences because, not being amenable to experimental manipulation, it was one of the few nonempirical sciences in the era of the Scientific Revolution.

Its major advances were therefore accomplished by mathematically inclined scholars. But before Copernicus and even before Ptolemy and Aristotle, the long prehistory of astronomy was significantly shaped by the daily practice of mariners.

Seamen were less stationary than most people in the ancient world. Their occupation required them to move great distances—and relatively rapidly—which gave them opportunities to notice systematic changes in the night sky as a function of where they were on Earth. The sailors of the Red Sea four thousand years ago "must have played an important part in advancing knowledge of astronomy" because their north-south voyages "involved a change of latitude, and so of Sun and star, of at least 20°, to which must be added the 5° or 6° sailed northwards by Egyptian mariners, who are known to have trafficked with Byblos in Syria from very remote times."[105] The Roman encyclopedist Pliny, noting that the motions of the heavens "are most clearly discovered by the voyages of those at sea," observed that with the passing of the night hours "the stars that were hidden behind the curve of the ball suddenly become visible, as it were rising out of the sea."[106]

Observations of this kind led to the knowledge that the earth is round rather than flat. Most of us were taught in our earliest school years that before "Columbus sailed the ocean blue in fourteen-hundred-ninety-two," he was virtually alone in his insistence on the roundness of the earth. That myth (initiated by Washington Irving's biography of Columbus) is now routinely debunked in standard history textbooks, but it has been replaced by another false story. Now students are taught that Columbus learned of the earth's spherical shape from the university scholars of his era, who in turn learned it from Aristotle and other ancient authors. But because the sailors of that century were allegedly not as enlightened as the scholars, Columbus's crew is said to have feared sailing over the edge of the earth.

It is evident from the arguments of the ancient authors, however, that their knowledge of the earth's roundness was drawn from the experience of seafarers, and there is no reason to believe that sailors of subsequent ages were any less aware of it. "The spherical shape of the earth is seen," Strabo wrote, "in the phenomena of the sea and of the heaven":

> For instance, it is obviously the curvature of the sea that prevents sailors from seeing distant lights at an elevation equal to that of the eye; however, if they are at a higher elevation than that of the eye, they become visible, even though they be at a greater distance from the eyes; and similarly if the eyes themselves are elevated, they see what was before invisible. . . . So, also, when sailors are approaching land, the different parts of the shore become revealed progressively, more and more, and what at first appears to be low-lying land grows gradually higher and higher.[107]

Aristotle likewise drew on the experience of mariners when he deduced the sphericity of the earth from the fact that "the stars above our heads change their position considerably, and we do not see the same stars as we move to the North or South."[108] It is doubtful, then, that the seamen of Columbus's time were afflicted with flat-earth ignorance. As for Columbus himself, he was already an accomplished sailor before he could read, so it is more likely that he learned the world was round from his nautical mentors than from books by scholars.

It is true that Columbus had a hard time selling the idea that he could sail due west and wind up in Asia, but it was not because anybody thought the world was flat. The problem was that most people believed the world to be much bigger than Columbus thought it was. Sailors were not afraid of falling off a flat earth, but they were understandably reluctant to set off around a globe so large that they would run out of provisions long before they made land. As it turned out, Columbus was wrong and all of those who said the world was much larger than he thought were right. But what neither Columbus nor any of his contemporaries knew was that there was another land mass between Europe and Asia.

DIRECTION AND LATITUDE

THE MOST FUNDAMENTAL benefit that astronomical knowledge provided to early European navigators was the ability to know their direction in the middle of a vast sea. With no landmarks by which they could orient themselves, they turned to the sun and the stars. During daylight hours, they had the sun's east-to-west path to go by, and at night they became familiar with certain constellations that always occupied the northern sky. Steering toward those star groups meant heading north, keeping them on their right meant they were westbound, and an eastward trip required keeping them on their left. Odysseus, Homer tells us, "never closed his eyes, but kept them fixed on the Pleiads, on late-setting Bootes and on the [Great] Bear," because the goddess Calypso had told him to keep those stars to his left as he sailed across the sea.[109]

But better than the Great Bear as a guide to steer by was the Little Bear, which Strabo said "did not become known as such to the Greeks until the Phoenicians so designated it and used it for purposes of navigation."[110] According to the poet Callimachus, it was none other than Thales who "measured out the little stars of the Wain, by which the Phoenicians sail."[111] There is some confusion here, because the Wain is usually another name for the Great Bear, but its "little stars" apparently constituted the Little Bear, and Thales is being credited with alerting the Greeks to the distinction. In any event, these acknowledgments constitute powerful testimony to the seminal influence of Phoenician seamen on Greek astronomy. Sailors on night watch also noted the regularity with which the two Bears revolved around a fixed central point, and learned to use them as a celestial clock for telling time.

A second very important discovery was that it was possible to estimate how far north or south a ship had traveled from its port of origin by measuring how much the apparent pattern of the stars overhead had shifted during the voyage. From that came the recognition that it was possible to measure one's latitude by means of astronomical observations. This was important for more than simply knowing a ship's immediate position; it was the key to a principal technique of transoceanic navigation called "running down the latitude." A captain would steer north or south to the known latitude of the port of destination, and when it was reached, turn east or west and maintain that latitude until landfall. The way from the Iberian peninsula, for example, to Santo Domingo was to sail in a southerly direction until arriving at about latitude eighteen degrees north and then to head due west.

The most straightforward method of ascertaining one's latitude is to measure the angular distance of Polaris, the polestar, above the horizon: the larger the angle, the farther north one is, and the proportion is exact. Other stars and the sun can also be used (and south of the equator must be used) to determine latitude, but the observations and calculations are more difficult because other variables are involved.

Polaris (which the Greeks called "the Phoenician star" due to its maritime associations) was the reference point of choice because it seems almost stationary in the night sky. As the earth rotates, all the other stars appear to revolve around Polaris because it is almost aligned with the earth's axis. *Almost*, but not quite, which means that as navigators required more precise determinations of their latitude, their astronomical prowess had to grow to include knowledge of how to correct for the small but significant displacement of Polaris from the axis.[112] Increased precision also demanded the introduction of measuring instruments and mathematical tables. The sailors' initial contri-butions to astronomy were enduring: "the only scientific contributions to the subject made in the next 2000 years were improvements" in instruments and tables.[113] This is an important corrective to claims that Henry "the Navigator" was responsible for initiating astronomical navigation and giving mariners the ability to find their latitude at sea. Systematic observations made at Henry's headquarters at Sagres improved the accuracy of the tables, but the basic techniques had long been known.

The sophisticated quadrants, sextants, and octants of the eighteenth century evolved from the ancient astrolabe and the primitive cross-staff and represented the collective creativity of innumerable sailors and other artisans. Although most were anonymous, one particular innovation can be credited to an English seaman named John Davis, "an ingenious mariner with a practical turn of mind who got tired of peering into the eye of the sun in order to find his latitude."[114] In 1607, he published an account of his invention, the back-staff, which solved the centuries-old problem of navigators damaging their eyesight while making solar observations.[115]

Back-staff. 1676, American colonies.

Another major improvement, the reflecting quadrant, was independently invented by at least three people, one of whom was "a poor glazier of Philadelphia" named Thomas Godfrey.[116]

THE LONGITUDE PROBLEM

SAILORS COULD TELL by the stars where they were on a north-south meridian but not how far east or west they were. To locate themselves precisely during a transoceanic voyage, they would have to know their longitude as well as their latitude. Navigators attempted to estimate distances by dead reckoning— that is, by multiplying the length of time they had been at sea by the average speed of the ship. The problem was that there was no reliable way of measuring a ship's speed as it moved through moving waters. Experienced pilots could make reasonably good guesses at how fast they were going by spitting in the water and timing (by saying Hail Marys) how quickly the spittle was carried away, but that was obviously not a high-precision method.

More sophisticated dead-reckoning techniques were devised over the centuries, but truly accurate measurement of longitude at sea remained a virtual impossibility. It was the central scientific problem of the so-called Age of Exploration; the Spanish, French, Dutch, and British governments mobilized their resources in an attempt to solve it, and the leading figures of the Scientific Revolution gave it their best efforts. But in spite of the cumulative brainpower of the likes of Galileo, Newton, William Gilbert, Christian Huygens, and Edmund Halley,[117] it was not a member of the scientific elite but a skilled craftsman—a clockmaker named John Harrison—who finally, well into the eighteenth century, provided the solution to the longitude problem.[118]

Throughout the sixteenth and seventeenth centuries, numerous methods were suggested and extensively tested. One that proved workable in principle was Galileo's proposal to utilize the four moons of Jupiter that he had discovered with his telescope. The predictability and frequency with which they eclipsed their mother planet made it possible to draw up tables that could tell an observer with a telescope the exact time at some other place that could be used as a prime meridian, or "zero point" for longitude. Then stellar measurements could determine the observer's exact local time at any location, even in the middle of an ocean. Subtract one time from the other, and *voilà*, the longitude has been calculated, because clock time is a function of the earth's rotation and the time differential is equivalent to a specific east-west distance. The circumference of the earth is by convention divided into 360 degrees of longitude; in 24 hours the earth rotates through 360 degrees; therefore each hour of time differential equals fifteen degrees of longitude. If the time at the prime meridian is noon, and your local time is 3:00 PM, then your longitude is 45 degrees east of the prime meridian.

Galileo's method was eventually found to be useable on land but not at sea, because precise telescopic observations of the moons of Jupiter proved impossible to perform on a rolling, pitching ship. An analogous method based on measuring the position of the earth's moon was thoroughly investigated by Isaac Newton. Newton's intensive analysis of "lunar distances" turned out to be scientifically fruitful in serendipitous ways—it led him to his formulation of the law of universal gravitation and contributed to his development of the calculus—but he was unable to predict adequately the moon's erratic motions. It was, he declared, the only problem that made his head ache.[119] Useful tables of lunar positions were developed in the mid-eighteenth century, but they were never considered fully satisfactory as a basis of longitude calculation at sea.

A third method, championed by Edmund Halley (for whom the famous comet was named), had nothing to do with astronomy, but relied instead on characteristics of the earth's magnetic field. Navigators had long been aware that their compass needles did not point precisely to true north as determined by the stars. (This difference, called the angle of declination, was known in China by the ninth century C.E. The earliest definite evidence that it was known in Europe dates to about 1450, which suffices to disprove the tradition that it was discovered by Columbus.[120]) Sailors also noticed that the angle of declination tended to vary according to how far across the Atlantic they were. That observation suggested that perhaps a mathematical formula might be found that could relate the compass's deviations to longitude. Halley enlisted the efforts of mariners to make careful records of their compass readings in hopes of compiling data that would yield to analysis. It was a logical and worthwhile course to pursue, but it did not give the hoped-for answer because the earth's magnetic field proved to be insufficiently uniform and unpredictably variable. That conclusion was based on observations made by a craftsman named Robert Norman.

Norman, an instrument-maker by trade, also discovered that magnetic compass needles rotate and point around not only one axis, but two.[121] In addition to pointing approximately toward the north, the needle also tends to "dip" and point slightly downward. The needle's dip varied with longitude, which suggested another measurable variable that might serve to determine longitude at sea. William Gilbert brought Norman's suggestion to the attention of the scientific elite,[122] and it, too, was systematically investigated with the collaboration of working seamen, but again the instability of the earth's magnetic field rendered it futile. Although the two attempts to determine longitude by means of the compass needle were unsuccessful, they stimulated knowledge of geomagnetism.

Theoretically, the most straightforward way to measure longitude at sea would be to carry aboard ship an accurate clock set to the time of the prime meridian. A simple comparison with local solar-measured time would then give the longitude. But the operative word is "accurate": no existing clocks were even remotely accurate enough for that purpose. Even on land, the best portable clocks tended to gain or lose several minutes a day, but the ship's motions and the variability of temperature and humidity at sea made timekeeping all the more difficult. A clock that lost or gained only one minute a day would have yielded errors of hundreds of miles after just a few days at sea—and crossing the Atlantic typically took two months.

This challenging problem stimulated Christiaan Huygens and other savants to study the theoretical mechanics of timekeeping, and their work resulted in the development of pendulum-controlled clocks.[123] But alas, pendulums are particularly unsuitable for conditions aboard a rocking ship, so scholars looked elsewhere for a solution to the longitude problem. It would be found, Isaac Newton said with hauteur, "not by Watchmakers or teachers of Navigation . . . but by the ablest Astronomers."[124] He was wrong—it was a persistent "watchmaker," John Harrison, who in the 1760s eventually succeeded. Harrison was not, in fact, a clockmaker by trade but a carpenter who had taught himself how to make clocks as a sideline. His first precision clocks were made of the carpenter's material, wood, rather than metal.

The theoretical predispositions of elite scientists like Huygens hampered them in their attempts to design more accurate clocks. "For one thing, they were not always ready to accept the fact of temperature effects on solids. . . . Craftsmen, unhampered by theory, knew better. A number of branches of industry were based on the empirical observation that heated metal contracts on cooling." As timepieces increased in accuracy, "clockmakers and watchmakers quickly became aware that temperature made a difference."[125]

John Harrison's "Chronometer no. 4," 1760.

After creating a timepiece adequate to the task (a clock that lost only five seconds on an eighty-one-day test voyage to Jamaica!), Harrison encountered a great deal of resistance on the part of the gentlemen scientists, who refused to acknowledge his accomplishment. Eventually, after repeated demonstrations made his triumph undeniable, he was awarded the large cash prize that the English Parliament had promised for more than half a century to anyone who could find a way to measure longitude accurately at sea. But Harrison's chronometer was a very expensive instrument, so even after it gained general acceptance, it was many years before precise longitude measurement became routine at sea. The cumbersome and unreliable method of lunar positions continued to be used into the early twentieth century.

Meanwhile, the quest for finding longitude at sea had stimulated investigations that had made it possible to measure longitude on land, and that in turn gave a great impetus to progress in mapmaking. With the ability to pinpoint specific locations on the earth's surface by both latitude and longitude, scientific cartography as dreamed of by Ptolemy many centuries earlier at last became a reality. But that depended on two pieces of apparatus that had been created by artisans: the sufficiently accurate clock and the telescope. Although credit for inventing the telescope is often ascribed to Galileo, Galileo himself knew that was untrue. "Indeed, we know," he wrote,

> the Hollander who was first to invent the telescope was a simple maker of ordinary spectacles who in casually handling pieces of glass of various sorts happened to look through two at once, one convex and the other concave, and placed at different distances from the eye. In this way, he observed the resulting effect, and thus discovered the instrument.[126]

It is also worth noting that the original motivation for developing the two lenses into an instrument for seeing distant objects was not astronomical but nautical—not to look at the stars, but to spot and identify ships far out at sea. The commercial and military advantages of that capability are obvious. Galileo himself originally promoted his improved telescope to his patrons by stressing this aspect of their utility. In a letter of August 29, 1609, he told of demonstrating it for "the entire Senate" of Venice. "Numerous gentlemen and senators," he wrote,

> though old, have more than once climbed the stairs of the highest bell towers of Venice to observe at sea sails and vessels so far away that, coming under full sail to port, two hours and more were required before they could be seen without my spyglass.[127]

That the telescope was created from the lenses of a spectacles-maker suggests a further question. Convex-lens eyeglasses are known to have been invented more than three centuries earlier, in the 1280s, but how? Some authors have speculated that they resulted from the conscious application of scientific principles articulated by Oxford University scholars Robert Grosseteste and Roger Bacon. The evidence suggests, however, "that eyeglasses emerged not by scientific inspiration but from the world of glass-makers and the cutters of glass, gems, and crystal" (that is, in the *empirical* science of craftsmen rather than the *theoretical* science of schoolmen).[128]

THE MAGNETIC COMPASS

THINK OF NAVIGATIONAL instruments and you will undoubtedly think first of all of the magnetic compass. As we have seen, seamen had for millennia been able to successfully traverse the open seas without it, but its introduction nonetheless amounted to a major technological revolution in seafaring, which stimulated the historic expansion of global commerce and paved the way for European world domination. In the thirteenth century, after Mediterranean sailors began routinely using magnetized needles for direction-finding, they "no longer had to waste time ashore waiting for winter to pass." The ever-cloudy winter skies of the Mediterranean rendered navigating by the stars impracticable, but with a compass to guide them the Venetian fleets, could "make two round-trips a year instead of one—and neither fleet had to winter overseas."[129]

The origins of the magnetic compass are obscure, but, not surprisingly, a heroic tradition exists that credits it to a lone genius. If you travel to Amalfi, Italy, you will find in the center of town a large bronze statue of one Flavio Gioia, with a plaque asserting that it was he who invented the compass in 1302. But as an Italian historian wrote more than a century ago, "Flavio Gioia never existed. He represents only a kind of myth, created late after his presumed lifetime, and hence suspect. He is a fantasy produced by the fertile southern imagination of the people of Amalfi and elsewhere."[130]

The fact that certain elongated rocks ("lodestones") exhibit a north-south directional property was originally discovered in China more than two thousand years ago. Magnetic compasses were first used in China for divination and religious ritual, but documentary evidence establishes that Chinese sailors began steering their ships by "the south-pointing needle" no later than 1117 C.E.[131] It is likely that knowledge of these strange devices, with their seemingly supernatural powers, was transmitted from China to the Mediterranean world by merchants— anonymous predecessors of Marco Polo. The first recorded reference in Europe to a magnetized needle being used by sailors for direction-finding appeared in a book by an Englishman, Alexander Neckam, in 1187 C.E., but the casualness of its mention suggests that by then its use may already have become a matter of common, everyday practice.[132]

The earliest navigational compasses were simply magnetized needles floating on water or suspended by threads in order to give a gross estimate of direction. Over time, these primitive instruments developed into the apparatus familiar to us: a pivoting magnetized needle attached to a round printed card featuring a "wind rose" divided into 360 degrees. This transformation represented a series of innovations over many years; the "invention" of the magnetic compass was the collective achievement of several generations of mariners and instrument-makers.

OCEANOGRAPHY: WHAT LIES BELOW?

OF ALL THE earth's territory, the most foreign to everyday human experience is the vast, three-dimensional area beneath the surface of the seas. The depth of their waters and what lies on the ocean floor were mysteries investigated first of all by fishermen and other mariners. These were not matters of idle curiosity to sailors; it was of the utmost importance to them to know, as they approached land, the exact depth of the water separating the bottom of their vessel from solid earth. And because the materials that constituted the ocean floor varied significantly from place to place, identifying them provided experienced pilots with valuable clues as to their location.

Both of these bodies of knowledge were gained by the sailors' practice of taking soundings with the lead and line. The official hydrographer of the British Navy wrote in 1955,

> There is no navigator afloat today who will not approve the words, written some hundreds of years ago, "navigating is not by chart and (magnetic) compass, but by the sounding lead!" Though the officer of the watch may be surrounded by all the accoutrements of a scientific age it still remains a basic fact, and one he will be wise not to forget for a moment, that if the draught of a ship exceeds the depth of water he is most assuredly aground![133]

The impossibility of documenting the origins and development of this important source of scientific information is evident in the chronological distance separating the earliest mention of the practice by Herodotus in the fifth century B.C.E. from the first descriptions of the equipment two millennia later. Herodotus, describing the approach to Egypt by sea, wrote that "when you are still a day's sail from the land, if you let down a sounding-line you will bring up mud, and find yourself in eleven fathoms' water."[134] There is no reason to assume that the Greek sailors of his day were the first to take soundings; the Phoenicians surely did so, and their Minoan predecessors probably did as well.

There are other ancient literary accounts of sailors taking soundings—in the New Testament, for example[135]—but details about the leads and lines themselves are not found recorded before the late sixteenth century C.E. The standard apparatus, which remained very much the same in later centuries, was simple: a fourteen-pound lead weight attached to a line two hundred fathoms long.[136] The line was marked at ten- and twenty-fathom intervals for measuring depths. By this method the existence and dimensions of the continental shelves were discovered. When a pilot is "within soundings" the lead hits bottom, but far out at sea, beyond the edge of a continental shelf, it does not. The cumulative knowledge this practice produced made it possible to map the contours of much of what lay hidden beneath the ocean's surface.

The lead was designed to pick up a sample of the material on the ocean floor; Herodotus's sailors found mud attached to theirs. Whatever it brought up—mud, sand, silt, coral, seaweed, shell fragments, rocks of varied composition—tended to remain constant at specific locations over long periods of time, so pilots learned to recognize where they were by sampling the ocean floor. The information they collected and recorded in their pilot-books could have been obtained in no other way.

KIDNAPPING AND THE ORIGINS OF ETHNOLOGY

IN ADDITION TO knowledge about the physical characteristics of the earth, travelers and mariners also gained and transmitted information about the human societies they encountered in far-away places. Their observations of the unfamiliar customs of foreign cultures constituted the beginnings of ethnology and ethnography. The novelties they witnessed were intrinsically interesting, but their primary motive for gathering ethnological data was trade. Merchants wanted most of all to find out what commodities the foreigners might fancy and what valuable things they might be willing to give in exchange.

Europeans of the era of Henry "the Navigator" particularly craved information about the black people of Guinea from whom Arab traders were known to obtain gold. Prince Henry was not the first to utilize kidnapping as a routine method of knowledge confiscation, but he was certainly one of its most relentless practitioners. According to Peter Russell, Henry was "eager to have any information his people could pick up about the economic, ethnographical and political situation" in Guinea, and he "early realized that a better source for this material than anything Portuguese eyewitnesses could supply was . . . the local inhabitants of these regions." For that reason, the orders he gave his sailors "included instructions that, when a new country was discovered, one or two local people must be secured by force or deception and brought back to Portugal so that he or his officials could interrogate them at leisure about the land whence they had been kidnapped."[137]

In one such case, captives described the mouth of the Senegal River to him in such detail that Portuguese sailors were able to recognize it the first time they saw it. Henry also gained access to the linguistic knowledge of African men and women by having them kidnapped and forcing them to serve as translators in Portuguese dealings with other Africans.

As an adjunct to kidnapping peaceful natives to extract information from them, the Portuguese utilized their own least fortunate citizens—convicts sentenced to death or banishment—as a shield against not-so-peaceful natives: "Cabral took twenty among his fleet, da Gama ten or twelve. They were used as guinea-pigs to test the temper of possibly hostile natives, or put ashore in the hope that if the ships had to put in there again they would find water and food."[138]

A Venetian seaman in Henry's employ, Alvise da Cá da Mosto (Cadamosto for short), wrote an account of two voyages to Guinea he made under Henry's auspices in 1455 and 1456. Cadamosto's *Navigazioni* recorded valuable botanical and zoological data, including his observations of the habits of elephants and hippopotamuses, that are "notable for the accuracy and completeness with which eyewitness observation of these and other animals and birds of the region had equipped him." But what Henry was really interested in was Cadamosto's investigations into the peculiar way in which business was conducted between Muslim gold merchants and the blacks who mined it. He described the system of nocturnal silent barter that allowed the gold miners and merchants to carry out their exchanges without using language or any other form of face-to-face communication. Cadamosto augmented his direct experience by interrogating nomadic tribesmen and others who had been brought to Portugal as slaves.[139]

Henry's imperialistic ventures, which included pioneering the maritime trade in African slaves, were underwritten by the ideological support of the Church. The recognition that knowledge was an essential element of power and that black people were as capable of using it as whites was implicit in papal bulls that "specifically forbade any Portuguese attempt to teach Africans about navigation. To do so, the curia feared, might be to undermine the position of the Europeans."[140]

Kidnapping was not Henry's only method of knowledge acquisition. One of his squires, João Fernandes, happened to have talents that allowed the gathering of ethnological data in a more humane way. Because he could speak Arabic and was already somewhat familiar with Islamic customs, Fernandes was able to go directly to the Saharan interior and live among the people there. For seven months this "discerning and objective observer" gathered intelligence for Henry by traveling around the western Sahara and asking questions of anyone who would talk to him. He learned that camel caravans used magnetic compasses to guide them across the Sahara. He reported that the nomads' staple food was camel's milk but that they also placed a high value on wheat, a commodity that the Portuguese could supply at a very good profit. This latter fact was the kind of information that Henry was especially interested in acquiring.[141]

Henry's program of intelligence-gathering was continued by Portugal's rulers after his death in 1460. An expedition to Sierra Leone in 1461 or 1462 was ordered by King Afonso V to bring back, by force if necessary, "one of the blacks of that land so that he could give an account of his country, either through the many black interpreters to be found in Portugal, or because, with time, he himself became able to make himself understood [in Portuguese]."[142] Thanks to Cadamosto's account, the outcome of this affair is known: an African was kidnapped, and after an extensive search, a female slave in Lisbon was found who could communicate with him. This is a telling example of how "knowledge transfer" actually occurred in real life.

MONTAIGNE'S SERVANT

THE WAY ETHNOLOGICAL information reached and affected Europe's intellectual elite is beautifully illustrated by Montaigne's influential ruminations on the native peoples of the Americas and what their lifestyles revealed about human nature. He related in considerable detail what he had learned from one of his servants, a former sailor "who had lived for ten or twelve years in that other world which was discovered in our century," among Amerindians in what is now Brazil. To explain to his peers why he would accept the testimony of a man of such low social status, Montaigne inverted the usual correlation of nobility with trustworthiness. His informant was believable, he asserted, *precisely because* he "was a simple and ignorant fellow." That meant that he was "the more fit to give true evidence; for your sophisticated men . . . cannot help altering their story a little. They never describe things as they really are."[143]

His servant, Montaigne continued, was "so simple that he has not the art of building up and giving an air of probability to fictions, and is wedded to no theory." Besides, the man's descriptions were corroborated by "several sailors and traders" who had been on the same voyage. Leaving aside Montaigne's condescending depiction of his servant, it is significant that he considered this "ignorant" seaman's knowledge to be superior to that of the scholars: "I shall content myself with his information, without troubling myself about what the cosmographers may say about it."[144]

This is a small indication of the general significance of mariners' knowledge as a historical prerequisite of modern science. When Columbus and his crew returned to Europe with the news of their success in reaching the Indies, their tales set off a mad rush, and within a few years, transatlantic voyages were commonplace. An explosive growth of natural knowledge resulted from the discovery of peoples, places, and species of plants and animals that had been unknown to Aristotle and Ptolemy. European scholars were compelled to face as facts numerous phenomena the ancients had been quite sure could not possibly be observed because they were bound not to exist. Examples are Aristotle's denial that the tropics could be inhabited; Ptolemy's mathematically derived conviction that all dry land is confined to part of the Northern Hemisphere, and so on.[145]

The ancient natural philosophers' "narrow world" based on "all-too-rational speculations was now being blown to pieces. And this was not spontaneously being done by fellow natural philosophers, but rather at the urging of scarcely literate sailors!"[146] Those seamen "made an important contribution to the rise of modern science by unintentionally undermining the belief in scientific authorities and by strengthening the confidence in an empirical, natural historical, method."[147]

> The great change (not only in astronomy or physics, but in all scientific disciplines) occurred when, not incidentally but in principle and practice, the scientists definitively recognized the priority of Experience. The change of attitude caused by the voyages of discovery is a landmark affecting not only geography and cartography, but the whole of "natural history."[148]

But although the authority of the ancient authors as the arbiters of all scientific knowledge had obviously been severely weakened, it did not immediately crumble. Too many professorial, medical, ecclesiastical, and legal careers were founded on that authority for it to simply disappear without a struggle. The scientific elite resisted the infusion of new natural knowledge with all its might, but in the long run, its rearguard efforts were futile. Montaigne's servant ultimately proved more credible than the cosmographers. The common sense of working people prevailed and brought about the changes in worldview that have come to be known as the Scientific Revolution.

NOTES

1. See Virginia de Castro e Almeida, ed., *Conquests and Discoveries of Henry the Navigator; Being the Chronicles of Azurara.*
2. See Peter Russell, *Prince Henry "the Navigator,"* p. 374, n. 15. The German geographer was J. E. Wappäus.
3. Russell, *Prince Henry,* pp. 6–7.
4. R. H. Major and others made this claim. See Edward Gaylord Bourne, "Prince Henry the Navigator," p. 185.
5. Ibid., p. 185. On Henry's will, see Russell, *Prince Henry,* pp. 346–353.
6. Lloyd Brown, *The Story of Maps,* p. 109 (emphasis added).
7. "Deserving of special mention . . . are the three massive volumes entitled *Descobrimentos Portugueses* edited by J. M. Silva Marques (1944–71) and the fifteen volumes of *Monumenta Henricina* (1960–74)." Russell, *Prince Henry,* p. 9.
8. Ibid., pp. 6–8.
9. Clements R. Markham, *Sea Fathers,* p. 6.
10. The alleged flat-earth ignorance of Columbus's sailors is discussed later in this chapter.
11. K. St. B. Collins, "Introduction," to E. G. R. Taylor, *The Haven-Finding Art,* pp. x–xi.
12. Markham, *Sea Fathers,* p. 56.
13. Claudius Ptolemy, *The Geography,* book 1, chap. XI (Dover ed., p. 33). Ptolemy knew them as the Fortunate Isles and chose their position as the prime meridian from which all other locations were measured.
14. J. R. Hale, *Renaissance Exploration,* p. 19. These nine expeditions are those for which archival evidence has survived; undoubtedly many more occurred. When the Portuguese "discovered" the Canaries in 1336, they found the islands inhabited by a farming and herding people they called the Guanche. "By 1496 the Guanche had ceased to exist, the first indigenous people to become extinct as a consequence of European maritime expansion." Judith Ann Carney, *Black Rice,* p. 9.
15. R. A. Skelton, Thomas E. Marston, and George D. Painter, *The Vinland Map and the Tartar Relation,* p. 168.
16. E. G. R. Taylor, *The Haven-Finding Art,* p. 65.
17. See Herodotus, *The History,* book IV, 42–43; and Lloyd Brown's comment on that passage in *Story of Maps,* p. 119.
18. See J. H. Parry, *The Age of Reconnaissance,* p. 84.
19. Brown, *Story of Maps,* p. 114.
20. Taylor, *The Haven-Finding Art,* p. 38.
21. Hale, *Renaissance Exploration,* p. 99.

22. Taylor, *The Haven-Finding Art*, p. 3.
23. Brown, *Story of Maps*, p. 114.
24. Alfred W. Crosby, *Ecological Imperialism*, p. 114.
25. In fact, a Spanish explorer, Vicente Yáñez Pinzón, most likely preceded Cabral to Brazil by a few months, but as Samuel Eliot Morison explained, "the progressive opening up of [Brazil] stemmed from Cabral's rather than Pinzón's discovery." Morison, *The European Discovery of America: The Southern Voyages*, p. 224.
26. Armando Cortesão, *The Mystery of Vasco da Gama*. Ironically, Cortesão attempted to use this lack of documentation as a blank check to enhance the legend of da Gama's "genius." But, as Morison pointed out, this argument "is perhaps the most preposterous in modern history, because it uses absence of evidence as positive evidence that the Portuguese discovered everything." Morison, *The European Discovery of America: The Northern Voyages*, p. 110.
27. Anonymous, *Calcoen: A Dutch Narrative of the Second Voyage of Vasco da Gama to Calicut*. The pages are unnumbered; these incidents are described on the seventh and eighth pages of the narrative.
28. Markham, *Sea Fathers*, p. 46. This sailor's name has been variously rendered in the historical literature as Del Cano, Elcano, and de Elcano. Markham's negative assessment of his "genius" should be discounted.
29. Crosby, *Ecological Imperialism*, p. 122.
30. Laurence Bergreen, *Over the Edge of the World*, pp. 242–243. See also Morison, *European Discovery of America: Southern Voyages*, p. 435.
31. Bergreen, *Over the Edge of the World*, pp. 278, 286.
32. Ibid., p. 394.
33. Ibid., p. 394.
34. Lionel Casson, ed. and trans., *The Periplus Maris Erythraei*.
35. Ibid., p. 87.
36. Strabo, *The Geography*, vol. I, pp. 377–385.
37. For speculation concerning the connection between Eudoxus and Hippalos, see J. H. Thiel, "Eudoxus of Cyzicus"; and Lionel Casson, *The Ancient Mariners*, p. 187.
38. Casson, *Periplus Maris Erythraei*, translator's notes, pp. 12, 224.
39. From a fragment of Poseidonius quoted in Thiel, "Eudoxus of Cyzicus," p. 13. See also Strabo, *Geography*, vol. I, pp. 377–379.
40. The monsoon winds were accurately described by Marco Polo in the thirteenth century (Marco Polo, *Travels*, p. 210), but that information seems not to have helped Vasco da Gama.
41. The expert whose aid da Gama secured was possibly Ahmad Ibn Majid, by that time an elderly man, who was well known in the Arab world as the leading author of navigational treatises and guidebooks. Exactly how da Gama secured his cooperation—if indeed it was Majid who guided him—is not known.

42. Crosby, *Ecological Imperialism*, p. 120.
43. Jack Beeching, "Introduction" to Richard Hakluyt, *Voyages and Discoveries*, p. 10.
44. Hale, *Renaissance Exploration*, p. 42 (emphasis added).
45. Benjamin Franklin, *The Ingenious Dr. Franklin: Selected Scientific Letters of Benjamin Franklin*, p. 129.
46. Ibid., p. 131 (emphasis added).
47. Ibid., p. 131.
48. Ibid., p. 133.
49. American Philosophical Society, *Transactions* (Philadelphia, 1786), vol. 2, opposite p. 315.
50. Franklin, *Ingenious Dr. Franklin*, p. 132.
51. E. C. Krupp, *Skywatchers, Shamans & Kings*, p. 51.
52. "This same observation, however, was elsewhere attributed to Euthymenes, one of Pytheas's fellow Massaliots living a century or so earlier." Barry Cunliffe, *The Extraordinary Voyage of Pytheas the Greek*, p. 102.
53. Strabo, *Geography*, vol II, pp. 149, 153.
54. Ibid., p. 149.
55. George Sarton, *A History of Science*, vol. 1, p. 524.
56. Parry, *Age of Reconnaissance*, p. 86.
57. Taylor, *The Haven-Finding Art*, pp. 136–137.
58. Stillman Drake, *Cause, Experiment and Science*, p. 210.
59. Galileo Galilei, *Dialogue Concerning the Two Chief World Systems*, pp. 419, 445.
60. Ibid., pp. 446, 454–455, 460.
61. Casson, *Ancient Mariners*, p. 77.
62. Brown, *Story of Maps*, p. 12. The origins of cartography are now considerably less obscure, thanks to a multivolume work of cooperative scholarship, J. B. Harley and David Woodward, eds., *The History of Cartography*. Nonetheless, Lloyd Brown's point remains essentially valid; the original mapmakers were anonymous and predated recorded history.
63. Urban Wråkberg, "The Northern Space."
64. Massalia was where Marseilles, France, is today.
65. All of the fragments of Pytheas's book that exist in the works of other authors have been assembled and published as *Pytheas of Massalia, On the Ocean*, C. H. Roseman, ed.
66. Casson, *Ancient Mariners*, p. 139.
67. *See* Strabo, *Geography*, vol. I, pp. 399–401.
68. Cunliffe, *Extraordinary Voyage of Pytheas the Greek*, pp. 168, 173.
69. Ptolemy, *Geography*, book 1, chap. II, p. 26.
70. Ibid., book 1, chaps. XI and XVII, pp. 33, 37. It should also be noted that one of Ptolemy's main sources was Marinus of Tyre, and that "Marinus's method was simply to employ the various records of travelers and merchants." O. A. W. Dilke, "The Culmination of Greek Cartography in Ptolemy," p. 179.

71. Strabo, *Geography*, vol. I, p. 267.

72. With Ptolemy, Strabo, Hipparchus, and Eratosthenes in mind, the editors of *The History of Cartography* wrote, "On first inspection, the sources for the development of cartography in Hellenistic Greece strongly convey the impression that its knowledge and practice were confined to relatively few in an educated elite. Certainly the names associated with the history of mapping are largely drawn from a handful of outstanding thinkers traditionally associated with the history of Greek science in general. From other sources, however, albeit fragmentary, a broader picture can be drawn." Harley and Woodward, eds., *History of Cartography*, vol. I, p. 157.

73. Aurelio Peretti, *Il Periplo di Scilace: Studio sul primo portolano del Mediterraneo.*

74. A. E. Nordenskiöld, *Periplus*, p. 7.

75. Ibid., pp. 11–12.

76. Quoted in Taylor, *The Haven-Finding Art*, p. 85.

77. Brown, *Story of Maps*, p. 114.

78. Ibid., p. 121.

79. Ibid., p. 9 (emphasis added).

80. Quoted in Brown, *Story of Maps*, p. 143.

81. Brown, *Story of Maps*, p. 144.

82. J. B. Harley and David Woodward, "The Growth of an Empirical Cartography in Hellenistic Greece," p. 149.

83. Strabo, *Geography*, vol. I, pp. 453–455.

84. Brown, *Story of Maps*, p. 139.

85. Louise Levathes, *When China Ruled the Seas.*

86. Taylor, *The Haven-Finding Art*, pp. 184–185.

87. Bergreen, *Over the Edge of the World*, p. 11.

88. Taylor, *The Haven-Finding Art*, p. 103.

89. Ibid., p. 104. The "single coherent whole" into which "scattered sailing directions" of the Mediterranean and Black Sea were combined was the late-thirteenth-century *Compasso da navigare*. See B. R. Motzo, ed., *Il Compasso da navigare.*

90. Taylor, *The Haven-Finding Art*, p. 111.

91. See J. A. Bennett, "The Challenge of Practical Mathematics."

92. Anonymous, *An Essay on the Usefulness of Mathematical Learning*. This essay is dated November 25, 1700, and has been attributed variously to Martin Strong, John Arbuthnot, and John Keill. It should be noted that the author hoped to upgrade the respectability of mathematics by arguing that it should be removed from the province of seamen and other lowly types.

93. Brown, *Story of Maps*, p. 113.

94. Zvi Dor-Net, *Columbus and the Age of Discovery*, p. 62.

95. Taylor, *The Haven-Finding Art*, p. 112.

96. Brown, *Story of Maps*, p. 144.

97. Ibid., p. 74.

98. Tony Campbell, "Portolan Charts from the Late Thirteenth Century to 1500," p. 434.

99. David S. Landes, *Revolution in Time*, p. 111.

100. Quoted in Campbell, "Portolan Charts from the Late Thirteenth Century to 1500," pp. 427–428. Without this testimony, Campbell added, "we should have been left to guess how most of the new information must have reached the chartmakers."

101. O. A. W. Dilke, "Roman Large-Scale Mapping in the Early Empire," pp. 212, 226. Dilke's principal source is the *Corpus agrimensorum*, "an extant collection of short works in Latin of quite varied dates that have been recognized as the primary written records of Roman land surveying." Dilke, "Roman Large-Scale Mapping," p. 217.

102. Derek J. de Solla Price, *Science since Babylon*, p. 64.

103. Silvio A. Bedini, *Thinkers and Tinkers*, pp. xv–xvii.

104. See "Archaeoastronomy" in chapter 2.

105. Taylor, *The Haven-Finding Art*, p. 40. Pharaoh Snefru was said to have trafficked with Byblos c. 3200 B.C.E.

106. Pliny the Elder, *Natural History*, book 2, chap. 71 (Loeb Classical Library, vol. 1, p. 313).

107. Strabo, *Geography*, vol. I, pp. 41–43.

108. Aristotle, *On the Heavens*, book 2, chap. 14, 298a.

109. Homer, *The Odyssey*, book V, 262. "Bootes" is Arcturus.

110. Strabo, *Geography*, vol. I, p. 9. The Great Bear and Little Bear are also known as the Big and Little Dipper, respectively.

111. Callimachus, *Iambus*, trans. by and quoted in Kirk, Raven, and Schofield, *The Presocratic Philosophers*, p. 84.

112. That displacement was more pronounced for the ancient mariners than it is today. At the beginning of the Common Era it was enough to result in a 4° error in latitude measurement, compared with 1° at present. The change is a result of the precession of the equinoxes.

113. Brown, *Story of Maps*, p. 180.

114. Ibid., p. 184.

115. John Davis, *The Seaman's Secrets*.

116. Two others were Isaac Newton and "a country gentleman" named John Hadley. Godfrey and Hadley announced their inventions in 1730/31; Newton's came to light posthumously, in 1742. Brown, *Story of Maps*, pp. 192–193.

117. Other recognizable members of the scientific elite who could be added to this list are Leibniz, Pascal, Hooke, Bernoulli, and Euler.

118. Dava Sobel, *Longitude*, provides a very readable account of the longitude problem and its solution. See also David W. Waters, "Nautical Astronomy and the Problem of Longitude"; and Landes, *Revolution in Time*, chap. 9.

119. Richard S. Westfall, *Never at Rest: A Biography of Isaac Newton*, p. 544.

120. Robert Temple, *The Genius of China*, pp. 153–155.

121. Robert Norman, *The Newe Attractive*. The axis discovered by Norman is perpendicular to the more familiar one.

122. William Gilbert, *De magnete magneticisque corporibus et de magno magneto tellure physiologia nova*. For more on the relation between the work of Gilbert and that of Norman, see chapter 5.

123. See Christiaan Huygens, *Horologium oscillatorium*.

124. Quoted in Landes, *Revolution in Time*, p. 146.

125. Landes, *Revolution in Time*, pp. 133–134.

126. Galileo Galilei, *Il Saggiatore*, p. 212. See also Galileo Galilei, *Sidereus nuncius*, pp. 36–37. The question of the precise identity of the "simple spectacles-maker" is considered further in chapter 5.

127. Galileo Galilei, *Sidereus nuncius*, p. 7.

128. Lynn White, Jr., "Pumps and Pendula," p. 104.

129. Amir D. Aczel, *The Riddle of the Compass*, p. 103. See also Lionel Casson, *Ships and Seamanship in the Ancient World*, pp. 270–271.

130. Timoteo Bertelli, *Discussione della legenda di Flavio Gioia, inventore della bussola* (Pavia, 1901), quoted in and trans. by Aczel, *Riddle of the Compass*, p. 7.

131. Chu Yü, *P'ingchow Table Talk*, 1117; cited in Temple, *Genius of China*, p. 150.

132. Alexander Neckam, *De naturis rerum libri duo*, p. 183.

133. Collins, "Introduction," p. x.

134. Herodotus, *The History*, book II, 5. It should be noted that the eleven-fathom figure is almost surely a copyist's error.

135. Acts 27:28: "[The shipmen] sounded, and found it twenty fathoms: and when they had gone a little further, they sounded again, and found it fifteen fathoms."

136. See Parry, *Age of Reconnaissance*, p. 80. A fathom equals six feet.

137. Russell, *Prince Henry*, p. 131.

138. Hale, *Renaissance Exploration*, p. 87.

139. Russell, *Prince Henry*, pp. 210, 301–302.

140. Ibid., p. 232. Russell cited as an example Pope Nicholas V's bull *Romanus pontifex*, January 8, 1455.

141. Russell, *Prince Henry*, pp. 204–205.

142. Cadamosto, *Navigazioni*, quoted in Russell, *Prince Henry*, pp. 340–341.

143. Michel Eyquem de Montaigne, *The Essays*, book I, no. 30, "Of Cannibals," pp. 173, 175.

144. Ibid., p. 176.

145. H. Floris Cohen, *The Scientific Revolution*, p. 355.

146. Ibid., p. 355.

147. R. Hooykaas, "The Portuguese Discoveries and the Rise of Modern Science," p. 580.

148. R. Hooykaas, "The Rise of Modern Science," p. 472 (emphasis in original).

5

WHO WERE THE REVOLUTIONARIES IN THE SCIENTIFIC REVOLUTION?

THE FIFTEENTH THROUGH SEVENTEENTH CENTURIES

THE SCIENTIFIC REVOLUTION was the most important "event" in Western history.
—RICHARD S. WESTFALL, "The Scientific Revolution"

IT OUTSHINES EVERYTHING since the rise of Christianity and reduces the Renaissance and Reformation to the rank of mere episodes, mere internal displacements, within the system of medieval Christendom.
—HERBERT BUTTERFIELD, The Origins of Modern Science

OF ALL THE kinds of knowledge that the West has given to the world, the most valuable is a method of acquiring new knowledge. Called "scientific method," it was invented by a series of European thinkers from about 1550 to 1700.
—CHARLES VAN DOREN, A History of Knowledge

I T SEEMS THE only thing historians agree on about the Scientific Revolution—including its very existence—is that whatever it was or was not, it was very, very important. The paradox-lovers who exclaim "there was no such thing" while writing books about it[1] at least make clear that *something* momentous happened in Europe between, say, 1450 and 1700 that resulted in the emergence of modern science. Perhaps the process was too drawn out to be properly considered a *revolution*, but it utterly transformed the way we human beings understand the world around us.[2]

The assertion by Charles Van Doren in one of the epigraphs is a faithful rendition of the familiar textbook explanation of what was important about the Scientific Revolution. The "series of European thinkers" he credits are Francis Bacon, Nicolas Copernicus, Tycho Brahe, William Gilbert, Johannes Kepler, Galileo Galilei, René Descartes, and Isaac Newton.[3] The activities and ideas of these men dominate the traditional narrative, whereas the more fundamental contributions of innumerable anonymous artisans and tradespeople are overlooked.

Historians with a more expansive perspective, however, insist that to come to grips with the Scientific Revolution, "we must exorcise from our mythology all the great men,"[4] and that "any discussion of the 'foundations' of the Scientific Revolution must consider a much broader base for it than historians of science have so far attempted."[5] The purpose of this chapter is to consider that broader base. It will then be possible to appreciate Blaise Pascal's description of his era as one in which "simple workmen were capable of convicting of error all great men who are called philosophers."[6]

Although the *concept* has been around since the seventeenth century, the term "Scientific Revolution" is relatively recent; it was coined by Alexandre Koyré in the 1930s. There is much in Koyré's work to admire, but from the standpoint of people's history, its great influence on subsequent generations of historians is unfortunate. Koyré based his analysis on a narrow definition of science that focused only on its purely theoretical aspects. He saw the Scientific Revolution as the advent and triumph of what he called the "mathematization of nature." At the same time, he downplayed experimentalism as a relatively unimportant aspect of the new science.

Koyré's exaltation of the "Platonic and Pythagorean" elements of the Scientific Revolution, however, was based on a demonstrably false understanding of how Galileo reached his conclusions. Koyré asserted that Galileo merely used experiments as a check on the theories he devised by mathematical reasoning. But later research has definitively established that Galileo's experiments *preceded* his attempts to give a mathematical account of their results.[7] Whereas Koyré assumed Galileo had used his brain first and his hands and eyes only secondarily, it turns out to have been the other way around. Koyré's denial of the priority of the empirical side of the Scientific Revolution has thus been refuted to the satisfaction of all but those who insist that only theoretical endeavors are worthy of the name "science."

There is another great, gaping hole in Koyré's picture of the Scientific Revolution. H. Floris Cohen, in his extensive *Historiographical Inquiry* into the subject, explained that "the preponderant part inevitably assigned to astronomy and mechanics by the mathematical view of the origins of early modern science left wide open the question of what, if any, role nonmathematical physics, chemistry, and the life sciences had played in the birth of early modern science." Koyré, he said, "ducked the issue."[8] By avoiding consideration of nonmathematical sciences, Koyré reduced the Scientific Revolution to the ideas of Copernicus, Kepler, Galileo, and Newton. To resolve this difficulty, Thomas Kuhn proposed a dualistic interpretation, portraying the Scientific Revolution as a tale of two distinct kinds of science: "Baconian" and "classical physical."[9] Kuhn echoed Koyré, however, in giving pride of place to the mathematized disciplines.

The "Baconian" sciences were the kind Francis Bacon had in mind when he issued a call to revitalize science by basing it on craftsmen's knowledge of nature. Bacon is remembered as the most effective critic of the traditional learning promulgated by the elite institutions of his day. The university-based sciences, he said, "stand like statues, worshipped and celebrated, but not moved or advanced," while "the mechanical arts . . . having in them some breath of life, are continually growing."[10]

Accordingly, Bacon advocated compiling a "history of arts," or encyclopedia of craft knowledge. The "particular arts to be preferred" in such a project, he declared, are those that "exhibit, alter and prepare natural bodies and the materials of things such as agriculture, cookery, chemistry, dyeing, the manufacture of glass, enamel, sugar, gunpowder, artificial fires, paper, and the like." He also categorized as less useful (but by no means to be neglected) "weaving, carpentry, architecture, manufacture of mills, clocks, and the like."[11] A later prominent Baconian scientist, Robert Hooke, enthusiastically extended the list of "histories," indicating how vast he thought the contributions of artisans to science could be. Among the hundreds of types of craftsmen he enumerated were:

> surveyors, miners, potters, tobacco pipe-makers, glass makers, glaziers, glass grinders, looking-glass makers, spectacle makers, optick glass makers, makers of counterfeit pearls and precious stones, bugle makers, lamp-blowers, colour makers, colour grinders, glass painters, enamellers, varnishers, colour sellers, painters, limners, picture drawers, makers of bowling stones or marbles, brick makers, tile makers, lime burners, plasterers, furnace makers, china potters, crucible makers, masons, stone-cutters, sculptors, architects, crystal cutters, engravers in stones, jewelers, locksmiths, gun smiths, edge-tool makers, grinders and forgers, armourers, needle makers, tool makers, spring makers, cross-bow makers, plumbers, type founders, printers, copper smiths and founders, clock makers, mathematick instrument makers, smelters and refiners, sugar planters, tobacco planters, flax makers, lace makers, weavers, malters, millers, brewers, bakers, vintners, distillers.[12]

In France, meanwhile, René Descartes likewise called for the systematization of craft knowledge. The first to be investigated, he suggested, were "those arts of less importance"; that is, "those which are easiest and simplest, and those above all in which order most prevails. Such are the arts of the craftsmen who weave webs and tapestry, or of women who embroider or use in the same work threads with infinite modification of texture."[13]

The Baconian program of cataloguing all empirical knowledge culminated more than a century later in the *Grande*

Encyclopédie of the French *philosophes*. But the movement to base science on artisanal knowledge had already been in full swing well before Bacon came on the scene: "he was, in truth, only the most publicized preacher of a method that had been growing for decades before him."[14] A Londoner named Hugh Plat, to cite but one example, was energetically gathering information from craftspeople in the 1570s and publishing books based on what he learned from them.[15] Bacon could not have been unaware of Plat's endeavors.

He was surely also influenced by the presence in London of Cornelius Drebbel, a Dutch immigrant mechanic and alchemist whose experimental activities had attracted a great deal of notice. Drebbel's best-known public demonstration took place in 1620, when he immersed a submarine of his invention in the Thames for three hours. The people aboard the vessel were able to breathe because Drebbel had provided them with oxygen in bottles that could be opened as needed. Although the concept and word "oxygen" lay almost two centuries in the future, Drebbel had empirically learned how to generate it by heating saltpeter.[16] Robert Boyle later credited Drebbel with recognizing that the air we breathe is a mixture of various "airs," one of which is essential for sustaining life.[17]

Drebbel also added to physical and chemical knowledge with regard to optics, mechanical systems, heat, explosives, and much more. As a former engraver's apprentice with no university education, he was representative of a significant social trend summarized by William Eamon:

> Nonacademics, amateurs, and craftsmen made important contributions to the development of the Baconian sciences. ... Robert Boyle, in devising his chemical experiments, drew extensively from the empirical information accumulated by metallurgists, dyers, and distillers. The development of these sciences depended directly upon the dissemination of information from such occupational groups, whose activities had been irrelevant, or unknown, to academic scientists.[18]

Koyré's conception of the mathematization of nature is clearly inapplicable to much of the science of the Scientific Revolution. While giving the nonmathematical "Baconian" sciences the emphasis they deserve, however, let us not concede the "classical physical" sciences to Koyré's interpretation; craft knowledge and practices played an important role in their development as well. Galileo, for example, credited workers at the Arsenal of Venice with inspiring his investigations into the science of mechanics.[19] Leonardo Olschki observed that

> what enabled Galileo to transcend the amassed erudition of his predecessors in science was *a recently emerged tradition of applying mathematical notions to practical matters of a technological nature,* which he adopted from the preceding literature in the vernacular. That is to say, matters of perspective, mining, fortification, ballistics, and so on provided the impulse for *the turn toward the empirical without which the decisive renewal of science in the 17th century would have been inconceivable.*[20]

The vernacular literature to which Olschki alluded were the writings of craftsmen.[21]

HOW WAS NATURE "MATHEMATIZED"?

KOYRÉ'S GLORIFICATION OF Galileo's mathematics not only leads to a one-sided picture of the Scientific Revolution but also idealizes the history of mathematics itself. The mathematical procedures Galileo applied to natural philosophy were not immaculate conceptions of contemplative thinkers but were developed over many centuries by people who found quantitative and geometric techniques beneficial to the pursuit of their various trades and occupations. In general,

> Renaissance developments in *practical* mathematics *predated* the intellectual shifts in natural philosophy [i.e., the theoretical contributions of Copernicus, Kepler, and Galileo]. . . . On

the strength of [practical mathematicians'] successes in navigation, cartography and surveying, they asserted its importance and widespread relevance. In time these claims impinged on natural philosophy, and it is significant that the reformed natural philosophy adopted practical mathematical techniques in its new methodology.[22]

The occupational groups that most stimulated the development of mathematics were merchants, instrument-makers, sailors, miners, surveyors, engineers, architects, and graphic artists. First of all, mercantile pursuits "encouraged an interest in exactitude and a concern for the accuracy of measurements of time, distance, and capacity." Merchants'

> quest for answers to problems from numerical data elevated mere computation to the status of an empirical science. Thus these traditions were well-established by nonscientists well before the time of Galileo, Copernicus, and Descartes and their mechanistic world view. Further, the quest for new markets and greater profits also gave rise to studies in geography and astronomy and led to advances in cartography, navigation, and marine architecture.[23]

Although "the role of merchants in medieval and early Renaissance society as instigators of . . . scientific innovation is seldom fully appreciated," their substantial contribution to mathematical progress is undeniable.[24] In a previous chapter it was noted that the adoption of modern numerals was a necessary precondition of further mathematical advance. "It is no mere coincidence that Fibonacci, one of the principal conveyors of a knowledge of the Hindu-Arabic numeral system to Europe, was also a merchant."[25]

> In trade contacts around the Mediterranean and Barbary coasts, Italian merchants became exposed to the Hindu-Arabic numeral system and its methods of computation. Raised in a Pisan trading colony in Bugia (Bougie), in what is

now Algeria, Leonardo of Pisa (1180–1250) studied this new system of arithmetic under the guidance of an Arab master. He became convinced that the new numerals and their methods were vastly superior to the Roman numerals commonly employed in Europe. Leonardo, also known as Fibonacci . . . became the evangelist of the new knowledge and published his impressions in a book, *Liber abaci* (1202). . . . The book and its message were well-received in the . . . merchant houses of Pisa, Genoa, and Venice, and soon the Hindu-Arabic symbols were replacing Roman numerals in account books and the use of the abacus was giving way to computations performed with pen and ink.[26]

Standard histories of mathematics often depict the early Renaissance as a period of stagnation except for the isolated contributions of a few great geniuses: "a large featureless plain broken up by such occasional peaks as Cardano, Copernicus and Galileo," as Paul Rose described it.[27] Other historians have asserted that "no significant mathematics appeared in Europe between the time of the death of Fibonacci (1250) and the beginnings of the sixteenth century." But, a perceptive critic asked, what does "significant" mean in this context?

Significant for what and *for whom?* . . . There is more to mathematics than theory, for theory by itself, without a mode of expression or articulation, remains impotent. While not dramatic in the sense of new theories, the mathematical accomplishments taking place in the fourteenth and fifteenth centuries are profound in a subtler way.[28]

By the time of Fibonacci's death in the mid-thirteenth century, "arithmetic was being taught as a science at European universities," but the "teaching was often theoretical and devoid of practical applications." For that reason, a student with a serious interest in practical mathematics "usually did not go to the university, but sought out a reckoning master, a man skilled in the arts of commercial computation, with whom to study." The explosive growth of European commerce led to a rapid increase in the numbers of reckoning masters—called *maestri d'abbaco* in Italian, *maistres d'algorisme* in French, and *Rechenmeister* in German—and their abacus schools.[29]

The mercantile example-problems contained in Fibonacci's *Liber abaci* "formed the basis of a whole descendent genre devoted, primarily, to the mathematical needs of merchants"— a genre that "extended from the thirteenth through the sixteenth centuries."[30] With the advent of printing in the fifteenth century, the proliferation of vernacular arithmetic texts ensured "that this knowledge was disseminated to the 'common man.'"[31] The first mathematics book known to have been printed in Europe, the *Treviso Arithmetic*, appeared in 1478, four years before the first printed edition of Euclid's geometry—a fact that "tells much about the real mathematics climate of this time." Its anonymous author was a *maestro d'abbaco*, and it was written in the Venetian dialect, a reflection of "its egalitarian mission of communicating knowledge to a broad audience." The example was followed elsewhere. "The first printed, dated, arithmetic in Germany appeared in 1482; in France and Spain, 1512; in Portugal, 1519; and in England, 1537. All of these arithmetics were of a commercial type and many were written by reckoning masters."[32] The knowledge they contained had not trickled down from university sources; quite the opposite:

> *It is to the elementary textbooks of the abacus schools that one must trace the rise of algebra to become a part of learned mathematics* in the sixteenth century. Moreover, the use of mathematics in the crafts seems to have encouraged thinking in three dimensions, as was required for many practical problems, rather than in the two-dimensional terms that characterize almost all the theorems of Euclid. Such habits eventually led to important changes within the learned tradition of geometry.[33]

A boom in the mining industry expanded the demand for practical mathematics. "Alligation," the process of alloying metals:

as a topic of mathematical importance had first made its appearance in the arithmetic books of the fifteenth century in connection with metallurgy. . . . As metallurgy, stimulated by the demands of bell and cannon casting and minting, began to be recognized as a science, consideration of its quantification techniques moved from the manuals of alchemists to those of the reckoning master.[34]

INSTRUMENT MAKERS
AND PRACTICAL MATHEMATICS

IN THE SIXTEENTH century, "mathematician-craftsmen" revolutionized surveying by supplanting traditional linear land-measuring procedures with the method of triangularization that requires measuring angles and applying the principles of trigonometry. Gemma Frisius, who established a workshop in Louvain that became a center for practical mathematics, was among the leading innovators. "In charge of the Louvain workshop for more than twenty years was the instrument maker and cartographer Gerard Mercator . . . and he was succeeded by Gemma's nephew, the celebrated instrument maker Walter Arsenius."[35] Their workshop acquired an international reputation that stimulated the development of practical mathematics far and wide:

> John Dee complained that in the 1540s, finding no appropriate expertise in England, he had had to learn his mathematics on the Continent, spending time particularly with Gemma Frisius and Gerard Mercator. It was partly due to his enthusiastic promotion that by the end of the century English mathematical science was thriving, with many books published in the vernacular and a new instrument making centre established in London.[36]

As John Dee discovered, "the Englishman lagged far behind his Continental rivals in scientific knowledge," and especially in the mathematical arts. "The reason was simple," E. G. R.

Taylor explained; "the Italians, the French, the Germans, possessed a scientific language in the mother tongue, but as soon as an English boy had his A B C he was put to Latin grammar and his sole reading was in the classics." In the middle of the sixteenth century "the universities at large appeared indifferent or even hostile to any mathematics that that went beyond the meager medieval curriculum, upon which a single lecture course was provided at Cambridge." Meanwhile, "there was still but scanty evidence of any application of mathematics either to civil, military or nautical practices in England."[37]

But growing demand "for instruction in the geometry and astronomy necessary for improved techniques in navigation, surveying, horology, cartography, gunnery and fortification" led some technicians to set themselves up as "professors" of mathematics:

> While a few of them were men with a university background, the majority were not. They were almanack-makers, astrologers, retired seamen, surveyors, gunners, gaugers—in fact they were themselves mathematical practitioners who simply handed on their art. But, as might be expected, they all worked in close association with the instrument-makers . . . the handling of instruments was the very badge of this new profession.[38]

Some of the "professors" produced textbooks. Leonard Digges "became famous as a pioneer writer on geometrical practice for the common man"; his "avowed aim was to bring within reach of the artisan and the master-craftsman a knowledge of the mathematical arts." About 1571, William Bourne published a manual of navigation entitled *A Regiment for the Sea*. Bourne, responding to "scornful criticism he had been subjected to for trespassing on the scholars' field," acknowledged "that he is indeed an utterly unlearned man, but he has not presumed to write for the educated but only for the simple and ignorant." John Blagrave's *Mathematical Jewell* (1585), a description of an astronomer's astrolabe, sought to spread mathematical knowledge "unto everie ingenious practiser . . . whence many inventions yet unthought of may spring from the common sort of handicraftmen and workmen."[39]

Astrolabe, dated 1603. (Reputedly Samuel de Champlain's)

The astrolabe was "not a terribly important instrument itself—it is only a rotating star map, after all," but it "spawned the craft" of instrument-making: "It is the instrument that was the chief product, chief training ground, and the masterpiece of work for a continuous tradition of craftspeople who preserved very complicated techniques including scientific engraving."[40]

Crafts characterized by precision metalwork were the main progenitors of the new profession. "It was from the engravers, who knew the meaning of the phrase 'to a hair's breadth,' that the mechanical-instrument makers took their first rise." They were joined by "clock-makers, themselves derived from black-smiths and locksmiths, while optical-instrument makers, the last on the scene, emerged from among the glass-grinders and spectacle-makers."[41]

An astrolabe produced in 1552 by engraver Thomas Lambritt, alias Gemini, "affords the first definite evidence of an instrument-maker's workshop in London."[42] Gemini, a Belgian immigrant, may have trained the first generation of English-born instrument-makers. One who may have been apprenticed to Gemini, Humphrey Cole (also an engraver), is generally credited with being "the first great instrument-maker of England."[43]

John Aubrey, writing in 1690, attributed the continued propagation of "mathematicks" in England in large part to a vernacular book published earlier in the century by a "mathematicall instrument maker" named Edmund Gunter. Gunter, he claimed,

> was the first that brought Mathematicall Instruments to perfection. His *Booke of the Quadrant, Sector and Crosse-staffe* did open men's understandings and made young men in love with that Studie. Before, the Mathematicall Sciences were lock't-up in the Greeke and Latin tongues; and so lay untoucht, kept safe in some Libraries. After Mr. Gunter published his Booke, these sciences sprang up amain, more and more to that height it is at now.[44]

It is evident that from Louvain to London, instrument-makers were in the forefront of the "mathematization of nature."[45] The mathematized sciences were greatly dependent upon habits of precision and refinements in measurement, which were in turn dependent upon improvements in instrumentation. The mathematical practitioners—"the early mass movement in science"—were not well rewarded for their contributions: "the record shows that most of them lived in acute poverty and died of starvation."[46]

Even Galileo and Newton, who claimed to have built their own instruments, could not have done so without having learned a great deal from skilled artisans. It was a partnership between the "greatest scientists" and the "best of the craftsmen," David Landes said, "that finally gave us precise timekeepers to the measure of the stars and the sea; and toward the end, when the scientists thought they had done as much as they could, it was the craftsmen who persisted and completed the task." Those artisans, he added, "possessed surprising theoretical knowledge and conceptual power."[47]

The mathematized science at the core of traditional histories of the Scientific Revolution is astronomy, often presented as the story of one array of mathematical abstractions (Copernicus's) superseding another (Ptolemy's). But the path to Copernicus was cleared by Renaissance astronomers such as Georg Peurbach and Johann Regiomontanus:

> The practical aspect of the mathematical sciences formed part of this Renaissance movement from its inception. Both Peurbach and Regiomontanus were involved with the design of instruments . . . indeed Regiomontanus was a maker and established, along with his printing press, a workshop for the manufacture of instruments. He was naturally attracted to Nuremberg as the most important centre in Europe for metalwork and other crafts, and he cited the availability of instruments as a reason for his settling there in 1471.[48]

The "pretelescope" tradition of astronomical observation culminated in achievements associated with the name of Tycho Brahe. Although historians have frequently portrayed Brahe's enterprise as "almost miraculous in its originality," it must be "placed in perspective by the activities of earlier practical astronomers. In particular, Wilhelm IV, Landgrave of Hesse, founded a well-equipped observatory at Kassel with the aim of providing a new star catalogue, and here he employed the instrument maker and designer Joost Bürgi."[49] Tycho Brahe's project and its relevance to a people's history of science will be discussed in more detail below. Meanwhile, another major contribution by a particular group of superior artisans to the "mathematization of nature" deserves consideration.

THE DISCOVERY OF PERSPECTIVE

Renaissance painters, sculptors, and architects are ordinarily thought of as representatives of "high culture" rather than as manual workers practicing a craft to make a living, and their activities are usually considered the province of art historians rather than historians of science. Both of these assumptions require modification. First it must be recognized that the artists emerged from the ranks of manual laborers: "During the fifteenth century Italian painters, sculptors and architects had slowly separated from whitewashers, stone-dressers and masons. As the division of labor was still only slightly developed, the same artist usually worked in several fields of art, and often in engineering too."[50]

It is necessary to recognize that the architects of that era, who immeasurably advanced the science of mechanics, were not generally accorded social status above the ranks of skilled artisans:

> Today we are so used to celebrating the brilliance of architects like Michaelangelo, Andrea Palladio, and Sir Christopher Wren that it is hard to imagine a time when architects and architecture were not esteemed. But the great architects of the Middle Ages had been virtually anonymous. . . . Part of the reason for this anonymity was a prejudice against manual labor on the part of both ancient and medieval authors, who assigned architecture a low place in human achievement, regarding it as an occupation unfit for an educated man. Cicero claimed that architecture was a manual art on the same level as farming, tailoring, and metalworking, while in his *Moral Letters* Seneca mired it in the lowest of the four categories of art, those which he classified as *volgares et sordidae*, "common and low." Such arts were mere handiwork, he claimed, and had no pretense to beauty or honor.[51]

In the established university curriculum of the late Middle Ages, "painting and sculpture are nowhere. The figurative arts are left to a purely servile status. They are not supposed to be a freeman's activity." At the beginning of the fifteenth century, "what we call science was still a hole-and-corner affair without a character of its own," and painters and sculptors were a "group of men who had no connection with the universities, little access to books, who hardly even dressed as burghers and went around girt in the leather apron of their trade."[52]

Leonardo da Vinci is today hailed as the supreme example of a Renaissance man who mastered many diverse fields of knowledge and combined achievements in the *beaux arts* and in science to a degree attained by no individual before or since. Ironically, however, in his own time he was denied the prestige of a fully learned man because he lacked a classical education and was not literate in Latin. His originality, one leading historian of science has written, "was partly due to his ignorance and his lack of academic inhibitions."[53]

Leonardo blasted "certain presumptuous persons" who "scorn me as an inventor" and go about "alleging that I am not a man of letters." He responded:

> If indeed I have no power to quote from authors as they have, it is a far bigger and more worthy thing to read by the light of experience, which is the instructress of their masters. They strut about puffed up and pompous, decked out and adorned not with their own labours but by those of others. [54]

The artisan's resentment at being looked down on as a social inferior is evident in Leonardo's angry retort to university scholars who categorized painters like himself as manual workers:

> You have set painting among the mechanical arts! Truly were painters as ready equipped as you are to praise their own works in writing, I doubt whether it would endure the reproach of so vile a name. If you call it mechanical because it is by manual work that the hands represent what the imagination creates, your writers are setting down with the pen by manual work what originates in the mind. If you call it mechanical because it is done for money, who fall into this error—if indeed it can be called an error—more than you yourselves? If you lecture for the Schools do you not go to whoever pays you the most?[55]

The artisanal status of painters, sculptors, and architects is clearly depicted in Giorgio Vasari's *Lives of the Artists,* a series of biographical sketches first published in 1550 that has long served as a primary source for historians of Renaissance art. He classified Donatello and Brunelleschi as "two distinguished craftsmen" and designates Michaelangelo as the "wisest of all the craftsmen."[56] Artists generally practiced their trades in workshops where a customer could place orders for the commodities produced there. And it was by the method of apprenticeship that artists transmitted their craft knowledge from generation to generation. Michaelangelo's father apprenticed him to "one of the finest living masters," Domenico Ghirlandaio, for three years, and Vasari himself was later apprenticed in turn to Michaelangelo.[57]

Sandro Botticelli, Lorenzo Ghiberti, Filippo Brunelleschi, Paolo Uccello, Andrea del Verrocchio, and Leonardo da Vinci were among the many prominent artists who began their careers as apprentices to goldsmiths. Botticelli took his famous last name not from his father but from the master craftsman to whom he was apprenticed. His father, Mariano Filipepi,

> apprenticed him as a goldsmith to a close companion of his own called Botticelli, who was a very competent craftsman. Now at that time there was a very close connection—almost a constant intercourse—between the goldsmiths and the painters, and so Sandro . . . became entranced by painting and determined to devote himself to it.[58]

Although goldsmiths were sometimes able to gain a measure of prestige for their artistry, the generally unpleasant nature of their working conditions set them apart from the genteel classes. The furnaces in which they melted gold and other metals

> had to burn for days on end, even in the heat of summer, polluting the air with smoke and bringing the danger of explosions and fire. Noxious substances such as sulfur and lead were used to engrave silver, and the clay molds in which

metals were cast required supplies of both cow dung and charred ox horn.

It is not surprising, then, that

> the workshops of most goldsmiths were found in Florence's most notorious slum, Santa Croce, a marshy and flood-prone area on the north bank of the Arno. This was the workers' district, home to dyers, wool combers and prostitutes, all of whom lived and worked in a clutter of ramshackle wooden houses.[59]

Brunelleschi's father likewise apprenticed his son "to the goldsmith's art with a friend of his so that he might study design. . . . [A]fter he had started to learn the art not many years passed before he was setting the precious stones better than experienced craftsmen." On a later occasion, Brunelleschi "ran short of money and had to meet his wants by setting jewels for some friends of his who were goldsmiths."[60]

Like many of his contemporaries, Brunelleschi "tried his hand at several crafts." One of them was surveying, a well-developed branch of practical mathematics that stimulated his thinking about perspective: "Perspective drawing is, after all, similar to surveying in that both involve determining the relative positions of three-dimensional objects for the purpose of protracting them on paper and canvas."[61] Among other accomplishments, Brunelleschi "made with his own hand some very splendid and very beautiful clocks," and for fifteen years made his living as a clockmaker.[62] His later prominence as an architect and engineer would win him powerful patrons and allow him to move in elite social circles, but he remained in essence a "superior manual worker."[63]

As in any other field of human endeavor, the lives of a few famous painters, sculptors, and architects are comparatively well documented, whereas those of thousands of their anonymous colleagues remain unknown. But the Renaissance artist-craftsmen tended to work collectively. Many of their works cannot be attributed to any single person but must be assigned to one "school" of art or another. Even those that are credited to a famous individual often combined the work of many. The celebrated doors of the Baptistry in Florence—Michaelangelo named them the "Doors of Paradise" and they were praised by Vasari as "the finest masterpiece ever created, either in ancient or modern times"—though always ascribed solely to their principal designer, Lorenzo Ghiberti, were to a great extent a cooperative venture. According to Vasari,

> Lorenzo was assisted in finishing and polishing the work after it had been cast by many young men who afterwards became accomplished artists in their own right. . . . By *working in close collaboration* on the doors and *conferring with each other as a team* they benefited themselves as much as Lorenzo.[64]

In particular, Ghiberti's grandson Bonaccorso "himself very diligently completed the frieze and ornamentation, which, I maintain, constitute the rarest, most marvelous work in bronze anywhere to be seen."[65]

For another example, when Raphael was engaged in his great work at the Vatican, Pope Leo X

> asked Raphael to make the designs for the stucco ornaments and the scenes that were to be painted there, and also for the various compartments. Raphael put Giovanni da Udine in charge of stuccoes and the grotesques, and (although he worked there only a little) in charge of the figures. He also employed . . . . many other painters who supplied various of the scenes and figures and other details that were wanted.[66]

In one instance, Raphael "had his pupil Giovanni da Udine (who is unrivaled as a painter of animals) depict all the animals owned by Pope Leo."[67]

Vasari apparently did not consider it unethical for Raphael systematically to borrow ideas from designs produced by other artists. "Such was Raphael's stature," he offhandedly remarked, "that he had draughtsmen working for him throughout all Italy, at Pozzuolo and even in Greece; and he was always looking for good designs which he could use in his work."[68]

Ghiberti, in the first case, and Raphael, in the second, certainly deserve a great deal of credit for the works they designed, organized, and supervised, but they were not solitary creators. And the same is true of the Renaissance artists' invention of mathematical perspective. It was their collective determination to depict nature realistically that put them in the vanguard of the quest for knowledge of nature:

> At a time when what *we* mean by science was still beyond the horizon, when the *name* of science was monopolized by scholastic officials, who officially denied to mathematics any link with physical reality, these men had conceived of an original prototype of science based on mathematics, which was to provide them with a creative knowledge of reality.[69]

Their predecessors, the postclassical artists of the Middle Ages, had been more concerned with religious symbolism than with realism and therefore had felt little need to introduce three-dimensionality into their work. By the fifteenth century, however, artists were actively seeking ways to translate the three-dimensional objects of physical space onto flat, two-dimensional surfaces. In doing so, they developed the mathematics of linear perspective. "Linear perspective," da Vinci explained in treatises addressed to aspiring young painters and draftsmen, "has to do with the function of the lines of sight, proving by measurement how much smaller is the second object than the first and the third than the second, and so on continually until the limit of things seen." The images of objects, he added, are transmitted "by pyramidal lines" to the eye, but those sight lines "are all intersected at a uniform boundary when they reach the surface of the painting."[70]

Theory-oriented mathematicians were slow to respond to the

artists' innovations. "A mathematician however great who does not understand drawing," the painter Ludovico Cigoli declared, "is not only half a mathematician, but indeed a man deprived of sight."[71] It was not until the mid-sixteenth century that learned mathematicians began to pay attention to perspective, and only in the late seventeenth century did it finally "become a part of mathematics, for mathematicians."[72] (In this context, it is worth recalling John Wallis's comment that well into the seventeenth century "Mathematicks . . . were scarce looked upon as *Accademical* studies, but rather *Mechanical*.")[73]

Although development of mathematical perspective was due to broad artistic *movements* throughout Europe—"to Florentine artists of the fourteenth century and also, in a smaller degree, to Flemings and Germans"—its invention is frequently attributed to a small number of those who achieved fame and fortune in the following century. Among the Italians, Filippo Brunelleschi "is said to have been the first to make a systematic study of perspective"; Leon Battista Alberti, Paolo Uccello, and Piero della Francesca wrote widely discussed treatises on the subject; and Leonardo da Vinci is credited with having "brought these ideas to their climax." Albrecht Dürer is representative of the many northern artists who contributed —for the most part independently of the Florentines, it should be noted—to the creation of mathematical perspective.[74] Justice would be better served if we could focus directly on the collective contributions of thousands of anonymous painters, sculptors, and architects, but the documentary record on which we depend was left by their better-known colleagues.

The treatises of Leon Battista Alberti and Piero della Francesca eloquently testify to the craft origins of perspective. Alberti reveals his general approach to gaining knowledge by telling us that he "would learn from all, questioning smiths, builders, shipwrights, and even shoemakers lest any might have some uncommon or secret knowledge of his craft, and often he would feign ignorance in order to discover the excellence of others."[75] As for Piero, his "work as a painter lies in the province of the craftsman (and the style of his mathematics . . . is that associated with the artisan)." His treatise *De prospettiva pingendi*

> was intended to be instructional, addressed to an apprentice: the reader is called "tu" and is given detailed instructions on how to draw the diagrams. Almost throughout, the style is exactly that of the vernacular textbooks used in abacus schools. . . . A similar style was employed in the training of apprentices in artists' workshops. . . . All subsequent treatises on perspective addressed to painters follow more or less faithfully the pattern laid down by Piero.[76]

Brunelleschi's previously mentioned background as a clockmaker was not unrelated to his contributions to the theory of perspective, which were based on experiments with apparatus of his own design and construction. He has been credited with inventing "the earliest optical instrument after the eyeglasses," a perspective device that was a forerunner of the *camera obscura*.[77] The *camera obscura*, a room entirely darkened except for sunlight admitted by a single small hole in a window shade, generates a dramatic optical effect: if the sun is shining brightly outdoors, an upside-down image of the scene outside the window is projected onto the wall opposite the hole. No single artist can be credited with inventing it, but da Vinci and Alberti were among the first to exploit its potential.

Later generations of experimenters improved the *camera obscura* by placing a small convex glass lens at the hole where the light entered, allowing them to focus the projected images more sharply. More than two centuries after its appearance in Italy, Johannes Vermeer (1632–1675) apparently utilized a *camera obscura* to execute his masterpieces.[78]

"Brunelleschi's perspective device [circa 1413] and then Alberti's *camera obscura* of about 1430, both produced not as scientific instruments but rather as craftsmen's tools, initiated novel scientific speculations on the nature of light and of vision."[79] In their impact on the science of optics, they were "in every way as significant as the achievement of the telescope."

The optical phenomena exhibited by the *camera obscura*

> went a long way towards establishing in natural philosophy a
> new idea concerning the nature of light. . . . By showing the
> passive character of vision, it cut the ground from under a vast
> set of theories, primitive and also Platonic in origin, which
> assume vision to be an "active function," a reaching out, as it
> were, of the soul.[80]

The invention of perspective by the Renaissance artists, in
summary, amounted to a "new science of light and space."[81] By
demonstrating that mathematics could be usefully applied to
physical space itself, a momentous step was taken toward the
general representation of physical phenomena in mathematical
terms. Galileo is traditionally hailed as the great pioneer of that
movement, but what was the source of his mathematical
knowledge?

> During his student days there was no mathematical instruc-
> tion at Pisa. He learned mathematics privately, his tutor,
> Ostilio Ricci, being an architect and teacher at the Accademia
> del disegno which had been founded in 1562 by the painter
> Vasari as something between a modern academy of arts and
> a technical college. Thus Galileo's first mathematical educa-
> tion was directed by persons who were artist-engineers.[82]

It should be clear that the "mathematization of nature" was
not, as Alexandre Koyré maintained, a product of "pure unadulterated thought."[83] The "classical physical" sciences owed a
great deal to the practical mathematics of commerce and the
crafts.

In the nonmathematical "Baconian" sciences that have traditionally been given short shrift by Koyré and his followers, the
empirical rather than the mathematical side of scientific innovation comes to the fore. The Renaissance artists' drive to
achieve an accurate depiction of nature stimulated the growth
of medical and botanical as well as mathematical knowledge. "It

was really the painters, beginning with Pollaiuolo, and not the doctors," who advanced the science of anatomy by performing dissections on human cadavers "for purposes of exploration rather than demonstration."[84] Michaelangelo was following a tradition established by previous generations of artists when "in order to achieve perfection he made endless anatomical studies, dissecting corpses in order to discover the principles of their construction and the concatenation of the bones, muscles, nerves, and veins, and all the various movements and postures of the human body."[85]

The publication of Andreas Vesalius's anatomical master-piece, *De humani corporis fabrica*, in 1543 is often cited as one of the landmark events of the Scientific Revolution, but its impact owed less to Vesalius's Latin text than to the striking anatomi-cal drawings of one or more anonymous draftsmen presumed to

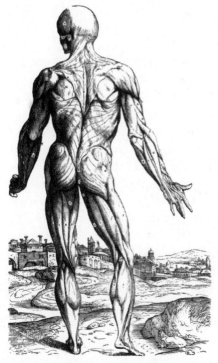

Illustration from Vesalius.

have been affiliated with the workshop of Titian.[86] With regard to the 420 illustrations in Vesalius's book, historian Vivian Nutton observed: "so much have they been seen as the very essence of the *Fabrica* that they have diverted attention away from the text that surrounds them and have thereby contributed to making the *Fabrica* an unknown document." Whereas the illustrations have remained in print for more than four centuries, Vesalius's text "has never yet been fully translated into a modern Western language." The artists who drew the illustrations are unknown because Vesalius did not bother to give them credit:

> Vesalius reveals the name of the manager of the Venetian branch of the merchant house of Bomberg who had helped him pack the blocks, Nicolaus Stopius, and that of the Milanese merchants, the Danoni, who took them over the Alps, but on the name of his artist or his block-cutter he is silent.

"Whoever he was," Nutton concluded, the anonymous artist "produced a masterpiece that, almost at a stroke, solved most of the technical questions involved in making the two-dimensional image on the printed page serve as a substitute for three-dimensional reality."[87] But the anatomical illustrations could not have been printed without highly skilled engravers who could transfer them to copper plates; one such craftsman was the Belgian engraver Thomas Gemini whose seminal role in English instrument-making was noted previously.[88]

As for botany, George Sarton considered it "one of the most exhilarating chapters in the history of science" when artists began to execute finely detailed realistic drawings of plants based on their own careful observations. "As some naturalists realized the need of illustrations made directly from nature, there grew up a class of draftsmen and woodcut makers who learned to do this and do it well. Art and science came together and great was the result." The *Herbarum vivae eicones* of 1530, "the first herbal with illustrations made from nature," resulted from the collaboration between a naturalist, Otto Brunfels, and a draftsman and blockcutter, Hans Weiditz.[89] Weiditz was a pupil of Albrecht

Dürer; his "detailed record of observation and his realistic woodcuts in this herbal evince an entirely new attitude toward nature, in contrast to the scholar Otto Brunfels's text, which is borrowed from classical authorities."[90]

Leonhard Fuchs's *De historia stirpium* (1542) was also based on ancient texts, but it included drawn-from-life illustrations of some four hundred wild and one hundred domesticated plant species. In this case the illustrators, Heinrich Füllmauer and Albert Meyer, and the blockcutter, Veit Rudolf Speckle, received credit for their work. Their portraits appear at the end of the book.

Portraits of the three illustrators from Leonhard Fuchs's herbal
De Historia Stirpium *(1542).*

In general, the early herbals' texts were in Latin, which "put them out of reach of all but the learned doctors," but their illustrations made them accessible to a much wider audience. "Love of plants and interest in herbs and roots was not by any means confined to scholars; even women might want to know more about them and to be able to recognize them in the fields."[91]

"*Even women*"? Women, in fact, played an essential part in the origins of botanical science. But as Pamela Smith has pointed out, "The story of . . . the 'old women' and 'herbalists' mentioned by almost every early modern botanist as the basis of his specimens and local plant knowledge has yet to be written."[92] Thomas Sydenham, the seventeenth-century doctor acclaimed as "the English Hippocrates," said of academic botany: "Nonsense! Sir, I know an old woman in Covent Garden who understands botany better." ("And as for anatomy," he added, "my butcher can dissect a joint full and well.")[93]

Although anatomy and botany are empirical sciences, their methodology is primarily observational—examining the "book of nature" as it presents itself directly to our senses. Experimentation is active rather than passive and therefore represents a higher level of empirical practice. Performing experiments involves creating artificial situations in order to observe natural phenomena that do not exhibit themselves in everyday settings. Knowing where the habit of experimenting came from is therefore essential to understanding the origins of modern science.

THE ORIGIN OF "EXPERIMENTAL PHILOSOPHY"

HISTORIOGRAPHER COHEN LAMPOONED what he called a "simpleminded picture of the Scientific Revolution":

> Before Galileo, if one wished to acquire an understanding of nature, the best thing to do was to follow the sole authority in these matters, Aristotle. Then came Galileo, who once and for all taught us to think for ourselves, thus inaugurating early modern science and clearing the path toward its further improvement.[94]

This is, in every respect, an intentional caricature, but it is fair to say that before the Renaissance the dominant trend among European scholars was the belief that the best way to gain knowledge of nature was through scrutiny of Aristotle's writings. It is also reasonable to assert that by Galileo's era, a new way of seeking knowledge—by means of experiment—had come into favor in some elite intellectual circles. Although Galileo was a leading propagandist for this important change, it certainly was not brought about by him.

A well-known morality tale drawn from the "simpleminded picture" has Galileo dropping two objects of unequal weight from the top of the leaning tower of Pisa. As the story goes, Galileo's scholastic opponents insisted that a heavier object must fall faster than a lighter object—because Aristotle had said so. Galileo supposedly suggested simultaneously releasing a heavy and a light ball from the top of a tower to see whether the heavier one would in fact hit the ground first. The philosophy professors rejected his proposal as a waste of time. But Galileo went ahead and, sure enough, proved Aristotle wrong. Both balls hit the ground at the same time, which meant that they had been falling at the same speed.

This apocryphal story, "perhaps false in all essentials," is one of the foundation myths of modern science.[95] It serves to illustrate the rise of experimentalism in seventeenth-century science but supports the erroneous notion that it was "invented by a series of European thinkers." In addition to Galileo, others who have routinely been credited with initiating empirical practices in science are Francis Bacon (for his championing of inductive as opposed to Aristotelian deductive logic) and William Gilbert (for publishing the first scholarly book on experimental physics, *De Magnete*, in 1600—a detailed account in Latin of his experiments with magnets).

Although most accounts of the Scientific Revolution continue to retail some version of the "great thinkers" story, it has not gone without challenge. More than a half-century ago, one dissenting historian, Edgar Zilsel, offered an alternative point of view: "The experimental method did not and could not have descended from the metaphysical ideas of the natural philosophers. We have to look elsewhere and in other social ranks for its immediate predecessors."[96] Experimentalism, Zilsel explained, had been developing for a long time before a few scholars took note of it and began adopting it for their own purposes. The writings of Galileo, Bacon, and Gilbert themselves all clearly reveal that their inspiration came from miners, sailors, blacksmiths, foundry-men, mechanics, lens-grinders, glassblowers, clockmakers, and shipwrights—the manual workers of that era.

In all of the crafts, the leading practitioners had developed their skills through long experience, and in many they had worked out numerical rules based on repeated measurements to guide their productive activities. "These quantitative rules of the artisans of early capitalism are, though they are never called so, the forerunners of modern physical laws," Zilsel wrote.[97] William Gilbert's achievement was not that he created the experimental method but that he was "the first academically trained scholar who dared to adopt the experimental method from the superior craftsmen and to communicate the results in a book . . . to the learned public."[98]

SCHOLARS VERSUS CRAFTSMEN?

Zilsel's challenge, as might be expected, met with stiff opposition: "A high proportion of scientists and historians of science recoiled, with all the quivering primness of a virgin insulted, from the notion that society, through technology, dictates the shape of science." A "counterrevolution of the purists" assembled behind the banner of Alexandre Koyré, whose outlook became "so fashionable as to be orthodox . . . among historians of science, especially in America."[99]

The Koyréan fashion was not a fleeting fad. In the late 1950s, a conference was held to discuss "Critical Problems in the History of Science," and the first point on the agenda was "Scholars versus Craftsmen."[100] Although the relative weight of academics and artisans in the making of modern science had been identified as a controversial issue, it was treated in an entirely one-sided manner at the conference. Rupert Hall summarized his assessment in these words:

> The roles of the scholar and the craftsman in the scientific revolution are complementary ones. . . . [The scholar] holds the prime place in its story. . . . The scholar's function was active, to transform science; the craftsman's was passive, to provide some of the raw material with which the transformation was to be effected.[101]

Hall's pronouncement echoes Aristotle's view of human reproduction. Aristotle believed that the female role was a passive one—merely "providing the raw material"—while the active, truly creative contribution was made by the male: "transforming" the raw material provided by the woman into a living organism. And just as Aristotle's outlook was a product of male bias, Hall's position as an academic no doubt inclined him to be partial to his scholarly forebears. To portray contemplative types as active and productive workers as passive might seem paradoxical to most people, but not to a professional intellectual who believed that "the genesis of the scientific revolution is in the mind."[102]

A. C. Crombie responded to Hall's paper with a declaration of wholehearted agreement:

> Dr. Hall, I think rightly, has strongly emphasized the highly intellectual and theoretical character of the Scientific Revolution, and it is not surprising that he has identified the architects of the revolution as "scholars" who alone were capable of thinking at that level, and who drew into their highly sophisticated system of scientific thought both the problems and *ad hoc* techniques of the craftsmen.[103]

Crombie's only qualification was with regard to timing; he believed that certain scholars of the thirteenth and fourteenth centuries had been as important to the process as the better-known scholars of the sixteenth and seventeenth.[104] Another participant, Francis R. Johnson, felt that Hall's presentation had not been proscholar *enough*. Hall, he said, had "often done less than justice to the 'scholar' and to the institution that represented him—the university."[105] Meanwhile, no voices were raised in defense of the artisans' priority.

With all of the proscholar arguments going unopposed, outsiders might well have wondered how this issue came to be treated as a matter of controversy in the first place. Late in the discussion, however, Crombie finally mentioned Edgar Zilsel, the historian who had called attention to the essential role of artisans in the Scientific Revolution.[106] Zilsel's arguments had been too powerful for the traditionalists to ignore, but his untimely death in 1944 meant that they did not have to confront him in person. His side of the debate was not heard at the Critical Problems conference.

After Zilsel's death, no other historians immediately stepped forward to take up where he had left off in his pioneering analysis of the Scientific Revolution. He had raised his challenge to traditional history of science in the 1930s, an era of political radicalism in which radically new ideas could receive a hearing in American universities. Shortly after he died, however, the onset of the Cold War created a political atmosphere in academia that anathematized any ideas with Marxist associations. Zilsel, a refugee from Nazi persecution, was a self-professed Marxist historian, and his emphasis on the contributions of working people to science was clearly of that provenance.

It was not until after the McCarthy witch-hunts ended that echoes of Zilsel's ideas began to be heard again.[107] Although the Critical Problems conference was completely stacked in favor of the "scholars," the effort devoted to refuting Zilsel's conclusions was a backhanded tribute to the power of his challenge. The symposium had been designed to confront "interpretations that have made changes in social circumstances enhancing the social importance of the craftsman and of technology the primary factor in the Scientific Revolution."[108] But Roy Porter more bluntly described it as an exercise in which conservative historians "militantly denied, faced with the claims of Marxism and sociology, that 'external influences' shaped the intellectual history of science, which they tended to see as Objective Rational Truths, discovered by Great Minds."[109]

Those who placed scholars in the vanguard of the Scientific Revolution did so by treating it, in the spirit of Alexandre Koyré, as "primarily a revolution in theory." Hall confined it to "the innovations and criticisms in the academic sciences—astronomy, physics, anatomy."[110] With only the *academic* sciences taken into account, the contributions of artisans were guaranteed to be minimized. In the course of the discussion, however, both Hall and Crombie were obliged to make important concessions with regard to the scientific contributions of craftsmen.

Hall, for example, acknowledged the existence of *nonacademic* sciences—the "Baconian" ones that "had no regular place in academic studies," such as chemistry, metallurgy, botany, zoology, and experimental (as opposed to theoretical) physics. In chemistry,

> the influence of craft-empiricism was strong. It can hardly be doubted that the range of chemical phenomena known to craftsmen about 1550 was much greater than that known to scholars, and that, as Professor C. S. Smith has pointed out, craftsmen had developed both qualitative and quantitative techniques of vital necessity to the growth of chemistry as an exact science.[111]

In general, Hall stated, "the technological progress of Europe during the Middle Ages was due to transmission" that "occurred at the level of craftsmanship rather than scholarship." If a natural philosopher of that era paid more attention to the processes of artisans "and less to his own consciousness and limited academic horizon, he could learn much of what the world is like. As the Middle Ages verged on the Renaissance, an increasingly rich technological experience offered ample problems for enquiry, and besides, much knowledge of facts and techniques."[112]

Among many important technological advances—most ultimately of Chinese origin—were "improved methods of harnessing both saddle-horse and draft-horse, the watermill, the windmill, the mechanical saw, glazed windows, spectacles, the wheeled plow, the true rudder, lock-gates, the grandfather-clock, and finally print-ing."[113] These new techniques were indispensable stimuli of the "mechanical philosophy" that characterized the Scientific Revolution.[114] For example:

> Wind and water-mills needed to be made and serviced, a task beyond the skill of most village smiths. So there grew up a trade of millwrights who went about the country making and mending mills. These men were the first mechanics in the modern sense of the word. . . . They were the reposito-ries of ingenuity from which the Renaissance, and even more the Industrial Revolution which followed it, drew the crafts-men who alone could have put into practice the ideas of the new philosophy.[115]

From the fourteenth century, professional clockmakers and watchmakers were "to become for science what the millwright was to be for industry—a fruitful source of ingenuity and work-manship." Similarly, "the need for compasses and other navi-gating instruments brought into being a new skilled industry, that of the card and dials makers, whose subsequent influence on science, particularly in setting higher and higher standards for accurate measurement, was enormous."[116]

As for scientific methodology, Hall acknowledged that "the early exploitation of observation and experiment as methods of scientific enquiry drew heavily on straightforward workshop practice." Nevertheless, he credited the scholars with creating the experimental method on the grounds that "the initiative for this borrowing seems to be with scholars rather than crafts-men."[117] Leaving aside for the moment whether it was borrowed or stolen, in either case it clearly originated with the craftsmen, a fact sufficiently attested to by the published works of artisan-authors from the fifteenth century onward.[118] And because the craftsmen possessed knowledge that the scholars wanted and needed, it was obviously incumbent on the scholars to take the initiative by going to the artisans' workshops.

THE ZILSEL THESIS

NO CONTRIBUTION TO a people's history of science is more important than the one made by Edgar Zilsel. It was the theory he formulated—the "Zilsel thesis"—that first indicated the indispensable part played by artisans in the creation of modern science. "The pioneering work of Zilsel's research," his editor-biographers noted, "can today more easily be appreciated than in his own times. Historians, sociologists, and philosophers of science have vastly demonstrated the impact of experimentation, intervention, instrumentation, or in one word, of practice on scientific knowledge formation."[119] Unfortunately, Zilsel's writing on the subject remained unfinished at the time of his death; all he left were a few short articles and outlines of work that he had planned to undertake.

The Zilsel thesis did not exalt the craftsman to the detriment of the scholar; it was not a counterposition of "scholars *versus* craftsmen." He would have agreed with Hall that their roles were complementary but not that the craftsmen were merely passive participants. According to the Zilsel thesis, modern science arose in early modern Europe through the interaction of artisans and elite intellectuals. Both elements were essential to the process. The scholars' role will receive less attention here because traditional histories of science have done a more-than-adequate job of explicating it; the purpose of a people's history of science is to direct the spotlight toward the generally under-appreciated artisans, for a change.

The craftsmen did indeed provide the "raw material"—the factual data that constituted the fundamental building blocks of scientific knowledge—but, Zilsel insisted, that was not all. "The artisans, the mariners, shipbuilders, carpenters, foundry-men, and miners" were also "the real pioneers of empirical observation, experimentation, and causal research." Systematic logical and mathematical thought, on the other hand, "was preserved for upper-class learned people, for university scholars, and for humanists." But "eventually the social barrier between the two components of the scientific method broke down, and the methods of the superior craftsmen were adopted by academically trained scholars." And that is when "real science was born."[120]

In recent years the Zilsel thesis has been most eloquently defended—and refined—by Pamela Smith. Her research has strengthened the view of the Scientific Revolution as an "intellectual revolution from the bottom up" wherein artisans "laid the foundations for a new epistemology, a new *scientia* based on nature." The essential premises of this "artisanal epistemology" were that "knowledge of nature is gained through direct observation of particular objects and that nature is known through the hands and the senses rather than through texts and the mind."[121] The most effective exponent of this new way of knowing nature was Paracelsus, whose writings and career are considered more thoroughly in the following pages.

It is important to remember that the early modern scholars who sought to tap the artisans' knowledge were a small minority of the educated elite. By far the greatest number fiercely resisted the "reform of learning" promoted by Francis Bacon and a few others who were frustrated by the conservatism of the universities which, as Hall acknowledged, "prevented the recognition and implementation of the victories of the revolution in each science until long after they were universally applauded by thoughtful men outside."[122]

In summary, the experimental method that characterizes modern science originated not in the minds of a few elite scholars in universities but in the daily practice of thousands of anonymous craftsmen who were continuously utilizing trial-and-error procedures with materials and tools in their quest to perfect their crafts. The "Experimental Philosophy," Baconian philosopher Joseph Glanvill declared in 1668, was founded upon "those things which have been found out by illiterate tradesmen."[123]

SEEKING WISDOM IN THE STREETS
AND THE WORKSHOPS

ALTHOUGH BACON IS traditionally credited with first recognizing that real knowledge and therefore "true command" of nature was to be found in the artisans' workshops, he was but one of a number of early modern philosophers who had begun to think along those lines. In about 1450, a book by the German philosopher Nicolas of Cusa rejected the learning of the orator and suggested instead "that wisdom can be found in the streets and the marketplace, where ordinary weighing and measuring occur."[124] In 1531, Juan Luis Vives told his scholarly colleagues they should not "be ashamed to enter into shops and factories, and to ask questions from craftsmen, and to get to know about the details of their work."[125] Martin Luther's typically blunt assessment of the value of Aristotle's works was that "any potter has more knowledge of nature than these books."[126] Bacon, representing the culmination of this tradition, did not introduce a new approach to gaining scientific knowledge; he was describing a social phenomenon that had begun centuries earlier and was advocating that it be systematically exploited.

The fertile "association of technology and learning," Crombie pointed out, originated in the twelfth and thirteenth centuries, well before the Renaissance. The writings of Hugh of St. Victor and Domingo Gundisalvo in the twelfth century promoted "the study of technical subjects that give control over nature," especially "practical geometry, which includes the work of land-surveyors, carpenters, masons and black-smiths." Also significant was "the interest of thirteenth-century scholars like Albertus Magnus, Vincent of Beauvais and Raymond Lulle in the activities of artisans such as peasants, fishermen and miners." The upshot was that "contacts between scholars and craftsmen [were] already bearing fruit by the middle of the thirteenth century."[127]

Crombie also called attention to the "precursor Bacon," Roger, a thirteenth-century Franciscan friar and Oxford scholar who made a "strong plea for experiment and mathematics, as he understood them, as means of obtaining useful power over nature."[128] Roger Bacon has often been portrayed as having anticipated Francis Bacon by three hundred years in recognizing the scientific value of craft practices, because in his *Opus majus* he declared,

> More secrets of knowledge have always been discovered by plain and neglected men than by men of popular fame, and I have learned more useful and excellent things without comparison from very plain people unknown to fame in letters, than from all my famous teachers.[129]

At about the same time, Nicholas of Poland, another university-educated friar (but of the Dominican order), was expressing similar sentiments with regard to medical knowledge:

> God loves the humble, Nicholas proclaimed. He chose to reveal his deepest secrets to ordinary people, just as He had conferred the most marvelous medical virtues on the meanest beings in nature. Hence the common people of the villages had deeper insights into the secrets of nature than did the learned physicians: "The people love empirical things," Nicholas declared, "because none of them are harmful; but the physicians are ashamed because great works prefer the villages, where the marketplaces resound in their praises of empirical remedies."[130]

Crombie's general point is well taken, but his focus was almost entirely on the intellectual elite. He and like-minded historians have attempted to locate the genesis of the Scientific Revolution in the work of medieval natural philosophers such as Roger Bacon, Robert Grosseteste, Thomas Bradwardine, and Jean Buridan.[131] Although it is true that these scholars began to question Aristotelian dogma in the realm of theoretical physics, their science "was a very bookish business . . . and its odor remained that of ink and parchment."[132] Their challenges were intellectual exercises solidly within the tradition of scholasticism—more metaphysics than physics—rather than serious attempts to understand the actual workings of nature. Genuine advances in knowledge of nature remained in the hands of working people, who, until the fifteenth century, did not ordinarily commit what they knew to writing.

GALILEO AND THE ARTISANS

GALILEO DID NOT need Bacon's writings to tell him that he could profit greatly from interactions with artisans. In his most influential work, the *Dialogues Concerning Two New Sciences*, he wrote that the "constant activity" at Venice's weapons factory, the "famous arsenal":

> suggests to the studious mind a large field for investigation, *especially that part of the work which involves mechanics;* for in this department all types of instruments and machines are constantly being constructed by many artisans, among whom there must be some who, partly by inherited experience and partly by their own observations, have become highly expert and clever in explanation.[133]

He had his interlocutor respond:

> You are quite right. Indeed, I myself, being curious by nature, frequently visit this place for the mere pleasure of observing the work of those who, on account of their superiority over other artisans, we call "first rank men." Conference with them has often helped me in the investigation of certain effects including not only those which are striking, but also those which are recondite and almost incredible.[134]

Was all of this "only a rhetorical device," as suggested by those who portray Galileo as "a vast intellect palpitating in a realm of pure Ideas unsullied by . . . practical considerations, unaffected by the outer world which pays the bills"? That interpretation is ruled out by the documentable fact that during his early career, his "environment and interests were largely technological." In Florence he studied engineering and "mastered machine design, canal construction, dyking and fortification"; his professorship at Padua was due to the sponsorship of a military engineer named Guidobaldo del Monte; in Venice he patented a mechanical device designed to raise water.[135]

The famous story that Galileo discovered the isochronism of the pendulum by watching an oscillating chandelier in a church during a boring sermon is yet another of the myths that have attached to him. It was "the rapidly expanding mechanic arts of Galileo's age" that make his contributions to science "historically intelligible." His "environment of technical innovations like suction pumps and pendula," a historian of technology observed, "provided novel controlled situations, almost laboratory situations, in which he could be among the first to observe natural phenomena, like isochronism or the breaking of a column of water, which are not easily perceived in a pure state of nature."[136]

Galileo's scientific reputation owes a great deal to his investigations of projectile motion, a subject with important potential military applications. He demonstrated mathematically that "if projectiles are fired . . . all having the same speed, but each having a different elevation, the maximum range . . . will be obtained when the elevation is 45°: the other shots, fired at angles greater or less will have a shorter range." But in recounting how he arrived at that conclusion, he revealed that his initial inspiration came from discussions at the Arsenal: "From accounts given by gunners, I was already aware of the fact that in the use of cannons and mortars, the maximum range, that is the one in which the shot goes the farthest, is obtained when the elevation is 45°."[137] Although Galileo's mathematical analysis of the problem was a valuable original contribution, it did not tell the workers at the Arsenal anything they had not previously learned by empirical tests, and had little effect on the practical art of gunnery.[138]

WILLIAM GILBERT AND THE MAGNET

SOME HISTORIANS OF science have carelessly identified William Gilbert as Bacon's "disciple"[139] or characterized his experimental work as "in accordance with the teaching of Francis Bacon."[140] In fact, however, Gilbert was performing his experiments while Bacon was still a child. Gilbert's modern reputation as the first experimental scientist is based on the book he published in 1600, *De Magnete:* "the first great work of modern science in England";[141] "a landmark in the Scientific Revolution";[142] its "form and content . . . stand out like a watershed in the country of the mind."[143]

Gilbert, one of the physicians to Queen Elizabeth I, took it on himself to investigate and demystify the apparently occult power of the lodestone to cause iron objects to move without touching them. *De Magnete* is the report of his results, showing with great care exactly how he performed his experiments so that the reader could replicate them. It might be mistaken for a work of popular science had it not been written in Latin; nearly three centuries passed before the first English edition was published in 1893. He was obviously not writing for craftsmen and manual workers but was transmitting information, methods, and techniques learned from them into the realm of international scholarship—the "Republic of Letters."[144]

Taking into consideration the level of knowledge and instrumentation in the late sixteenth century, it is difficult to suggest anything that Gilbert might have done to improve his experiments. In other words, in *De Magnete* the practice of experimentation suddenly appears in fully developed form, with no primitive or intermediate stages. Gilbert's approach "is so exceptional in his period that the question arises where it originates."[145] The answer is to be found in a careful reading of *De Magnete* itself, which shows how heavily Gilbert drew on the knowledge of blacksmiths, miners, sailors, and instrument-makers. His experiments, Zilsel pointed out, often "simply repeated the working processes of contemporary iron manufacture." In short, Gilbert's "spirit of observing and experimenting was taken over not from scholars but from manual workers."[146]

Gilbert's description of a key experiment involving the tendency of magnetic rocks to be aligned with the earth's north-south axis is revealing:

> We once had chiselled and dug out of its vein a loadstone 20 pounds in weight, having first noted and marked its extremities; then, after it had been taken out of the earth, we placed it on a float in water so it could freely turn about; straightway that extremity of it which in the mine looked north turned to the north in water.[147]

Gilbert evidently left his cloistered study for this experiment. There can be no doubt that he actually descended into iron mines—the source of lodestones—and had close contact with the miners.

His discussion of the phenomenon of magnetism induced by a blacksmith's working of iron aligned in a north-south direction makes it clear that he had spent time observing in the smithies and gaining information from the smiths. The text describing some of the experiments is accompanied by a woodcut illustration of a blacksmith's shop and tools.[148] As important as they were to the history of science, "since the miners and foundrymen of the period belonged to the lower ranks of society and were uneducated we know neither their names nor their ideas."[149]

"Navigation and nautical instruments play an even greater part in *De Magnete* than mining and metallurgy."[150] In rejecting the hypothesis that variation in the magnetic compass is caused by large deposits of lodestone scattered irregularly over the earth's surface, Gilbert cites the evidence of sailors who reported that when they passed Elba, renowned as a source of powerful lodestones, there was no deflection of the needle toward the island.[151] As for how he was able to make the careful measurements that established his reputation for scientific accuracy, "all of his physical instruments are actually nautical instruments or are at least nearly related to the mariner's compass," Zilsel noted. "When the seamen of the sixteenth century went to sea, they laid the foundation-stone of the British Empire and when they retired and made compasses, of modern experimental science."[152]

An illustration from William Gilbert's De Magnete. *Blacksmiths were a prime source of information for Gilbert.*

Although Gilbert's book provides ample evidence of his debt to manual workers, almost none of his working-class collaborators are named. In a discussion of the range of latitudes throughout which the magnetic compass works, for example, he cited as his source "the most famous captains and also many of the more intelligent sailors," but recorded for posterity only the names of two gentlemen mariners, Thomas Cavendish and Sir Francis Drake, and did not identify any of the ordinary seamen.

Perhaps it is understandable that Gilbert would only include names of "famous captains" that his readers might recognize, but his failure to give full credit to one craftsman in particular, Robert Norman, is less excusable. Norman was a seaman-turned-compass-maker who wrote a book about magnetism entitled *The Newe Attractive*. It was published in 1581 in English—its author could not write Latin—and its purpose was mainly to convey practical information to other artisans. As noted in the previous chapter, its most important revelation was the discovery of the "dip" of the compass needle. In the book, Norman described himself as "an unlearned Mechanician," alluded to the "18 or 20 years that I have travelled the seas," and apologized for his "want of eloquence." But at the same time, he suggested that artisans like himself who have "the use of those artes at their fingers endes" may have more to contribute to mathematics than "the learned in those sciences beeying in their studies amongst their bookes."[153]

Gilbert did mention Norman by name several times and identified him as "a skilled navigator and ingenious artificer."[154] Nonetheless, in spite of Norman's crucial influence on Gilbert's investigations, "Gilbert himself does not emphasize it at all, but rather hides it. . . . If we wish to learn what Gilbert actually owes to him, we have to examine Norman's treatise."[155]

Many of Gilbert's most important conclusions about magnetism, much of the evidence supporting them, and the method he used to arrive at them appeared in *The Newe Attractive* nineteen years before the publication of *De Magnete*. Norman reported his experiments with magnets suspended by threads and by pieces of cork floating on water; these "new experimental devices were simply taken over by Gilbert." Zilsel cited "an experiment in every detail" that Gilbert "borrowed from Norman's book," as well as two other "outstanding and most carefully performed experiments" that also originated with Norman.[156] Gilbert's "best quantitative experiment" (establishing the weightlessness of magnetism) was taken directly, without attribution, from Norman.[157]

Norman's was not the only book that preceded Gilbert's in taking "an experimental and instrumental approach to geomagnetic questions"—others included William Borough's *A discovrs of the variation of the cumpas, or magneticall needle* (1581); Thomas Blundeville's *Exercises* (1594); and William Barlow's *Navigator's supply* (1597)—but Norman's was the vernacular source from which Gilbert most liberally borrowed.[158] Aside from "the Latin erudition, the quotations and polemics, and the metaphysical philosophy of nature," Norman's book has everything Gilbert's has.[159] It is somewhat unjust, then, that

"Norman is virtually unknown today, whereas Gilbert is counted among the pioneers of natural science."[160]

ARTISAN-AUTHORS

To say that scholars "went to the workshops" to learn from craftsmen is to speak figuratively. Although Galileo and Gilbert literally paid visits to artisans and directly observed their operations, a more common channel of communication was the printed word.

Robert Norman, as we have already noted, was by no means unique as an artisan-author. By the time he published his book on magnetism, European artisans had been disseminating scientific and technical information in written form for hundreds of years. The previously discussed arithmetic texts of the reckoning masters and the artists' treatises on perspective were representative of that trend. "For a century and a half prior to [1646]," William Eamon said, "Europe had been inundated with scores of treatises that professed to reveal the 'secrets of nature' to anyone who could read."[161] Another historian, Pamela Long, has demonstrated a significant circulation in manuscript form of artisans' books on the mechanical arts in "the early decades of the fifteenth century"—that is, even before the introduction of printing in Europe.[162] Although Zilsel noted that the rudiments of "the idea of science we usually regard as 'Baconian' . . . appear first in treatises of fifteenth-century craftsmen,"[163] this corpus of artisanal literature had for the most part been ignored by historians until Eamon and Long redirected attention to it.[164]

"Naturally only members of the most highly skilled crafts wrote treatises," Zilsel commented. "Even in the sixteenth century a considerable number of the manual workers, particularly outside Italy, were illiterate."[165] Those who could write in their "vulgar tongues," however, were often quite prolific, and "when apothecaries, potters, sailors, distillers, and midwives got into print along with scholars, humanists, and clerics, the Republic of Letters was permanently changed."[166] This change represents an especially important juncture for a people's history of science, because it means that it can begin, at long last, to be based on solid documentation.[167]

The moribund state of learned science—"lock't-up in the Greeke and Latin tongues," as John Aubrey put it[168]—could not have been superseded without the "living instrument" of vernacular language, "the language that one has learned from his mother's lips and uses in complete innocence and freedom." The allusion to mothers is not an empty figure of speech. "The fact is that no language can be truly alive that is not used by women." But women were denied advanced education, so "Latin was spoken only by men, and the number of men able to use it spontaneously was steadily decreasing."[169] The new science was thus dependent on a fundamental element of cultural transmission in which the role of women was vital.

Before Robert Norman's book inspired William Gilbert, the artisans' writings were largely disregarded by contemporary university-trained scholars and thus had had little direct impact on the course of elite science.[170] Their latent contribution to the creation of scientific knowledge was nonetheless crucial. "The textual and pictorial elaboration of the mechanical arts," Pamela Long explained, "allowed their transformation from 'know-how,' available for constructing things in the world, to 'knowledge' involving rational or mathematical principles."[171] The craftsmen's literary efforts, in other words, reveal the scientific content of what is frequently dismissed as "mere technology." But whereas the artisans' authorship "helped to transform some arts from the arena of skilled know-how to that of discursive knowledge, it did not change artisans into learned men. It is more accurate to say that it prepared certain of the mechanical arts for appropriation by learned cultures."[172]

Galileo was one of the "appropriators." He was not far behind Gilbert in recognizing the value of the vernacular scientific books; among other things, they posed the questions that formed his research agenda: "The problems of the economy of power and of how much machines can accomplish, of the accuracy of guns, of the resistance of fortifications, are the very same ones that *for two centuries* had found treatment in the technical literature."[173]

Eamon's study of the artisans' "books of secrets" describes them as "works of 'popular science'" that "played an instrumental role in disseminating craft information" to the elite *virtuosi*. These works "articulated a novel concept of experimentation" that later became identified with Baconianism. They

> were grounded upon a down-to-earth, experimental outlook.
> . . . What they revealed were recipes, formulas, and "experiments" associated with one of the crafts or with medicine: for example, instructions for making quenching waters to harden iron and steel, recipes for mixing dyes and pigments, "empirical" remedies, cooking recipes, and practical alchemical formulas such as a jeweler or tinsmith might use.[174]

Recipes, it should be noted, "are the record of trial-and-error experimentation. They are the accumulated experience of practitioners boiled down to a rule."[175]

THE "PRINT REVOLUTION" AND MODERN SCIENCE

With the advent of printing, "the world of learning, hitherto the domain of a tiny privileged elite, was suddenly made much more accessible to the common man."[176] The coming together of artisans and academics, which Zilsel identified as the key to the rise of modern science, was greatly accelerated. Elizabeth Eisenstein, who has most thoroughly investigated the historical impact of printing, argued that the collaboration of scholars and craftsmen would not have permanently borne fruit had it not become a continuous and cumulative process. And that in turn depended on the records of their cooperative efforts being stabilized by their reproduction in thousands upon thousands of standardized printed copies. According to Eisenstein's thesis, the "print revolution" was a *sine qua non* of scientific advance.[177]

If that is so, then yet another multitude of artisans must be recognized as indispensable to the history of science. First of all was the mechanic or mechanics directly responsible for the innovation. Unfortunately, "this stupendous achievement left but little record of its origin and development besides the mass of material that it produced," leaving "the identity of the great mechanic who invented printing shrouded in impenetrable mystery. . . . Whether he spelled his name Gutenberg, Fust, Schöffer, Coster or something else, has not been established."[178] Gutenberg alone is often credited, but would it not be better to recognize it as a collective accomplishment?

Furthermore, the rapid rise of the printing industry brought into existence a constellation of related crafts whose practitioners included "printers, typefounders, engravers, compositors, woodblock cutters, proofreaders, booksellers, and even peddlers." More than that, it can be thought of as having created a *new culture* that "brought together scholars, craftsmen, merchants, and humanists engaged in common pursuits."[179] The proofrooms of early printshops, where proofreaders, authors, and proofreader-authors exchanged ideas and criticized each other's works, became in sixteenth-century Europe far more important than the moribund universities as centers of creative intellectual activity—scientific and otherwise.

But although "on one level printing opened up new avenues of communication among scholars, craftsmen, and the general public," it did not completely "bridge the gap between town and gown." Academic disdain for manual labor remained strong, but "the explosion of vernacular publication" meant that "the literacy barrier no longer coincided with the Latin barrier," and the power of the ancient prejudices to retard science gradually began to break down.[180]

As previously noted, treatises of artisan-authors began to circulate in manuscript in the preprint era, but the literature they created was transformed when it began to be mass-produced in printed form. "Another scientific tradition emerged alongside academic science" when publication of scientific books for the "common man" became commonplace. "It was in large measure the printer's willingness to accept the risks associated with the competitive and uncertain vernacular book market that made sixteenth-century 'popular science' a reality." The social background of the new occupational category was "a mosaic of the Renaissance petite bourgeoisie." Many of the pioneer printers "had spent years working for wages as pressmen, typefounders, and proofreaders . . . some had previously been goldsmiths or woodblock-cutters, while others had been painters, bookbinders, barbers, or even tavern-keepers."[181]

Early printing entrepreneurs who emerged from the ranks retained their artisanal social connections, which provided them with a supply of texts to publish and a pool of customers for the finished products. A "barrage of how-to-do-it manuals and of technological treatises" appeared in tens of thousands of copies:

> The most widely circulated of these printed how-to-do-it books was a group of craft manuals known collectively as the *Kunstbüchlien* (Skills-booklets), which appeared in various German towns in the early 1530s. Originally issued as four separate pamphlets, the booklets became immediate bestsellers. . . . The *Kunstbüchlien*, all anonymous publications, were printers' compilations. Popular printers like Christian Egenolff compiled them from workshop notes, from word-of-mouth sources, and from various "experimental" treatises.[182]

WALTHER HERMAN RYFF

AT ABOUT THE same time, a new class of authors exemplified by Walther Herman Ryff (c. 1500–1548) "created a new kind of scientific literature." Ryff, who had been trained as an apothecary, became "by far Germany's most prolific and best-known scientific writer." Although not an elite scholar, he could read Latin and used that ability to translate classical scientific works into German. The sources he drew on for his own writings were "medical empirics and craftsmen-surgeons, distillers, oculists, and reckonmeisters."[183]

Ryff's books "played a key role in disseminating scientific information to the German people," but they were denounced as "vulgarizations" by the academic establishment.[184] He was not deterred. "I know very well," he wrote, "that this and my other works, which I have put into print in the German language for the simple man, have brought upon me the anger and scorn of many learned people."[185]

That Ryff was consciously attempting to create a literature of people's science is evident in the way he described his intended audience:

> I do not write this little book for the educated people, for they already know this art. Nor do I write for those ignorant blockheads whose brains you could make into pig's troughs. I write only for the simple, respectable and devout little people who have until now, through God, asked for my advice and help. Some of them have not reached me only because they are too distant, or because poverty makes the way too hard by which they might give themselves help or at least comfort.[186]

When Ryff stated, as he did repeatedly, that he was writing for the common people, we may presume, according to Eamon, that he meant to include "journeymen and masters in the trades, merchants, shopkeepers, and growing numbers of women. Ryff assumed the common man was literate, but not learned. He could read German, but not Latin." Ryff was not, however, directly addressing those at the very lowest socioeconomic levels. Although he provided extensive recommendations on medical treatment for the poor, "he did not imagine they were the main readers of his books, which they could rarely afford to buy. He was more concerned with providing medical advice to the apothecaries, surgeons, empirical healers . . . under whose care the poor might find themselves."[187]

PALISSY THE POTTER

Bernard Palissy, in George Sarton's estimation, was one of the two outstanding figures of Renaissance geology.[188] "Next to

Leonardo," he added, Palissy "is the best representative of the man of science who was not a scholar, whose knowledge was found not in books but in the very bosom of nature."[189] Palissy's own two scientific books rank among the most valuable of the artisan-authors' works.

Palissy was born to a French peasant family about 1510,[190] but as a young man he apprenticed himself to a master glazier and embarked on the trade of making stained-glass windows. By 1539, he had changed occupations and had become a land surveyor. At about the same time, a porcelain cup he saw inspired him with an obsessive ambition to learn how to create fine ceramic objects. Starting with virtually no knowledge of pottery-making, he began experimenting with various materials until, sixteen years later, he mastered the craft and eventually gained recognition as one of the outstanding ceramic artists of the Renaissance. His *rustiques figulines* attracted the attention of the queen mother, Catherine de Medici, who named him the official designer of the king's ceramicware.

An example of Bernard Palissy's ceramicware.

Palissy's experimentation in search of "the perfect enamel" provided him with a great deal of empirically derived chemical knowledge:

He had tried some 300 mixtures, combining various clays and sands with tin, lead, iron, steel, antimony, copper sulfate, ashes of tartar, litharge, stone of Périgord (manganese), etc. The art of pottery required physical experiment; he had to find out how much heat was needed in his furnaces and how to regulate the speed of firing his materials and the speed of cooling them. He was aware of the existence of a multitude of salts, and knew that salts in solution were very different from those in solid form.[191]

Although Palissy's fame and fortune derived from his craftsmanship as a potter, his scientific interests were much broader. He "observed not only the city artisans in their shops but also the farmers in the open country. Every aspect of nature drew his attention, fields and forests, mountains and valleys, springs and rivers." His first book, the *Recepte véritable* (1563), was "a medley of facts and ideas about agriculture and geology, mineralogy and chemistry, philosophy, theology, etc., which reveals his vast experience as well as his lack of formal education."[192] His second, the *Discours admirables* (1580), was even more impressive in the variety of scientific topics it treated: "philosophy, geology, paleontology, botany, zoölogy, engineering, hydrology, chemistry, physics, medicine, alchemy, metallurgy, agriculture, mineralogy, embalming, toxicology, meteorology, and ceramics."[193]

More noteworthy than the breadth of Palissy's knowledge was his outspoken defiance of the established scientific authorities of the day. "I well know," he wrote,

> that some will scoff, saying that it is impossible that a man without Latin could have knowledge of Nature; and they will say that it is very bold of me to write against the opinion of so many famous and ancient philosophers, who have written on natural things. . . . I know also that others will judge by appearances, saying that I am but a poor workman: and by such statements will try to make my writings appear harmful.[194]

He was certainly right about that. As his translator noted, at that time,

> when timid questioning of the authority of medieval science was just beginning, such an attitude struck the learned doctors of the University of Paris like a thunderbolt. Many a gray beard waggled in indignation at such irreverence and many a professor of philosophy fulminated *ex cathedra* at this impious and uncultured iconoclast.[195]

Undeterred, Palissy continued to denounce books "written from imagination by those who have practiced nothing" and to insist on the priority of practice over theory. "Be wary," he warned,

> of believing the opinions of those who say and maintain that theory begat practice. Those who teach such a doctrine argue wrongly, saying that the thing that one wants to do must be imagined and pictured in one's mind, before laying hand to one's task. . . . if the theory imagined by war leaders could be carried out, they would never lose a battle. I dare say, to the confusion of those who hold such opinion, that they could not make a shoe, not even the heel of a boot, even if they had all the theory in the world.[196]

Alas, it was religious rather than scientific heterodoxy that cost Palissy most dearly. As a Huguenot, he was on the wrong side when civil war erupted in France in 1588. He was imprisoned in the Bastille de Bussy and died there in 1590.

Palissy's scientific ideas had little immediate impact on scholarship, even in France, "because they were not published in scholarly books."[197] More than a hundred years would pass before their value would be recognized and appreciated by such eighteenth-century scientists as Antoine de Jussieu, Fontanelle, and Réamur.[198] But his career illustrates the scientific vitality of the crafts that was to make such a great impression on Francis Bacon in the next generation.[199]

HUGH PLAT

HUGH PLAT (1552–1608), son of a well-to-do London brewer, was a Cambridge-educated member of the urban elite, but he represents a transitional type between the pure artisan-author and the Baconian gentleman of science. Whereas Bacon and his successors sought to gain the knowledge of artisans without blurring the class distinctions between their informants and themselves, Plat did not hesitate to dirty his hands by personally engaging in artisanal practices.

Plat was trained to be a lawyer, but he developed a passionate interest in natural philosophy. Perceiving little of value in the learned, university-based sciences of his day—he was about ten years Bacon's senior—Plat "dedicated his entire leisured life to his experiments and inventions in agriculture, horticulture, and chemistry." But, aware of the limitations of his solitary efforts, he diligently sought the collaboration of working people. He learned metallurgy from blacksmiths and "corresponded with gardeners and farmers, gathering information about agricultural and horticultural practices from various parts of England."[200]

Plat published ten books based on the knowledge he had gleaned from artisans and his own experimental practice. In addition to his published works, Plat left behind a large volume of handwritten notes that have survived. His notebooks, which are in the British Library, provide a unique resource for social historians, because Plat, unlike Gilbert and most other gentleman-authors, gave credit to his working-class informants by name. One such historian, Deborah Harkness, has examined those notebooks and has determined that they record the names of some seventeen hundred practitioners from whom Plat received information.[201]

Plat's work should not be interpreted as a weak "anticipation" of Bacon but as a very strong stimulus to Bacon's philosophy of science. Harkness believes, with good reason, that Bacon's writings were intended in part as a response to authors of Plat's ilk. Bacon, sensing the potential power of the artisanal knowledge collected by Plat and others like him, was concerned that it not be unleashed without regard to its political consequences but controlled to forward the interests of the governing class he represented.[202] As the titles of Plat's books indicate (two of his best sellers were *Jewell House of Art and Nature* and *Delights for Ladies, to Adorn Their Persons, Tables, Closets, and Distillatories*), they were addressed to a popular readership rather than to the Republic of Letters. Bacon sought to rein in the power of the new knowledge by bringing it under the command of elite intellectuals.

BACON VERSUS PARACELSUS

BACON'S GENERAL ANTIPATHY to independent knowledge-seeking is perhaps most clearly revealed in his attitude toward the movement inaugurated by Paracelsus, which instigated revolutionary change in the chemical and medical sciences. Paracelsian influence rivaled Bacon's for the allegiance of intellectuals and the nonclerical, commercial elite. Although Baconianism eventually emerged as the main ideological current of the Scientific Revolution, the Paracelsian challenge—in many ways a "people's science" movement—represented a major countercurrent.

The movement's namesake, Paracelsus,[203] was a larger-than-life figure whose flair for the dramatic generated controversy wherever he went and made his challenge to medical tradition impossible to ignore. This "Luther of medicine" roiled long-stagnant theoretical waters, thereby stimulating and polarizing the study and practice of healing.

Both Bacon and Paracelsus aimed at a reform of knowledge, but their approaches to the problem derived from mutually exclusive motives and ideologies. Whereas Bacon coveted craft knowledge as raw material to be usurped by scholars, Paracelsus:

> took the methods of the artisan to be the ideal mode of pro-
> ceeding in the acquisition of all knowledge, for the artisan

worked directly with the objects of nature. Paracelsus considered this unmediated labor in and experience of nature as elevating the artisan both spiritually and intellectually above the scholar.[204]

Paracelsus.

Although not himself a scholar, Paracelsus was "the first in European culture to give scholarly voice to an artisanal understanding of the material world," Pamela Smith contended. "Bacon's contribution was to codify the artisanal construction of knowledge that was already going on around him." But "in his disregard of the artisan and mechanic as contributors to the reformed philosophy, Bacon provided a model for the self-styled new philosophers."[205]

Bacon perceived political danger in "a subculture of Puritan (Calvinist) gentry, yeomen, and artisans" inspired by "a Paracelsian medical literature insisting that healing knowledge does not come from the academically trained physicians and clergy but from the unmediated experience of nature. . . . Bacon found all this so subversive that he went on to construct his natural philosophy partly in answer to it."[206]

> The resonance between Paracelsus's ideas and the Radical Reformation was particularly strong. The radical reformers were self-professed saints, often known as Anabaptists, who were convinced of their salvation and claimed that, as God's chosen ones they were . . . without need of earthly interpreters. On that basis, they sometimes went on to . . . attack established authority in the name of social justice and equality. Thus, German peasants, inspired by such radical thinking, rose up against rural priests and landlords in the great Peasants' Revolt of 1525.[207]

Paracelsus took the side of the peasants in this momentous struggle. The radicalism of his approach to natural philosophy was evident in his harsh denunciation of the clerical and medical elites as

> vain and greedy, bankrupt in head and heart. They exploit the poor by pretending to knowledge they do not have. The common people are spiritually and intellectually superior to their social betters. If the notables would reform themselves, they would do well to go to peasants and artisans to . . . imbibe a genuine knowledge of nature.[208]

PARACELSUS AND THE PARACELSIANS

PARACELSUS SAW HIMSELF as a man of the people crusading against an entrenched oligarchy of wealthy, mammon-worshipping physicians who were defrauding the public. His bombastic,[209] confrontational style won him many influential enemies among the scientific elite, but the fierce opposition he provoked was not primarily motivated by dislike of his unorthodox ideas or his bad manners. His activities clearly represented a threat to the social and economic position of established medical scholars and practitioners.

Paracelsus briefly found himself in a position to wage war on the medical establishment from the inside. Due to a well-publicized therapeutic success (healing the influential printer-publisher Johann Frobin), he received and accepted an invitation in 1526 to be the city physician of Basle and a professor at its university. He used those appointments to launch violent attacks on learned medicine, culminating in 1527 in a public demonstration of his disdain for the classical medical authorities in which he threw a copy of the principal Galenic textbook, Avicenna's *Canon*, into a St. John's Day bonfire. The stunt greatly boosted his notoriety, but his academic career not surprisingly came to an end shortly thereafter, and he spent the rest of his life as an impoverished itinerant practitioner, preaching his dissident views far and wide.

The elite doctors of the medical faculties mobilized their political strength and succeeded in limiting Paracelsus's influence by excluding him from academic forums and blocking the publication of his writings. Although his attempt to overthrow the prevailing Galenic medical tradition met with little success during his lifetime, his charismatic crusade attracted a small number of devoted disciples, who carried his message forward posthumously.

Paracelsus's impact on history would have been minimal without the sustained collective efforts of those who took up his cause in the generations following his death. The history of the Paracelsians offers yet another demonstration that science—including alternative, or "people's science"—develops socially rather than as the product of solitary genius.

Paracelsianism began as a poor people's movement but transcended its origins by attracting educated and influential devotees. Within two or three decades of his death in 1541, Paracelsus's following had mushroomed. In addition to the itinerant alchemical healers who formed the movement's ranks, prominent physicians such as Adam von Bodenstein and Johannes Huser had been won to its banner. Von Bodenstein described himself as "the first doctor to graduate from a university and take up the wholesome and honest doctrines of Theophrastus [Paracelsus] and publicly defend them."[210] Peter Severinus, eminent physician to the King of Denmark, wholeheartedly embraced Paracelsus's teachings, as can be seen in the advice he gave in 1571 to those desiring genuine knowledge of nature:

> Sell your lands, your houses, your clothes and your jewelry; *burn up your books.* On the other hand, buy yourselves stout shoes, travel to the mountains, search the valleys, the deserts, the shores of the sea, and the deepest depressions of the earth; note with care the distinctions between animals, the differences of plants, the various kinds of minerals, the properties and mode of origin of everything that exists. *Be not ashamed to study diligently the astronomy and terrestrial philosophy of the peasantry.* Lastly, purchase coal, build furnaces, watch and operate with the fire without wearying. *In this way and no other,* you will arrive at a knowledge of things and their properties.[211]

PARACELSUS AND THE MINERS

THE PARACELSIANS' MEDICAL and social views were strongly conditioned by their hero's lifelong association with mines and miners. It was a two-way relationship: a great deal of Paracelsus's knowledge of nature was gained from miners, and in return he concerned himself with their ailments. One particular work, his *Von der Bergsucht und anderen Bergkrankheiten* (On the Miners' Sickness and Other Miners' Diseases) testifies to that concern on his part. It is of particular interest as the first example of what would later be called occupational medicine; "the first treatise in medical literature recognizing and systematically dealing with an occupational disease."[212]

Paracelsus was introduced both to medical practice and to the mines by his father, Wilhelm Bombast von Hohenheim, who was a doctor in a poor Swiss village. In 1502, when Paracelsus was nine years old, his family moved to Villach in the Austrian province of Carinthia, where his father reportedly practiced alchemy and taught at the mining school established by the Fuggers at Hutenberg.

Paracelsus himself began working in mines near Villach as a child, and then, in adolescence, he worked in the Fugger mines near Schwaz.[213] Later, as a wandering physician, he continued occasionally to work as a journeyman in the mines. In his *Von der Grossen Wundarzney* (The Great Surgery Book) he says that from the ages of twenty to twenty-five, he was employed in a smelting plant at Schwaz. "During his journeys through Denmark and Sweden, and later in Meissen and Hungary, he learned about the mines in these countries." In 1537 he once again "came into contact with the mining industry, when the management of the Fugger mines called him to Villach to take charge of the metallurgical work there."[214]

Paracelsus seems to have had some university training at Ferrara and possibly elsewhere, but the extent of his formal education is not known. "It is not clear where he studied medicine or where he obtained his M.D., if he did obtain one. He probably received his basic medical training from his father."[215] If he was proficient in Latin, there is no evidence of it. What is known is that Paracelsus "deployed the German language as part of a wholesale political attack upon the 'tyranny of Latin' exercised by the medical establishment."[216] During his stint as a university professor

he lectured not in Latin but in German, or rather in the Swiss dialect. It is possible that having been brought up in a mining environment instead of an academic one he could not do otherwise, but this was a final provocation to his colleagues. It was much as if a medical professor of today should lecture in thieves' argot. Moreover, it was a betrayal of professional secrecy. Latin was the esoteric language used to prevent the dissemination of learning to people who were deemed unworthy of it, or who might make a bad use of it.[217]

The time Paracelsus spent in the mines exposed him to the most advanced metallurgical knowledge of the day. The impact of this experience on his medical theory and practice is apparent: the hallmark of Paracelsian medicine was its use of metals as pharmaceutical agents. Orthodox medical practice depended almost exclusively on drugs produced from plants. Although Paracelsus was not the first to introduce chemical methods into medicine or metallic substances into the pharmacopoeia, in the debate over therapeutics between "herbalists" and "metalists," he certainly holds first place among the latter.

The metallurgical processes he learned in the mines and smelters provided the basis of the alchemical lore on which his medical theories were founded. To Paracelsus, the proper goal of alchemy was not the transmutation of base metals into gold but the transmutation of metals into useful medicines. Whereas the prescriptions of orthodox doctors combined numerous strange ingredients into a complex mess, Paracelsus insisted on an opposite approach: reducing a substance by alchemical means to its simplest level—the *quintessence*—in which only the medically active ingredient remains. According to the homeopathic principle ("like cures like"), an effective medicine for the whole range of diseases that afflict lead miners can be obtained by the reduction of lead, that for silver miners' diseases from silver, and so forth.

His advocacy of the homeopathic principle, a staple of folk healers, illustrates his willingness to learn from them and to incorporate important aspects of their knowledge into his system. Galenic medicine was founded on the opposite principle of "contraries cure." For a patient experiencing the "hot and dry" symptoms of a fever, for example, the Galenic doctor would prescribe medicines composed of "cold and moist" ingredients.

Paracelsus's debt to folk medicine was explicitly acknowledged. "A Physitian," he wrote, "ought not to rest only in that bare knowledge which their Schools teach, but to learn of old Women, Egyptians, and such-like persons; for they have greater experience in such things than all the Academicians."[218] He gathered information from "not only doctors, but also barbers, bath attendants, learned physicians, old wives, magicians (or *schwarzkünstlern*, as they call themselves), among the alchemists, in the cloisters, among the nobles, the common people, among the clever and the simple."[219]

From the standpoint of modern sensibilities, much of Paracelsus's alchemical and philosophical writings appear mystical and "unscientific." Some of that can be attributed to the need he felt to maintain craft secrecy and some can be attributed to the lack of a well-developed scientific vocabulary in sixteenth-century Swiss German that often left him no choice but to make up new terminology as he went along. Mostly it was a function of his premodern philosophical worldview, which is quite incompatible with our own. But however alien his motives and mode of expression may seem to readers today, he and his followers were undeniably engaged in a quest to extend their knowledge of nature.

Whereas much of the specific content of Paracelsian medicine has long been superseded, its insistence on the importance of chemistry to medicine was an enduring contribution. Most important, its frontal assault on Galenic orthodoxy played an indispensable role in clearing the way for the consideration of alternative medical views, without which the development of modern medicine could not have occurred.

PHYSICIANS, SURGEONS, APOTHECARIES, AND "QUACKS"

A KEY ISSUE that alienated Paracelsus from the medical establishment was his insistence that "there can be no surgeon who is not a physician," an attack on the medieval professional code separating surgeons from physicians.[220] "From the Middle Ages," Roy Porter explained, "medical practitioners organized themselves professionally in a pyramid with physicians at the top and surgeons and apothecaries nearer the base, and with other healers marginalized or vilified as 'quacks.'"[221]

Surgery had been looked down on as a manual art by physicians in Galen's time, but in 1163, when elite doctors were clergymen and the Church's word was law, the Council of Tours formalized the separation of surgery and medicine by proclaiming *Ecclesia abhorret a sanguine* ("the Church does not shed blood"). The alienation of surgery from medical practice—"so detrimental to both disciplines"—was to persist for another seven centuries.[222]

Paracelsus was not the only sixteenth-century critic of the socially imposed division of medical labor. His contemporary, the celebrated Andreas Vesalius, also lamented that "the technique of curing was so wretchedly torn apart," and he lambasted "doctors, prostituting themselves under the names of 'Physicians,'" for relegating most medical procedures "to those whom they call 'Chirurgians' and deem as if they were servants." The "elegant doctors," he charged, were "ashamed of working with their hands," and would therefore "prescribe to their servants what operations they should perform upon the sick [while] they merely stood alongside."[223]

In spite of these protests, medicine and surgery were to remain divided until the "Age of Revolution" in the late eighteenth century. "One of the great contributions of the French Revolution was the abolition of the separation between physicians and surgeons and the consequent creation of a united medical profession." Meanwhile, the American Revolution had had a similar effect: "The vicious separation between physician and surgeon never took root in [the United States]. . . . This is undoubtedly one of the reasons for the early excellence of American surgery."[224] Modern medical practice was thus formed in the crucible of social revolution.

In the era of the Scientific Revolution, however, the artisanal status of surgeons and apothecaries was evident in their forms of education and organization:

> Through most of Europe, surgery continued to be taught by apprenticeship and organized in guilds. In London a master surgeons' guild had been organized in 1368; the Mystery or Guild of the Barbers of London received its charter from Edward IV in 1462; and in 1540, by Act of Parliament, the

Guild of Surgeons merged with the Barbers to form the Barber-Surgeons Company.

Apothecaries in England were organized in 1607 as a subgroup within the Grocers' Guild; it was not until 1617 that they were allowed to form their own independent organization.[225]

And below the barber-surgeons and grocer-apothecaries in the medical establishment's pecking order,

> there existed a great variety of uncontrolled technicians who carried out procedures—sometimes highly skilled ones—not usually undertaken by the incorporated practitioners. Members of this group, which included eye doctors, tooth pullers, bladder-stone cutters, and quack salvers, had to petition the guilds and pay a fee for the right to practice.[226]

The lowly "empiricks" and folk healers, however, were widely perceived as being at least as therapeutically effective as the elite physicians. Thomas Hobbes's assessment was far from atypical: "I would rather have the advice or take physick from an experienced old woman that had been at many sick people's bedsides," he wrote, "than from the learnedest but unexperienced physician."[227] The social value of the irregular practitioners was explicitly acknowledged by a statute passed during the reign of Henry VIII extending legal protection to "divers honest persons, as well men as women, whom God hath endued with knowledge of the nature, kind, and operation of certain herbs, roots, and waters, and the using and administration of them to such as being pained with customable diseases."[228] Even as late as 1784, John Berkenhout could still declare, "a thousand indisputable facts convince me, that the present established practice of physic in England is infinitely destructive of the lives of his Majesty's subjects. I prefer the practice of old women, because they do not sport with edged tools; being unacquainted with the powerful articles of the *Materia Medica*."[229]

Berkenhout's mention of edged tools and powerful drugs alludes to elite physicians' heavy reliance on "heroic interventions"—shock treatments—that were usually more traumatic than beneficial. Massive blood-letting (phlebotomy) and purging by large doses of calomel (mercurous chloride) remained in the forefront of orthodox medicine's therapeutic arsenal even throughout the nineteenth century. (Physicians *prescribed* bloodletting by venesection or leeching; the procedures were actually *performed* by barbers and barber-surgeons.) But for all the pain and extreme discomfort these harsh treatments inflicted on patients, their power to heal or cure remained virtually nil. If the therapies offered by uneducated practitioners were superior to those of the university-trained doctors, it was because the former eschewed the harmful heroic measures and simply allowed nature to take its course. No wonder sick people of all social classes often sought out "old women," with their milder herbs and ministrations.

Surgeons—especially those who aspired to elite status—also practiced heroic interventions other than blood-letting. In sixteenth-century France, orthodox surgical practice was dictated by the scholars of the Surgeons' College of Saint Cosme. The standard treatment for wounds was cauterization, which required application of a red-hot cautery iron to the flesh. For gunshot wounds, a no less painful variant was used: boiling oil. The theory justifying these severe measures was the mistaken belief that they were necessary to drive out "poisons" and prevent the flesh around the wound from putrefying. By happy accident, a French military surgeon, Ambroise Paré, was able to disprove the theory.

Paré, born in 1510, was the son of a barber-surgeon and had been trained by apprenticeship to barber-surgeons. But his prowess on the battlefield as a military surgeon won him royal favor, and in 1552 he was appointed surgeon to Henri II. A few years later, "the elite surgeons of the College of St. Cosme were obliged to accept in their ranks this barber who did not even know Latin."[230]

The fortunate circumstance that led Paré to question the efficacy of cauterization occurred in 1536, well before his rise to prominence: on his first campaign with the French army, he ran out of oil to boil. Compelled to use a substitute, he treated some soldiers' wounds with "a digestive of yolkes of egges, oyle of Roses, and Turpentine." The next day he was astonished to find that those who had received the mild ointment were in much better shape than those he had subjected to the boiling oil. "And then," he said, "I resolved with my selfe never so cruelly to burne poore men wounded with gunshot."[231]

Paré extended this insight to challenge the orthodox use of cauterization to stop bleeding after amputations. It is difficult enough for us to contemplate the pain of having a limb sawed off without benefit of anesthesia, but imagine having it compounded by application of a red-hot iron immediately afterward. Paré was able to show that the latter step could be avoided by skillfully tying the blood vessels (vascular ligation).

Paré, like Paracelsus, felt there was much to be learned from folk medicine. Asked by a nobleman to treat the burns of one of his "kitchin boyes" who had fallen into a cauldron of nearly boiling oil, Paré went to an apothecary's shop, where he met "a certaine old countrey woman." The woman gave him a recipe for a dressing made from onions and salt, which Paré tried out on the boy and found remarkably effective in reducing the blistering.[232]

Another noteworthy sixteenth-century surgical innovation was accomplished by a practitioner in Switzerland far lower on the social scale than Paré. Jakob Nufer, whose skill with the surgeon's knife derived from castrating farm animals for a living, performed the first recorded cesarean section on a live mother in about 1500. The operation had routinely been performed on dead pregnant women in ancient times because Roman law required separate burial of the deceased mother and her fetus. But the Swiss pig-gelder extracted a living baby from a woman who is said to have later borne more children and lived to the age of 77.[233] Nufer's triumph could not have been replicated often, and then only with a great deal of luck, until antisepsis became standard in the twentieth century, but it was significant as a demonstration of what surgeons could hope to accomplish under ideal circumstances.

Almost three hundred years later, major advances in surgical knowledge were still proceeding from the most unexpected sources. In March 1793, Thomas Cruso and James Trindlay, two British colonial surgeons in Poona, India, witnessed a low-caste brickmaker performing successful plastic surgery on an unfortunate bullock driver whose nose had been deliberately cut off. Cruso and Trindlay published an account of the procedure, including diagrams. "The obscure brickmaker, reported the English surgeons, had performed a superb skin-graft and nose reconstruction using a technique superior to anything they had ever seen. It was taken up in Europe and became known as the 'Hindu method.'" The brickmaker was probably illiterate; his procedure "seems to have been passed down independently of the practice of educated physicians."[234]

HOMEOPATHS, HYDROPATHS, AND THOMSONIANS

THE TENSION BETWEEN establishment and alternative medical practice persisted well beyond the era of the Scientific Revolution. In the eighteenth and nineteenth centuries, the regular medical profession increasingly tried to differentiate itself from irregular healers by claiming a monopoly of scientific medicine. But "hardly any eighteenth-century scientific advance helped heal the sick directly," so "the net contribution of physicians to the relief and cure of the sick remained marginal." Late in the nineteenth century, science began to shed light on the causes of diseases but still could not cure them: "The half century after 1880 was marked by the doctor's ability to diagnose disease scientifically while remaining therapeutically powerless." The scientific pretensions of the orthodox physicians, coupled with their impotence as healers, "generated counter-trends—a populist, anti-elitist backlash."[235]

An early manifestation was the widespread popularity of inexpensive books such as *The Poor Man's Medicine Chest* (1791) that brought the principles of elite medicine to the masses. In England, there was William Buchan's *Domestic Medicine* (1769); in France, Samuel Tissot's *Avis au people sur la santé* (1761); and in the North American colonies, John Tennent's *Every Man His Own Doctor* (1730s). "Like many such works, Buchan's carried a radical message. Though himself a trained physician, he denounced the medical profession as oligarchic. Aiming to 'lay open' medicine to all, he espoused medical democracy as a fulfilment of the rights of man."[236]

A more radical development was the rise of social movements that—like the Paracelsians before them—challenged not only the practice of traditional medicine but its theoretical basis as well. Among the alternative healing sects, the "great inspiration and trail-blazer" was homeopathy. When Samuel Hahnemann first proclaimed his new medical doctrine in Leipzig, "the outraged Leipzig faculty and apothecaries had him banned from practice, and he moved on, everywhere meeting resistance from the elite."[237] But Hahnemann persisted until finally he found an appreciative public in Paris. In the late nineteenth century and into the twentieth, homeopathy enjoyed wide popularity throughout Europe and the United States. "The middle and upper classes flocked to this mild medicine, accepting it as a reprieve from the violent purges, poisonous mineral compounds and drastic blood-letting of medical orthodoxy."[238]

Other alternative medical movements offering gentler cures based on nature's healing powers followed in the wake of homeopathy's success. One was hydropathy, founded in the 1820s by Vincent Priessnitz, a farmer and folk healer in Austrian Silesia. Priessnitz opened a spa and preached the health-giving virtues of drinking large volumes of cold water, immersing the body in cold water, and taking cold-water douches. Although medical orthodoxy initially tried to dismiss hydropathy as unscientific, it eventually gained significant support among regular physicians and the favor of famous scientists, including Charles Darwin.

The European medical sects found their greatest success in the United States, so it was inevitable that indigenous American varieties would appear. The first was Thomsonianism, a "people's health movement" whose prophet, Samuel A. Thomson, lambasted "book doctors" and their megadoses of metallic drugs.

Thomson, who credited "an old woman" with teaching him what he knew about herbs, was an advocate of vegetable-based medicines.[239] Two other very influential American movements were osteopathy and chiropractic, founded in 1874 in Missouri by Dr. Andrew Taylor Still and in 1895 in Iowa by Daniel David Palmer, respectively.

A comprehensive people's history of medical science would take into account not only these well-known major challenges to orthodoxy but many all-but-forgotten minor ones as well. Taken together, the nineteenth-century alternative medicine campaigns transformed elite "allopathic" medical practice for the better. The allopaths could not simply ignore their immensely popular rivals, so a convergence occurred. Many individual orthodox physicians were won over to the new doctrines, while some of the innovators allowed themselves to be co-opted into the establishment in exchange for scientific respectability. When "scientific medicine" eventually abandoned the deleterious heroic interventions that had long been its stock in trade, it was not fresh scientific evidence that forced the change but competition from the alternative medical sects.

AGRICOLA, BIRINGUCCIO, AND THE MINERS

MEANWHILE, BACK IN the sixteenth century, others besides Paracelsus looked to the mines for inspiration in their quest for knowledge of nature. Georg Bauer's *De re metallica* has gained a reputation as "one of the great scientific classics."[240] Bauer, known to history by the Latinized form of his name, Agricola, was not an artisan-author; he was a medical doctor and university scholar who taught Latin and Greek.[241] But his famous work on metallurgy, published in Latin in 1556, owed its existence to information he gained from direct contact with miners and metalworkers and from the earlier vernacular writings of nonscholars.

One important source for Agricola was Vannoccio Biringuccio's *De la pirotechnia* (1540), "the first comprehensive textbook" on metallurgy. A historian of science commented that "considering the hoary antiquity of metallurgy," it is remarkable that a book like Biringuccio's "did not appear until 1540. The reason is simple: metallurgists were working men or at best craftsmen, who could not write or did not care to do so, while the learned men did not care about metallurgy."[242]

Biringuccio was not university educated; he was "a technician and industrialist, and there was no formal training for such men during the Renaissance. He had to train himself in the studios and workshops."[243] Nor was his Italian treatise completely unprecedented; at least two booklets on the subject had appeared earlier in German. *Eyn Nützlich Bergbüchlein*, "the first printed book in the field of mining," was published between 1505 and 1510. "It presented in simple terms the kind of knowledge a prospector would need, the tools he would require, the ores of the seven metals and their occurrence and associations, etc." The second was the *Probierbüchlein* (1524), which was written "for the benefit of all mintmasters, assay masters, goldsmiths, miners and dealers in metal." Its anonymous compiler "was obviously a man of experience but almost illiterate."[244]

Agricola's economic environment provided the inspiration for *De re metallica*. "The German cities where he spent most of his life, Joachimsthal, Chemnitz, Freiberg, etc., were all mining centers. This helped to develop his intense curiosity about geology, mineralogy, and related physical and chemical subjects." Having no direct metallurgical experience of his own, however, he was dependent on those who did for the information in his book:

> In reality he cribbed from Biringuccio as well as from the little German books printed during his lifetime. He also made some use of the abundant alchemical literature in spite of his distrust and contempt for it. His main source of information, however, was the manual tradition that he had observed with his own eyes in Germany and Italy and the oral tradition that he had collected with his own ears.[245]

A, B—Two furnaces. C—Tap-holes of furnaces. D—Forehearths. E—Their tap-holes. F—Dipping-pots. G—At the one furnace stands the smelter carrying a wicker basket full of charcoal. At the other furnace stands a smelter who with the third hooked bar breaks away the material which has frozen the tap-hole of the furnace. H—Hooked bar. I—Heap of charcoal. K—Barrow on which is a box made of wicker work in which the coals are measured. L—Iron spade.

Smelting methods. Illustration from Agricola's De re metallica.

Agricola acknowledged spending a great portion of his time "in visiting the mines and smelters" and "in association with the most learned among the mining folk."[246] He gave one "learned miner," Lorenz Berman, the title role in a dialogue he published *(Bermannus)* in 1530.[247] "Thanks to his association with workingmen," George Sarton said of Agricola, "he was an empiricist, stating as clearly as he could the knowledge available to contemporary miners, metallurgists, and smiths, describing their methods and their tricks."[248]

Subsequent contributions of René Descartes and his followers to knowledge of metals, which "a modern physical metallurgist" describes as "almost prescient," likewise derived from direct contact with artisans. "It seems probable that Descartes had spent some time in a smith's forge, for in a letter to Marin Mersenne on January 9, 1639, he discusses the hardening of steel."[249] Furthermore, "the description of the making of steel in his *Principia philosophiae* (1644) could only have been written by someone who had himself observed the granular nature of wrought iron coming to nature in the bath of molten cast iron in the hearth of the forge fire."[250]

The role of miners in the development of the materials sciences continued beyond the sixteenth and seventeenth centuries, as evidenced by their impact on the "new chemistry" associated with the name of Antoine-Laurent Lavoisier. "One of the most fundamental achievements of eighteenth-century chemistry, the enunciation of an analytic definition of simple substance," Theodore Porter explained, arose from "the longstanding practice of miners and assayers." Lavoisier and his colleagues translated "the practical, analytical assumptions of miners into a set of usable dicta for experimental and theoretical chemistry." They "simply adopted the practical notion of simple substance, which had been pioneered by assayers and mineralogists, as an empirical methodology, in order to establish secure foundations for the theory of chemistry."[251]

THE CRAFT ORIGINS OF MODERN SCIENTIFIC CULTURE

THOMAS SPRAT, A leading seventeenth-century publicist for the new science, recommended that scientists emulate the "Mathematical plainness" of the way artisans, merchants, and country people expressed themselves: a "close, naked, natural way of speaking."[252] But what these straight-talkers said was more important than how they said it. The books written by artisan-authors "were not merely passive vehicles for the transmission of 'raw data' to natural philosophers, but were bearers of attitudes and values that proved instrumental in shaping scientific culture in the early modern era."[253] The positive valuation of experimental practice was the most obvious, but there were other novel attitudes as well. Among those often erroneously credited to Francis Bacon were the notions that science embodies *progress*, that science is *useful* and can confer *public benefits*, and that science is a *collective, collaborative*, and *long-range* enterprise.

Just as the idea that scholars could gain knowledge from artisans and "low mechanicks" did not originate with Bacon, the same can be said about most of the philosophical innovations that are commonly attributed to him. Although his ideas were not entirely original, he was their most successful propagator to subsequent generations of elite scientists and in doing so laid the programmatic foundation of modern science.

But such essential "Baconian" ideals as the usefulness and public benefits of scientific discovery, which are today taken for granted, preceded Bacon. They were earlier

> expressed in several treatises composed by superior artisans, such as artists, instrument makers, and gun makers. Sometimes the authors even uttered the intention to further, through their treatises, the craftsmanship of their colleagues. Such statements reveal the social root of the modern ideal of progress. To modern ears they may sound rather trivial. We must not forget, however, that in classical, scholastic, and humanist literature . . . statements on the necessity of the gradual improvement of knowledge do not exist.[254]

The idea of science "as the product of a coöperation for nonpersonal ends, a coöperation in which all scientists of the past, the present, and the future have a part" is another of the key ideas generally attributed to Bacon. Today, Zilsel pointed out, this idea is considered almost self-evident. Yet no Brahmanic, Buddhistic, Moslem, or Catholic scholastic, no Confucian scholar or Renaissance humanist, no philosopher or rhetor of classical antiquity knew of this ideal. It is a specific characteristic of the scientific spirit and of modern Western civilization.[255]

Bacon's prescriptions for scientific cooperation, however, "were nothing but generalizations of [craftsmen's] practice." Although it was an idea that "appeared for the first time fully developed in the works of Francis Bacon," it ultimately stemmed, "like many other elements of modern scientific procedure, from the superior artisans of the fifteenth and sixteenth centuries."[256] Not surprisingly, Bacon gave it an elitist twist. His utopian treatise *The New Atlantis* described his vision of the ideal scientific institution of the future. His imaginative "Solomon's House" reveals that he clearly understood the collective nature of the scientific enterprise, but the kind of cooperation he envisioned reflected the social stratification that he assumed as an unalterable fact of life:

> Bacon's programme for a reformed natural philosophy emphasized a division of labour and a hierarchy of responsibility. Although he was convinced that the success of the work required the co-ordination of many helpers, these helpers were of two unequal sorts, a huge number of under-labourers and a handful of "Interpreters of Nature." . . . the organization of labour in "Solomon's House" clearly reflects Bacon's confidence in bureaucratic organization. Here, the work of the many assistants was directed by the elite "Brethren" who alone devised the "experiments," pondered their results, uncovered and possessed knowledge of the principles of Nature, and then devised useful technologies for the state (in the first instance) or for the people.[257]

Bacon's conception of a new science implied the existence of a new scientific elite. The artisans who produced the basic knowledge were to be excluded from any further role in the creation or control of science. Tycho Brahe's famous observatories on the Danish island of Hven bear witness that the organization of scientific research was proceeding along hierarchical lines in advance of Bacon's advocacy. Later, Robert Boyle's laboratory would exemplify the conscious development of genteel science in accord with Baconian prescriptions.

TYCHO BRAHE AND HIS "ASSISTANTS"

THE CENTRAL SCIENCE of the Scientific Revolution, according to those who insist that its essence was mathematical reasoning, was astronomy. Although the replacement of an Earth-centered mathematical schema by a sun-centered one was its most salient feature, the ultimate triumph of the Copernican view depended on a solid empirical foundation: prolonged, careful observation of the heavenly bodies.

Johannes Kepler's celebrated "laws" mathematically describing planetary orbits were derived from painstaking measurements of the planets' positions. Credit for producing the data Kepler used is traditionally ascribed to Tycho Brahe, founder of the most advanced astronomical research institute in the pre-telescope era, but Brahe had plenty of help. Although Brahe was the chief organizer of the project, dozens of highly skilled artisans were involved in constructing the precision instruments and taking the measurements that underlay Kepler's triumph. It is fortunate, from the standpoint of people's history, that although Brahe himself "never mentioned the names or even the principal competences of his artisans,"[258] historian John Robert Christianson has recorded the contributions of many of the men and women whom Brahe employed. Christianson set out "to focus upon the lives of Tycho Brahe's coworkers," he said, in order to show that "teamwork . . . was essential to the birth of modern science."[259]

Brahe's historical prominence is partially a function of the elevated social standing to which he was born:

Among the mass of detail that constitutes the personal, social, cultural, and intellectual background of Tycho Brahe's scientific achievement, the one indispensable fact is that he was born a Brahe, that is, born not merely into the Danish nobility but also into the small fraction of the noble class that had historically played significant roles in the administration, governance, and defense of the realm.[260]

The establishment of Brahe's research institute depended on a monumental gift from his royal patron, King Frederick II of Denmark. On May 23, 1578, Frederick granted to him in perpetuity "the whole island of Hven in fee, with all its resident peasants and servants of the crown and all royal incomes and rights, quit and free," making Tycho the feudal lord of the 1,850-acre island.[261] In addition, the king gave him a large cash grant to cover the costs of constructing buildings to house observatories.

"Tycho Brahe," Christianson said, "used his powerful position to bend the lives of hundreds of others toward a goal that he deemed important: a new understanding of the cosmos." First of all, "seigneurial lordship allowed him to utilize the labor of Hven peasants without pay." That meant that "two hundred peasants on the island of Hven were being *harnessed to the service of science,* along with hundreds of others on Tycho's fiefs in Skåne, Sjælland, and Nordfjord, and on his patrimonial lands under Knutstorp." The occupants of Hven were transformed "from freehold farmers into tenants and villeins, and they naturally resisted with all the means at their disposal." As for land on which to build "his manor, with its mansion, garden, grounds, barns, and fields . . . the method used by Tycho Brahe was to expropriate part of the village commons." It was not a small part; it amounted to two-thirds of the island.[262]

How many and what kind of people were required to keep his estate operating smoothly? Although a precise listing of his staff is not available, a comparable household established by his niece Sophie Axelsdatter Brahe included a chaplain, tutors, clerks, wetnurses, maids, footmen, a cook, cooks' helpers, bakers, brewers, gardeners, a tailor, a buttonmaker, watchmen, coachmen, bailiff, overseers, milkmaids, fishermen, a hops man, smith, and other peasants and workers on the estates. In addition, Sophie Axelsdatter Brahe and her husband often hired the services of goldsmiths, clockmakers, engravers, bookbinders, painters, cabinetmakers, gunsmiths, apothecaries, weavers, cobblers, saddlers, netbinders, ropemakers, wheelwrights, potters, brickmakers, roofers, sawyers, lime burners, charcoal burners, and many others.[263]

The castles that housed Brahe's observatories and alchemical laboratories—he named them Uraniborg and Stjerneborg—required a great amount of skilled labor to construct and maintain. "Masons, glaziers, cabinetmakers, painters, gilders, and other master craftsmen came with their journeymen and apprentices to work on Tycho's island."[264]

Brahe's scientific reputation rested in large part on a collection of innovative and unprecedentedly large astronomical instruments:

> First came a large (155 cm radius) quadrant mounted on a ball-and-socket swivel to measure distances between any two points in the sky up to 90° apart. Next came an immense azimuth quadrant of steel, with a brass arc over six feet (194 cm) in radius. . . . Then came . . . a large (155 cm radius) but rather unsuccessful bifurcated sextant for two simultaneous observations, followed by an excellent trigonal sextant of the same size, a bipartite arc that proved accurate for simultaneous observations of close objects, a huge mural quadrant (194 cm radius) . . . and an even more immense triquetrum (330 cm long) . . . as well as a portable azimuth quadrant (58 cm radius) for mapping.[265]

All of these were constructed for Brahe by skilled artisans. "Master craftsmen like Steffen Brenner, Hans Knieper,

Christopher Schissler, Georg Labenwolf, and many others worked on contract in their own shops in Copenhagen, Elsinore, Nuremberg, and Augsburg." The value of Schissler's service to Brahe, for example, was considerable: "The most significant item Tycho acquired while he was setting up his shop . . . was his great one-and-a-half-meter globe. Begun in 1570 at Augsburg by Schissler, *it was not even seen (let alone overseen) by Tycho* until his return there in 1575."[266]

Brahe also established "his own instrument shop, where he could convert skilled craftsmen into specialists, by hiring them to devote their time exclusively to his projects."[267] One particular German goldsmith and instrument maker merits special mention:

> The man who seems to have been Tycho's chief technician, Hans Crol, was also one of his most trusted observers. It would have been hard to let him go even if Tycho could have been certain that none of his instruments would ever need further modification or repair. And Tycho never did. Crol died on Hven in November 1591, which happens to have been the last year in which Tycho registered any new instruments.[268]

The reference to Crol's role as observer suggests another aspect of the cooperative nature of Brahe's accomplishments. To use the large instruments in the observatories "required a seasoned observer, a team leader to read off the positions of sightings, a secretary to record observations, and sometimes a clockwatcher."[269]

The scientific work on Hven was greatly dependent on Brahe's "staff of talented assistants." Brahe himself was at first a direct participant, but as time went by, other responsibilities demanded ever more of his attention. He became less active as a scientist and functioned more as "administrator, project initiator, author, and supervisor," as well as the one who "ensured the continuation of financial support" for the observatories. As a result, "much of the day-to-day scientific work fell on the shoulders of his . . . scholars and craftsmen."[270] For example, one of the signal accomplishments of the Brahe observatories was the completion, in 1592, of "a catalog of 777 stars, the first independent stellar catalogue since ancient times." It was produced by a team of observers working under the direction of Christian Sørensen Longomontanus "with little supervision from Tycho except in the very beginning."[271]

Tycho Brahe's role in the Scientific Revolution was certainly not insignificant. He was far more than a mere patron of science, but he does not deserve all of the credit for the scientific accomplishments that issued from his observatories. Perhaps the most judicious assessment of his contribution would remember him primarily as "a master of the patronage system who unified many strands of social and cultural life, created a new organizational model for the pursuit of science on a grand scale, and brought large teams of scholars, scientists, and technicians into the enterprise."[272]

TELESCOPES AND MICROSCOPES

BRAHE'S TEAM RAISED observational astronomy to a level of accuracy that likely could never have been surpassed without the aid of magnifying lenses. The invention of instruments that extended the range of human vision into realms of nature too small or too far away to see with the naked eye was obviously a stimulus of incalculable importance to scientific investigation. Telescopes and microscopes originated in close association with each other, and both were products of the inventiveness of spectacle-makers—craftsmen who ground lenses for a living.

It is impossible positively to identify an individual inventor of the telescope, "for the idea seems to have germinated in several minds at once." A leading candidate, however, is Hans Lippershey, "an obscure spectacle-maker of Middleburg in Zeeland" who was uncharitably described by Holland's best-known scientist, Christian Huygens, as an "illiterate mechanick."

The story goes, and it has several versions, that two children were playing in Lippershey's shop with some lenses and noticed that, by holding two of them in a certain position, the weather-vane of the nearby church appeared much larger. Lippershey at once tried this out for himself and then improved it by mounting the lenses in a tube. Some accounts say that an apprentice held the lenses, others that Lippershey was alone at the time or that he copied the idea from another optician. Some say he used a convex lens together with a concave, others that both lenses were convex and that he saw the steeple upside-down. Suffice it to say, Lippershey made a telescope and lost no time exploiting its financial possibilities.[273]

Another Dutchman, Jacob Adriaanzoon (a.k.a. James Metius), also laid claim to having invented the telescope, but when he raised a legal challenge to Lippershey's priority, the official ruling went in Lippershey's favor. A document in the archives of the states-general, dated October 2, 1608, declared:

On the petition of Hans Lippershey, a native of Wesel, an inhabitant of Middleburg, spectacle-maker, inventor of an instrument for seeing at a distance, as was proved to the States, praying that the said instrument might be kept secret, and that a privilege for thirty years might be granted to him, by which everybody might be prohibited from imitating these instruments, or else grant to him an annual pension, in order to enable him to make these instruments for the utility of this country alone.[274]

Another Middleburg spectacle-maker, Zacharias Jansen, has also been proposed as the inventor of the telescope. He and his father, Hans Jansen, were known to have been experimenting with lenses in tubes as early as 1590. A better case can be made, however, that the optical devices they originated were not telescopes but microscopes. One William Boreel, who had visited the Jansens and tried one of their instruments, wrote that "the minute objects which we looked at from above enlarged almost miraculously."[275]

Elite scientists such as Thomas Harriot in England and Galileo in Venice immediately recognized the value of the telescope; within a year of its invention they were utilizing it to dramatic effect in their astronomical researches. Galileo's discovery of the earthlike nature of the moon and the presence of satellites circling Jupiter hastened the destruction of the Aristotelian world picture and, while not *proving* the Copernican hypothesis, certainly increased its credibility.

ANTONY VAN LEEUWENHOEK

THE SCIENTIFIC IMPORTANCE of the microscope, on the other hand, took much longer to emerge. Early microscopes had limited powers of magnification and served more as a source of curiosities rather than as tools for the production of significant new knowledge of nature. As with the telescope, elite scientists attempted to improve the microscope and turn it into a useful instrument, but they were for the most part unsuccessful.

In the second half of the seventeenth century, however, a linen draper in the Netherlands began to use magnifying lenses to examine the threads of his fabrics and went on from there to develop microscopy into a major scientific enterprise. The draper, Antony van Leeuwenhoek, "was neither a philosopher, a medical man, nor a gentleman. . . . He had been to no university, knew no Latin, French, or English, and little relevant natural history or philosophy."[276] It is remarkable that this "ordinary shopkeeper—self-taught but otherwise uneducated"[277] made some of the most significant contributions to the knowledge of nature in the era of the Scientific Revolution. A reasonable evaluation of his accomplishments concludes that "he was the first man who ever saw living protozoa and bacteria under a lens, and by correctly interpreting and describing his observations he created the modern disciplines of Protozoology and Bacteriology."[278]

Leeuwenhoek, "in his spare time, when he was not selling buttons and ribbon . . . had taught himself how to grind and polish and mount lenses of considerable magnifying power."[279] It is not known exactly when he began systematically to use his lenses to examine infinitesimally small natural phenomena of every description, but there is evidence that it was no later than 1668. The first definitive documentation of his observations was recorded five years later, in a letter to the Royal Society in London, dated April 28, 1673.[280] He apologized to his primary correspondent there, Henry Oldenburg, for his lack of scientific and literary sophistication. "I have no style, or pen, wherewith to express my thoughts properly," he wrote, "because I have not been brought up to languages or arts, but only to business."[281]

In spite of his modest self-appraisal, he eventually gained widespread fame as a scientist. Scholars, statesmen, and even royalty traveled to Delft to have a look though his lenses. As "a common man," he "naturally felt flattered when a King or Queen of England, an Emperor of Germany, or a Tsar of Russia called upon him."[282]

Leeuwenhoek's first letter to the Royal Society was accompanied by a cover letter from a prominent anatomist, Reginald de Graaf, whose testimonial to the simple tradesman's worthiness was thought necessary to prevent his submissions from being dismissed out of hand. For more than fifty years—to the end of his life in 1723—Leeuwenhoek continued to send reports to the Royal Society containing illustrated descriptions of his observations. Although his work opened up vast new realms of scientific investigation, he remained virtually alone in the fields he created. At the end of the seventeenth century,

> Leeuwenhoek was actually the only earnest microscopist in the whole world. It is a remarkable fact that in all his later life he had no rivals and hardly a single imitator. His observations excited the greatest interest—but that was all. Nobody seriously attempted to repeat or extend them. The superexcellence of his lenses, combined with the exceptional keenness of his eye, killed all competition. As early as 1692, Robert Hooke, discoursing on "the Fate of Microscopes," says that they "are now reduced almost to a single Votary, which is Mr.

Leeuwenhoek; besides whom, I hear of none that make any other
Use of that Instrument, but for Diversion and Pastime."[283]

Leeuwenhoek's accomplishments are all the more remarkable in that, strictly speaking, he did not use microscopes
(compound-lens instruments by definition) but single lenses of
very high magnifying power. The invention of the microscope,
which occurred many decades earlier, thus "has no bearing
whatsoever upon his own work or discoveries."[284] But his revelations provided the stimulus for the development, after his
death, of true microscopes capable of equaling and surpassing
the power of his lenses.

Leeuwenhoek's reports of his observations were greatly
enhanced by the illustrations that accompanied them. "As I
can't draw," he admitted to Oldenburg, "I have got them drawn
for me."[285] Just as Vesalius failed to give credit to his anatomical artists, Leeuwenhoek made frequent reference to his illustrators but never by name. A number of different draftsmen
must have collaborated with Leeuwenhoek over the course of
his half-century of observations, but only Willem van der Wilt,
who was responsible for many of the later illustrations, can be
identified with a fair degree of confidence. It is less certain, but
a reasonable guess, that Willem's father, Thomas van der Wilt,
may have drawn many of the earlier ones.[286]

Whoever his illustrators were, their collaboration went
beyond the mere recording of images; "the unnamed Delft
draughtsmen participated in Leeuwenhoek's exploratory observations, intervened to specify and clarify the detail as they
recorded on paper what they mutually agreed had been seen,
and were a critical voice at Leeuwenhoek's shoulder as he made
sense of his microscopic images."[287]

The story of Leeuwenhoek has carried us into the early
eighteenth century, but now let us return to the mid-
seventeenth century in order to take up the development of
another important component of scientific history. The origin of
alchemy in the work of miners, dyers, and distillers from antiquity through the Renaissance has been previously discussed,

but in the era of the Scientific Revolution, "chymistry" became an essential part of the new experimental science.

ROBERT BOYLE:
"MAKING EXPERIMENTS BY OTHERS' HANDS"

IN THE CANONICAL literature of the Scientific Revolution, Robert Boyle is awarded a dual claim as the primary hero of the "Baconian" sciences.[288] He is held to be first and foremost among those who put Bacon's ideas about experimental method into actual practice, and he is also identified as the original modern chemist—that is, the first person to engage in scientific chemistry, as opposed to alchemy. But as an iconic figure of modern science, Boyle has traditionally been portrayed in an idealized way that leaves the important contributions of many, many other people out of account.

Boyle was a wealthy aristocrat. His father, Richard Boyle, first Earl of Cork, "was a robber baron of heroic stature, using his position to defraud Irish landowners . . . of their existing titles and to pass title to himself at absurdly deflated prices. He then expelled the Irish tenants and replaced them with more pliable and profitable English settlers." His lands yielded him some £20,000 a year by the 1630s, "a rental income larger than that of any other Crown subject."[289] Although Richard Boyle had bought his aristocratic title, his children were nonetheless accorded the privileges of hereditary nobility.[290] Because Robert Boyle was not the family's oldest son, he was not burdened with the responsibility of managing the immense estate when his father died, but he did have the wherewithal and leisure to pursue his interests in natural philosophy.

Boyle was fully committed to the Baconian program of gaining scientific knowledge from the crafts. "I freely confess," he wrote, "that I learned more of the kinds, distinctions, properties, and consequently of the nature of stones, by conversing with two or three masons, and stone-cutters, than ever I did from Pliny, or Aristotle and his commentators." He who "scorns to converse even with mean persons," Boyle added, "deserves not the knowledge of nature." It is "oftentimes from those that have neither fine language nor fine cloaths" that "the naturalist may obtain informations, that may be very useful to his design."[291]

"It seems to me to be none of the least prejudices," Boyle lamented, "that learned and ingenious men have been kept such strangers to the shops and practices of tradesmen":

> The phaenomena afforded by trades, are (most of them) a part of the history of nature, and therefore may both challenge the naturalist's curiosity and add to his knowledge. Nor will it suffice to justify learned men in the neglect and contempt of this part of natural history, that *the men, from whom it must be learned, are illiterate mechanicks*. . . . [This] is indeed childish, and too unworthy of a philosopher, to be worthy of a solemn answer.[292]

Boyle illustrated his argument by listing a number of artisan-mediated chemical and botanical processes that warranted systematic investigation:

> There are very many things made by tradesmen, wherein nature appears manifestly to do the main part of the work: as in malting, brewing, baking, making of raisins, currans, and other dried fruits; as also hydromel, vinegar, lime, &c. and the tradesman does but bring visible bodies together after a gross manner, and then leaves them to act one upon another, according to their respective natures; as in making of green or coarse glass, the artificer puts together sand and ashes, and the colliquation and union is performed by the action of the fire upon each body, and by as natural a way, as the same fire, when it resolves wood into ashes, and smoak unites volatile salt, earth, and phlegm into soot; and scarce any man will think, that when a pear is grafted upon a white thorn, the fruit it bears is not a natural one, though it be produced by a coalition of two bodies of distant natures put together by the industry of man, and would not have been produced without the manual and artificial operation of the gardener.[293]

Boyle "made no scruples about his distaste for necessary philosophical dealings with the artisan and trading classes,"[294] but he repeatedly demanded that those gentlemen who aspired to wisdom set aside their squeamishness about having to consort with people of humble social origin. "Knowledge of many retired truths," he wrote, "cannot be attained without familiarity with meaner persons, and such other condescensions, as fond opinion, in great men disapproves and makes disgraceful."[295] For a gentleman to become a natural philosopher, Boyle insisted, requires him to go "to such a variety of mechanick people (as distillers, druggists, smiths, turners, &c.), that a great part of his time, and perhaps all his patience, shall be spent in waiting upon tradesmen . . . which is a drudgery greater than any, who has not tried it, will imagine."[296]

But Boyle's Baconianism took a leaf from Tycho Brahe's book: instead of going to the artisans' workshops, Boyle could afford to bring the artisans' workshops to himself and, as he put it, "to make experiments by others' hands."[297] He created his own laboratories and hired skilled craftsmen—machinists, glassblowers, lens-grinders, and alchemical adepts, among others—as technicians to staff them.[298] The pursuit of scientific knowledge was thus accomplished more systematically and efficiently (and the "drudgery" of actually performing the scientific work was left to paid technicians).

Boyle's reputation as the archetype of the seventeenth-century "experimental philosopher" rests in large part on voluminous reports he published about the procedures and results of work performed in his laboratory. Most historians have simply taken those accounts at face value and have assumed that when Boyle described an experimental procedure, he meant that he had carried it out himself. But Steven Shapin has carefully reexamined Boyle's practices and has concluded that "there is reason to believe that relatively little of the manipulative and representational work involved in Boyle's experiments was done by Boyle himself." A great deal seems to have been performed by "individuals whom Boyle remuneratively engaged to do technical work on his behalf."[299]

The cause of the discrepancy is not intentional dishonesty on Boyle's part but a social assumption held by Boyle and his peers that we no longer share—namely that experimental work performed by technicians in Boyle's employ "belonged" to him and could be unproblematically claimed as his own. As Shapin explained, "the collective nature of experimental knowledge-making" was not acknowledged in "the political and moral economy of scientific work in seventeenth-century England." Although "knowledge was made by many" in Boyle's laboratory, it came "to be vouched for by the testimony of one."[300]

Even that testimony (experimental reports published over Boyle's signature) was not necessarily written by him. Most eye-opening—because it has to do with the famous scientific proposition that bears his name—is the likelihood "that Boyle's law [that in gasses there is an inverse relation between pressure and volume], as historically presented, owed much to assistants' intellectual labor. . . . it is virtually certain that this material was composed by Boyle's then paid assistant [Denis] Papin." Furthermore, "[Robert] Hooke, who was then in Boyle's employ, had substantial responsibility for the way it was represented in Boyle's text."[301]

It is probable, according to Shapin,

> that Boyle himself was involved only in a very limited way in "his" experimental manipulations. The device which became known as the *machina Boyleana* [the vacuum chamber, or "air pump"] was almost certainly constructed for him by remunerated assistants Ralph Greatorex and Robert Hooke, and even the extent of Boyle's role in its evolving design remains unclear. The glass J-shaped tube that yielded "his" law of pressures and volumes was again almost certainly made for him and had to be manipulated by him in collaboration with assistants, if not solely by them. The furnaces in his laboratory, and the alembics in which long-term distillations were performed, were probably attended by assistants.[302]

Among Boyle's most celebrated endeavors was a series of investigations of air pressure, or what he called "the spring and weight of the air." He acknowledged in the preface to his report, however, that the experiments had actually been carried out by technician Denis Papin.[303] Moreover, Shapin added:

> At least some, and perhaps the greatest part, of the design of the experimental project was also owing to the technician. . . . The technician performed the skilled manipulations which produced experimental phenomena; he recorded the phenomena that his skilled work made manifest; he embedded these records in literary form, adding occasional inferential corollaries which were likewise the product of his own thinking.[304]

Papin and Hooke were perhaps the "only two currently employed skilled assistants" whom Boyle "ever fully named . . . in the whole corpus of his experimental reporting, extending from 1660 to posthumous publications of 1692."[305] In general, "it was exceptional in the extreme for a seventeenth-century technician to be so identified by those who engaged his services. Anonymity is almost a defining characteristic of the technician in that setting":

> Technicians are triply invisible. First, they have traditionally been invisible to historians and sociologists of science. . . . Second, they have been largely, if not entirely, invisible in the formal documentary record produced by scientific practitioners. Even when one is committed to doing so, it is extremely difficult to retrieve information about who they were and what they did. Third, technicians have arguably been invisible as relevant actors to those persons in control of the workplaces in which scientific knowledge is produced. . . . Technicians have been "not there" in roughly the same sense that servants were, and were supposed to be, "not there" with respect to the conversations of Victorian domestic employers.[306]

Marie Boas Hall, a defender of the traditional heroic portrayal of Boyle, asserted that "the names of most of Boyle's . . . assistants have been lost, because they were not men of independent scientific merit."[307] They do not deserve to be so cavalierly dismissed, however, because, as Shapin has shown, *"Boyle was dependent upon them for the empirical foundations of his knowledge."* His experimental reports "largely spoke for, and vouched for, what others had done, observed, and represented."[308]

Shapin limited his study to experimental technicians and acknowledged that much more could be said about the collective nature of Boyle's research. Shapin did not, for example, examine in detail "the important assistance offered by instrument-makers and apothecaries, such as the apothecaries John Cross (with whom he lived in Oxford for some years) and Thomas Smith (who lived in Boyle's Pall Mall house for many years and who was one of the beneficiaries of his will)." Nevertheless, "without the work of all such people, 'Boyle's' science could not have happened."[309]

Boyle's great wealth gave him the means to hire more assistants and to create grander laboratories than most of his fellow *virtuosi* could afford, but the difference was one of degree rather than kind. "While it may well be objected that Boyle's Pall Mall laboratory was an atypical instance," Shapin said, "I see little significant difference between the evidence from Boyle's laboratories and that coming from the (perhaps less densely populated) workplaces of other gentlemen-practitioners."[310]

There were echoes of the laboratory relationships described by Shapin even in the twentieth century, despite the qualitative increase in specialization and professionalization of science that had occurred in the meantime. "During the golden age of experimental physics" of the 1920s, Derek de Solla Price contended,

> all progress seemed to depend on a band of ingenious craftsmen, with brains in their fingertips, and a vast repertoire of little-known properties of materials and other tricks of the trade. It is these that made all the difference in what could or

could not be done in a laboratory, and that, to a large extent, determined what was to be discovered.

Among those craftsmen were the "almost anonymous and unsung lab assistants, such as Lord Rutherford's man, George Crowe, or J. J. Thomson's aides, Ebeneezer Everett and W. G. Pye. These three assistants went on to found the Cambridge Instrument Company, one of the first high-technology companies of Britain."[311]

The point of these observations is not to deny Robert Boyle, Lord Rutherford, or J. J. Thomson the credit due them for organizing research and designing experiments that generated valuable results, but simply to illustrate once again the central themes of this book: that scientific knowledge production is a *collective social activity*, that *essential* contributions have been made by working people engaged in earning their daily bread, and that elite theoreticians are often unjustly awarded all the credit for knowledge produced by many hands and brains.

Among the revolutionaries of the Scientific Revolution—that is, those who set it into motion—were craftsmen and tradesmen and other ordinary people. But with whom did it end? When the dust cleared, who were the masters and who were the servants in the new world of science? *Who won and who lost?*

NOTES

1. See esp. Steven Shapin, *The Scientific Revolution*, p. 1.
2. Whether the term "Scientific Revolution" should or should not be capitalized is more than a copyediting issue, as the dispute on that subject between B. J. T. Dobbs and Richard Westfall illustrates. Although my views are in general closer to those of Dobbs (who eschews capitalization) than to those of Westfall (who capitalizes), I have nonetheless chosen to capitalize. See Dobbs, "Newton as Final Cause and First Mover," and Westfall, "The Scientfic Revolution Reasserted" in Margaret J. Osler, ed., *Rethinking the Scientific Revolution*.

3. Charles Van Doren, *A History of Knowledge*, pp. 139–142, 184, 195–209.

4. Derek de Solla Price, *Science Since Babylon*, p. 47.

5. William Eamon, *Science and the Secrets of Nature*, p. 11.

6. Quoted in Reijer Hooykaas, "Science and Reformation," p. 59.

7. The literature on this subject is immense in volume. See esp. Stillman Drake, *Galileo at Work;* and James MacLachlan, "A Test of an 'Imaginary' Experiment of Galileo's."

8. H. Floris Cohen, *The Scientific Revolution*, p. 99.

9. Thomas Kuhn, "Alexandre Koyré and the History of Science," pp. 67–69.

10. Francis Bacon, *The Great Instauration,"* p. 8.

11. Bacon, "Aphorisms on the Composition of the Primary History," Aphorism 5, in *The New Organon*, p. 278.

12. Robert Hooke, *General Scheme or Idea of the Present State of Natural Philosophy* (1705), pp. 24–26.

13. Descartes, *Rules for the Direction of the Mind*, rule X.

14. Price, *Science Since Babylon*, p. 46.

15. See "Hugh Plat" in this chapter.

16. Anthony F. C. Wallace, *The Social Context of Innovation*, pp. 21–23. See also L. E. Harris, *The Two Netherlanders*, pp. 171–181.

17. Boyle wrote: "Drebbel conceived, that it is not the whole body of the air, but a certain quintessence (as the chymists speak) or spirituous part of it, that makes it fit for respiration." Boyle, *New Experiments Physico-Mechanical, Touching the Spring of the Air*, p. 107.

18. Eamon, *Science and the Secrets of Nature*, p. 8.

19. See the section "Galileo and the Artisans" in this chapter.

20. H. F. Cohen, *The Scientific Revolution*, p. 323 (emphasis added). This is Cohen's paraphrase of a central thesis of Olschki's three-volume classic, *Geschichte der neusprachlichen wissenschaftlichen Literatur* (1919–1927).

21. See "Artisan-Authors" in this chapter.

22. J. A. Bennett, "The Challenge of Practical Mathematics," p. 176 (emphasis added).

23. Frank J. Swetz, *Capitalism and Arithmetic*, p. 295.

24. Ibid., p. 291.

25. Ibid., p. 292.

26. Ibid., pp. 11–12.

27. Paul L. Rose, *The Italian Renaissance of Mathematics*; quoted in Swetz, *Capitalism and Arithmetic*, p. 289.

28. Swetz, *Capitalism and Arithmetic*, p. 289 (emphasis added).

29. Ibid., pp. 14–16.

30. Ibid., p. 292.

31. Ibid., p. 33.

32. Ibid., pp. 24–25, 33.

33. J. V. Field, "Mathematics and the Craft of Painting," p. 74 (emphasis added).

34. Swetz, *Capitalism and Arithmetic*, p. 240.
35. Bennett, "Challenge of Practical Mathematics," p. 178.
36. Ibid., pp. 184–185.
37. E. G. R. Taylor, *The Mathematical Practitioners of Tudor & Stuart England*, pp. 16–17. Taylor describes her book as "a chronicle of lesser men—teachers, text-book writers, technicians, craftsmen—but for whom great scientists would always remain sterile in their generation" (p. ix). In it she provides biographical sketches of 582 mathematical practitioners (pp. 165–307). Equally valuable is its sequel, *The Mathematical Practitioners of Hanoverian England, 1714–1840*.
38. Taylor, *Mathematical Practitioners of Tudor and Stuart England*, pp. 9–10.
39. Ibid., pp. 22–23, 33, 41–42.
40. Derek de Solla Price, "Proto-Astrolabes, Proto-Clocks and Proto-Calculators," p. 61.
41. Taylor, *Mathematical Practitioners of Tudor & Stuart England*, p. 162.
42. Ibid., p. 20.
43. Price, *Science Since Babylon*, p. 54. For more on Humphrey Cole, see R. T. Gunther, "The Great Astrolabe and Other Scientific Instruments of Humphrey Cole."
44. John Aubrey, *Aubrey's Brief Lives*, p. 116. For more on Gunter's contribution to surveying, see Andro Linklater, *Measuring America*, pp. 5, 13–20.
45. In addition to the cited books by E. G. R. Taylor, see Silvio A. Bedini, *Patrons, Artisans and Instruments of Science, 1600–1750*.
46. Price, *Science Since Babylon*, pp. 53–55.
47. David Landes, *Revolution in Time*, p. 113.
48. Bennett, "Challenge of Practical Mathematics," p. 177. See also Pamela H. Smith, *The Body of the Artisan*, pp. 65–66.
49. Bennett, "Challenge of Practical Mathematics," p. 178.
50. Edgar Zilsel, "The Origins of Gilbert's Scientific Method," p. 91.
51. Ross King, *Brunelleschi's Dome*, pp. 157–158.
52. Giorgio de Santillana, "The Role of Art in the Scientific Revolution," pp. 41–42.
53. George Sarton, *Six Wings*, p. 113.
54. Leonardo da Vinci, *The Notebooks of Leonardo da Vinci*, pp. 57–58.
55. Ibid., p. 853.
56. Giorgio Vasari, *The Lives of the Artists*, pp. 105, 393.
57. Ibid., pp. 327, 365.
58. Ibid., p. 224.
59. King, *Brunelleschi's Dome*, p. 13.
60. Vasari, *Lives of the Artists*, pp. 134, 140.
61. King, *Brunelleschi's Dome*, p. 35.
62. Vasari, *Lives of the Artists*, pp. 135, 140.
63. See Zilsel, "Origins of Gilbert's Scientific Method," p. 91.
64. Vasari, *Lives of the Artists*, p. 120 (emphasis added).

65. Ibid., p. 121.
66. Ibid., p. 311.
67. Ibid., p. 310.
68. Ibid., p. 310.
69. Santillana, "Role of Art in the Scientific Revolution," pp. 60–61 (emphasis in original).
70. Leonardo da Vinci, *Notebooks*, pp. 875, 867, 993. Leonardo's instructional treatises remained unpublished in his lifetime and long thereafter.
71. Quoted in Santillana, "Role of Art in the Scientific Revolution," p. 34.
72. William Barclay Parsons, *Engineers and Engineering in the Renaissance*, pp. 94–95.
73. Christoph J. Scriba, ed., "The Autobiography of John Wallis, F.R.S.," p. 27 (emphasis in original).
74. Sarton, *Six Wings*, pp. 24–25.
75. Alberti, *Trattato della pittura* (1434), quoted in J. D. Bernal, *Science in History*, vol. 2, p. 390.
76. Field, "Mathematics and the Craft of Painting," pp. 74, 82–83.
77. Santillana, "Role of Art in the Scientific Revolution," p. 35. For a description of Brunelleschi's perspective device, see King, *Brunelleschi's Dome*, pp. 35–36.
78. Although not all art historians are ready to concede that Vermeer used the *camera obscura*, it seems to me that Philip Steadman has put it beyond reasonable doubt in his recent book *Vermeer's Camera*. It is possible that Vermeer obtained lenses for the apparatus from the pioneer microscopist Antony van Leeuwenhoek. Vermeer and Leeuwenhoek were exact contemporaries (both born in 1632) in Delft, Holland, and when Vermeer died, Leeuwenhoek was the executor of his estate.
79. Lynn White, Jr., "Pumps and Pendula," p. 101.
80. Santillana, "Role of Art in the Scientific Revolution," p. 36.
81. Ibid., p. 56.
82. Edgar Zilsel, "Problems of Empiricism," p. 174.
83. Alexandre Koyré, *Metaphysics and Measurement*, p. 13.
84. Santillana, "Role of Art in the Scientific Revolution," p. 33.
85. Vasari, *Lives of the Artists*, p. 418.
86. Based on the testimony of Vasari, historians of medicine have frequently credited the illustrations to the Flemish artist Jan Stephen van Kalkar (or Calcar), a pupil of Titian. Vasari's attribution, however, is highly doubtful. See the "Introduction" by Saunders and O'Malley to Andreas Vesalius, *The Illustrations from the Works of Andreas Vesalius of Brussels*, pp. 25–29.
87. Vivian Nutton, "Historical Introduction" to Vesalius, *On the Fabric of the Human Body*. The artwork, Nutton said, should not be attributed to the artist alone but should be considered "the co-production of anatomist, artist, block-cutter and printer."
88. Price, *Science Since Babylon*, p. 54.

89. Sarton, *Six Wings*, pp. 130–132.

90. Smith, *Body of the Artisan*, p. 99.

91. Sarton, *Six Wings*, pp. 135.

92. Smith, *Body of the Artisan*, pp. 240–241.

93. Quoted in Roy Porter, *The Greatest Benefit to Mankind*, p. 229.

94. H. F. Cohen, *Scientific Revolution*, p. 157.

95. See Price, *Science Since Babylon*, p. 46; and Lane Cooper, *Aristotle, Galileo and the Tower of Pisa*. Clagett pointed out that the result (virtually equal falling time for unequal weights) had been previously published by Giambattista Benedetti in 1554. He also cited evidence that John Philoponus made the same observation by experiment in the sixth century C.E. Marshall Clagett, *The Science of Mechanics in the Middle Ages*, p. 665.

96. Zilsel, "Origins of Gilbert's Scientific Method," p. 79.

97. Zilsel, "The Genesis of the Concept of Physical Law," p. 110.

98. Zilsel, "Origins of Gilbert's Scientific Method," p. 92.

99. White, "Pumps and Pendula," pp. 97–98.

100. The conference was held at the University of Wisconsin in September 1957. The papers presented there were published as Marshall Clagett, ed., *Critical Problems in the History of Science*.

101. A. Rupert Hall, "The Scholar and the Craftsman in the Scientific Revolution," p. 21.

102. Ibid., p. 22.

103. A.C. Crombie, "Commentary on the Papers of Rupert Hall and Giorgio de Santillana," p. 67.

104. Generally speaking, Crombie was upholding the banner of Pierre Duhem, who in 1913 initiated a long-running debate by arguing that the Scientific Revolution had begun in the fourteenth rather than the sixteenth century. But whereas Duhem focused exclusively on University of Paris scholars, Crombie extended the scope of investigation to include Oxford University scholars as well.

105. Francis R. Johnson, "Commentary on the Paper of Rupert Hall," p. 30.

106. Crombie, "Commentary on the Papers of Rupert Hall and Giorgio de Santillana," p. 68.

107. J. D. Bernal's *Science in History* (1954) was an exception.

108. Crombie, "Commentary on the Papers of Rupert Hall and Giorgio de Santillana," p. 68.

109. Roy Porter, "Introduction" to Stephen Pumphrey, Paolo L. Rossi, and Maurice Slawinski, eds., *Science, Culture and Popular Belief in Renaissance Europe*, p. 2.

110. Hall, "Scholar and Craftsman in the Scientific Revolution," pp. 7, 21.

111. Ibid., pp. 18-19. Hall cited *Lazarus Ercker's Treatise on Ores and Assaying*, trans. Anneliese Grünhaldt Sisco and Cyril Stanley Smith (Chicago, 1951).

112. Hall, "Scholar and Craftsman in the Scientific Revolution," pp. 18, 23.

113. Benjamin Farrington, *Science in Antiquity*, p. 145.

114. For the role of mechanics and mathematical practitioners in creating the mechanical philosophy, see J. A. Bennett, "The Mechanics' Philosophy and the Mechanical Philosophy."

115. J. D. Bernal, *Science in History*, vol. 1, p. 315.

116. Ibid., pp. 317, 320.

117. Hall, "Scholar and Craftsman in the Scientific Revolution," p. 21.

118. See "Artisan-Authors" in this chapter.

119. Diederick Raven and Wolfgang Krohn, "Edgar Zilsel," p. lv.

120. Edgar Zilsel, "The Sociological Roots of Science," pp. 12–15.

121. Smith, *Body of the Artisan*, pp. 151, 239.

122. Hall, "Scholar and Craftsman in the Scientific Revolution," p. 7.

123. Joseph Glanvill, *Plus Ultra, or, The Progress and Advancement of Knowledge Since the Days of Aristotle*, p. 105.

124. Pamela O. Long, "Power, Patronage, and the Authorship of *Ars*," p. 419, commenting on Nicholas of Cusa's *Idiota: De sapientia, De mente, De staticis experimentis* (c. 1450).

125. Juan Luis Vives, *De tradendis disciplines*; quoted in Smith, *Body of the Artisan*, p. 66.

126. Quoted in James R. Jacob, *The Scientific Revolution*, p. 27.

127. Crombie, "Commentary on the Papers of Rupert Hall and Giorgio de Santillana," pp. 72–74.

128. Ibid., p. 73.

129. Roger Bacon, *Opus majus*; quoted in Eamon, *Science and the Secrets of Nature*, p. 86. In context, Eamon comments, it appears that this statement was not intended as "a paean to the mechanical arts" but as "a warning against intellectual pride."

130. Nicholas of Poland, *Antipocras*; quoted in Eamon, *Science and the Secrets of Nature*, p. 78.

131. See, for examples, A. C. Crombie, *Augustine to Galileo*; Crombie, *Robert Grosseteste and the Origins of Experimental Science*; and Marshall Clagett, *The Science of Mechanics in the Middle Ages*. Crombie and Clagett were building on the earlier work of Pierre Duhem (see n. 104).

132. White, "Pumps and Pendula," pp. 101–102.

133. Galileo Galilei, *Dialogues Concerning Two New Sciences*, p. 49 (emphasis added). The italicized phrase attests to the importance of artisanal input to the "classical physical" as well as the "Baconian" sciences.

134. Galileo, *Dialogues Concerning Two New Sciences*, p. 49.

135. White, "Pumps and Pendula," pp. 96–98. White cited Leonardo Olschki's *Galileo and His Time* (1927) as the pioneering work on the technological context of Galileo's science.

136. White, "Pumps and Pendula," p. 110.

137. Galileo, *Dialogues Concerning Two New Sciences*, pp. 275–276.

138. See A. Rupert Hall; "Gunnery, Science, and the Royal Society."

139. Jean Daujat, *Origines et formation de la théorie des phénomènes électriques et magnétiques*; cited in Duane H. D. Roller, *The De magnete of William Gilbert*, p. 98.

140. A. Wolf, *A History of Science, Technology, and Philosophy in the 16th and 17th Centuries*; cited in Roller, *The De magnete of William Gilbert*, p. 98.

141. Robert K. Merton, *Science, Technology and Society in Seventeenth-Century England*, p. 6.

142. Hugh Kearny, *Science and Change*, p. 108.

143. W. P. D. Wightman, *The Growth of Scientific Ideas*, p. 209.

144. William Gilbert, *De magnete magneticisque corporibus et de magno magneto tellure physiologia nova*, book III, chap. 13. The term "Republic of Letters," Roger Hahn points out, "has a long history running back to the ancients." Hahn, *The Anatomy of a Scientific Institution*, p. 38.

145. Zilsel, "Origins of Gilbert's Scientific Method," p. 75.

146. Ibid., p. 82.

147. Gilbert, *De magnete*, book III, chap. 2.

148. Ibid., book III, chap. 12.

149. Zilsel, "Origins of Gilbert's Scientific Method," p. 81.

150. Ibid., p. 82.

151. Gilbert, *De magnete*, book IV, chap. 5.

152. Zilsel, "Origins of Gilbert's Scientific Method," pp. 72, 88.

153. Robert Norman, *The Newe Attractive*.

154. Gilbert, *De magnete*, book I, chap. 1.

155. Zilsel, "Origins of Gilbert's Scientific Method," pp. 85–86.

156. Ibid., pp. 85–86.

157. Ibid., p. 72. Gilbert, *De magnete*, book III, chap. 3.

158. Bennett, "Challenge of Practical Mathematics," p. 187.

159. Zilsel, "Origins of Gilbert's Scientific Method," p. 88.

160. Ibid., p. 90.

161. Eamon, *Science and the Secrets of Nature*, p. 3.

162. Long, "Power, Patronage, and the Authorship of *Ars*," p. 410.

163. Edgar Zilsel, "The Genesis of the Concept of Scientific Progress," p. 160.

164. An important exception is Leonardo Olschki (see nn. 20 and 135 above), whose work was perhaps Zilsel's primary inspiration, but unfortunately, as H. F. Cohen has pointed out, Olschki's masterpiece "is rarely cited, let alone discussed in any detail." Cohen, *Scientific Revolution*, p. 324.

165. Zilsel, "Genesis of the Concept of Scientific Progress," p. 150.

166. Eamon, *Science and the Secrets of Nature*, p. 94.

167. Most of this crucial documentation would not have been available to me if it had not been the subject of William Eamon's *Science and the Secrets of Nature*. I am in great debt to Eamon's research and I refer those who want to pursue this subject further to his book.

168. See n. 44.

169. Sarton, *Six Wings*, pp. 20–21. It would be more accurate to say that Latin was spoken *almost* exclusively (rather than "only") by men, but Sarton's general point is nonetheless valid.

170. The "association of technology and learning" of the twelfth through fourteenth centuries that was discussed earlier in this chapter demonstrates that "Baconianism" was not without precedent but its effect on contemporary elite science was minimal.

171. Long, "Power, Patronage, and the Authorship of *Ars*," p. 398.

172. Ibid., p. 433.

173. Olschki, *Geschichte der neusprachlichen wissenschaftlichen Literatur;* quoted in and trans. by H. F. Cohen, *Scientific Revolution*, p. 323 (emphasis added).

174. Eamon, *Science and the Secrets of Nature*, pp. 5, 9, 4.

175. Ibid., p. 7.

176. Price, *Science Since Babylon*, p. 51.

177. Elizabeth L. Eisenstein, *The Printing Press as an Agent of Change*.

178. Parsons, *Engineers and Engineering in the Renaissance*, pp. 105–107.

179. Eamon, *Science and the Secrets of Nature*, p. 94.

180. Ibid., pp. 94–95.

181. Ibid., pp. 111, 106, 122.

182. Ibid., pp. 109, 113–114.

183. Ibid., pp. 96–97, 104.

184. Ibid., pp. 97–98.

185. W. H. Ryff, *Prakticir Büchlein der Leibartzeney;* quoted in Eamon, *Science and the Secrets of Nature*, p. 102.

186. W. H. Ryff, *Kurtz Handbüchlein und Experiment vieler Artzneien;* quoted in Eamon, *Science and the Secrets of Nature*, p. 99.

187. Eamon, *Science and the Secrets of Nature*, pp. 101–102.

188. Sarton, *Six Wings*, p. 164. The other was Agricola, who is discussed in the section "Agricola, Biringuccio, and the Miners" in this chapter.

189. Sarton, *Six Wings*, p. 164.

190. His year of birth is unknown; biographers have cited estimates ranging from 1499 to 1520.

191. Sarton, *Six Wings*, p. 168.

192. Ibid., pp. 164–166.

193. Aurèle La Roque, "Introduction" to Bernard Palissy, *Discours admirables*, p. 15.

194. Palissy, *Discours admirables*, p. 24.

195. La Roque, "Introduction" to Palissy, *Discours admirables*, p. 14.

196. Palissy, *Discours admirables*, p. 26.

197. Sarton, *Six Wings*, p. 170.

198. La Roque, "Introduction" to Palissy, *Discours admirables*, p. 5.

199. Although it cannot be definitively established, Bacon, as a teenager living

in Paris in the late 1570s, may have attended Palissy's public lectures on geology, mineralogy, and other sciences.

200. Eamon, *Science and the Secrets of Nature*, p. 311.

201. Deborah Harkness discussed her work-in-progress in "Interview with an Alchemist: Hugh Plat's Pursuit of Natural Knowledge in Early Modern London," presented at a meeting of the History of Science Society in November 2003. Her book on the subject, tentatively entitled *The Jewel House of Nature: Elizabethan London and the Social Foundations of the Scientific Revolution*, is scheduled for publication in 2006.

202. See Julian Martin, "Natural Philosophy and Its Public Concerns." This theme is explored further in the next chapter.

203. According to Walter Pagel, it cannot be demonstrated that Theophrastus Bombast von Hohenheim himself ever used the name "Paracelsus." Pagel, *Paracelsus*, p. 5. Nonetheless, that is the name by which he has become known to posterity and it would be confusing to refer to him in any other way.

204. Pamela H. Smith, "Vital Spirits," p. 127.

205. Smith, *Body of the Artisan*, pp. 25, 233, 239.

206. Jacob, *The Scientific Revolution*, p. 61.

207. Ibid., p. 27.

208. Ibid., p. 27.

209. Whether the adjective "bombastic" originated in ironic tribute to Paracelsus's family name or not is a matter of dispute. Allen G. Debus assumes that it did, but Walter Pagel states flatly that it did not. Debus, *The Chemical Philosophy*, vol. 1, p. 52; Pagel, *Paracelsus*, p. 6. The Oxford English Dictionary agrees with Pagel but offers no alternative etymology.

210. See von Bodenstein, "Foreword" to Paracelsus, *The Diseases that Deprive Man of His Reason*, p. 136.

211. Petrus Severinus, *Idea medicinae philosophicae*; quoted in Allen G. Debus, *The English Paracelsians*, p. 20 (emphasis added).

212. Pagel, *Paracelsus*, p. 25. *On the Miners' Sickness* was published posthumously in 1567.

213. Pagel, *Paracelsus*, p. 13.

214. George Rosen, "Introduction" to Paracelsus, *On the Miners' Sickness*, p. 26.

215. Sarton, *Six Wings*, p. 109.

216. Eamon, *Science and the Secrets of Nature*, p. 102.

217. Sarton, *Six Wings*, pp. 109–110.

218. Paracelsus, *Sämtliche Werke*, vol. 14, p. 541, quoted in Debus, *English Paracelsians*, p. 22.

219. Paracelsus, *Sämtliche Werke*, vol. 10, p. 19; quoted in Smith, *Body of the Artisan*, p. 84.

220. Quoted in Pagel, *Paracelsus*, p. 15.

221. Porter, *Greatest Benefit to Mankind*, p. 11.
222. Erwin H. Ackerknecht, *A Short History of Medicine*, pp. 89–90.
223. Vesalius, *De fabrica*; quoted in John Henry, "Doctors and Healers," pp. 193–194.
224. Ackerknecht, *Short History of Medicine*, pp. 153, 220.
225. Porter, *Greatest Benefit to Mankind*, pp. 186, 194.
226. Smith, *Body of the Artisan*, pp. 196–197.
227. Quoted in Porter, *Greatest Benefit to Mankind*, pp. 282–283.
228. Quoted in Henry, "Doctors and Healers," pp. 197–198. The phrase "as well men as women" is indirect testimony to the leading role of women as healers in this era.
229. John Berkenhout, *Symptomatology* (1784); quoted in Porter, *Greatest Benefit to Mankind*, p. 267.
230. Ackerknecht, *Short History of Medicine*, p. 110.
231. Ambroise Paré, *Apologie and Treatise*, pp. 23–24.
232. Ibid., p. 140.
233. Henry, "Doctors and Healers," p. 197. Henry said Nufer's feat seems to have been first recorded by Gaspard Bauhin in an appendix to François Rousset, *Isterotomotochia* (Basle, 1588).
234. Porter, *Greatest Benefit to Mankind*, pp. 141–142. Cruso and Trindlay's report originally appeared in the *Madras Gazette* and was reprinted in the *Gentleman's Magazine* of London in October 1794.
235. Porter, *Greatest Benefit to Mankind*, pp. 266, 396, 686.
236. Ibid., pp. 283, 290.
237. Ibid., pp. 391–391.
238. Roberta Bivins, "The Body in Balance," p. 116.
239. See Porter, *Greatest Benefit to Mankind*, p. 393, and John Crellin, "Herbalism," p. 88.
240. Sarton, *Six Wings*, p. 125. *De re metallica* was translated into English by Herbert Clark Hoover and Lou Henry Hoover. (Herbert Hoover was an engineer before he became president of the United States.)
241. H. Floris Cohen's inclusion of Agricola in the class of "literate craftsmen" alongside Bernard Palissy and Robert Norman seems to me inaccurate. Cohen, *Scientific Revolution*, p. 350.
242. Sarton, *Six Wings*, p. 120.
243. Ibid., p. 120.
244. Ibid., pp. 121–122. Also see Appendix B to Agricola, *De re metallica*, Hoover and Hoover, trans. pp. 609–613.
245. Sarton, *Six Wings*, pp. 123–124.
246. Hoover and Hoover, "Introduction" to *De re metallica*, p. vii. Their source is Agricola's preface to his *De Veteribus et Novis Metallis* (1546).
247. Ibid., p. vii. Also see Appendix A, pp. 596–597.

248. Sarton, *Six Wings*, p. 125.

249. Cyril Stanley Smith, "The Development of Ideas on the Structure of Metals," pp. 473–474. The letter to Mersenne is in Descartes, *Oeuvres*, vol. II, pp. 479–492.

250. Smith, "Development of Ideas on the Structure of Metals," p. 474. See Descartes, *Oeuvres*, vol. IX, pp. 276–278.

251. Theodore M. Porter, "The Promotion of Mining and the Advancement of Science," p. 544.

252. Thomas Sprat, *History of the Royal-Society of London*, p. 113.

253. Eamon, *Science and the Secrets of Nature*, p. 9.

254. Zilsel, "Genesis of the Concept of Scientific Progress," p. 150.

255. Ibid., p. 129.

256. Ibid., p. 129.

257. Julian Martin, "Natural Philosophy and Its Public Concerns," p. 111.

258. Victor E. Thoren, *The Lord of Uraniborg*, p. 150.

259. John Robert Christianson, *On Tycho's Island*, p. 3. Christianson was able to identify almost a hundred of Brahe's collaborators "who were active on Hven during 1576–97 or in Wandsburg, Wittenberg, and Bohemia during 1597–1601. Three out of five of them were university trained, and one-third were artists or master artisans" (p. 247).

260. Thoren, *Lord of Uraniborg*, p. 1.

261. Christianson, *On Tycho's Island*, p. 24. Although the grant was intended to be in perpetuity, political changes in Denmark intervened to end Brahe's tenure on Hven in 1597.

262. Ibid., pp. 3, 35, 43, 80 (Emphasis added).

263. Ibid., p. 79.

264. Ibid., p. 81.

265. Ibid., p. 72.

266. Thoren, *Lord of Uraniborg*, p. 159 (emphasis added). For a biographical sketch of Schissler, see Christianson, *On Tycho's Island*, pp. 295–296.

267. Thoren, *The Lord of Uraniborg*, p. 150.

268. Ibid., p. 180. For a biographical sketch of Crol, see Christianson, *On Tycho's Island*, pp. 262–263.

269. Christianson, *On Tycho's Island*, p. 80.

270. Ibid., p. 143.

271. Ibid., p. 183.

272. Ibid., pp. 246–247.

273. Henry C. King, *The History of the Telescope*, pp. 30–31.

274. Quoted in ibid., p. 31.

275. Quoted in ibid., p. 32.

276. Steven Shapin, *A Social History of Truth*, p. 307.

277. Clifford Dobell, *Antony van Leeuwenhoek and His "Little Animals,"* p. 5.

278. Ibid., p. 362.

279. Ibid., p. 40.

280. Leeuwenhoek's famous first letter to the Royal Society was translated into English and published in *Philosophical Transactions* (1673), vol. 8, no. 94, p. 6037.

281. Leeuwenhoek to Oldenburg, August 15, 1673. Quoted in Dobell, *Leeuwenhoek*, p. 42.

282. Dobell, *Leeuwenhoek*, p. 54.

283. Ibid., p. 52. The source of the quotation is Robert Hooke, *Phil. Expts. & Obss.* (1726).

284. Dobell, *Leeuwenhoek*, p. 363.

285. Leeuwenhoek to Oldenburg, August 15, 1673. Quoted in Dobell, *Leeuwenhoek*, p. 42.

286. Dobell, *Leeuwenhoek*, pp. 343–344.

287. Lisa Jardine, *Ingenious Pursuits*, p. 99.

288. See esp. Marie Boas Hall, *Robert Boyle and Seventeenth-Century Chemistry*.

289. Shapin, *Social History of Truth*, pp. 132–133. A standard historical currency-conversion rule of thumb would estimate £20,000 in the seventeenth century as roughly equivalent to £1,200,000 in the twentieth century.

290. For more on Richard Boyle, see Nicholas Canny, *The Upstart Earl*.

291. Robert Boyle, "That the Goods of Mankind May Be Much Increased by the Naturalist's Insight into Trades," in Boyle, *Works*, vol. 3, pp. 443, 444.

292. Ibid., p. 442 (emphasis added).

293. Ibid., p. 443.

294. Shapin, *Social History of Truth*, p. 395.

295. Boyle, "An Account of Philaretus, during His Minority," p. xiii.

296. Boyle, "The Excellency of Theology Compared with Natural Philosophy," p. 35–36.

297. Boyle, "Some Considerations Touching the Usefulness of Experimental Natural Philosophy," p. 14.

298. Boyle's contemporaries did not use the word "technician" to denote their paid laboratory assistants; the latter were referred to by various terms, including "laborants" and "laborators."

299. Shapin, *Social History of Truth*, pp. 374, 367.

300. Ibid., p. 359.

301. Ibid., p. 326.

302. Ibid., p. 379.

303. Robert Boyle, *Continuation of New Experiments Physico-Mechanical, Touching the Spring and Weight of the Air, Second Part*, pp. 505–508.

304. Shapin, *Social History of Truth*, pp. 356–358.

305. Ibid., p. 372.

306. Ibid., pp. 359–360.

307. Hall, *Boyle and Seventeenth-Century Chemistry*, p. 208.
308. Shapin, *Social History of Truth*, p. 381 (emphasis added).
309. Ibid., pp. 367–369.
310. Ibid., p. 367. See also Stephen Pumphrey, "Who Did the Work?"
311. Derek de Solla Price, *Little Science, Big Science . . . and Beyond*, pp. 237–238.

6

WHO WERE THE WINNERS IN THE SCIENTIFIC REVOLUTION?

THE SIXTEENTH THROUGH EIGHTEENTH CENTURIES

THE BENEFITS OF discoveries may extend to the whole race of man . . . . discoveries carry blessings with them, and confer benefits without causing harm or sorrow to anyone.

—FRANCIS BACON, *The New Organon*, aphorism CXXIX

EVERY REVOLUTION HAS its winners and losers, and the Scientific Revolution was no exception. Although Bacon promoted his reform of learning as a "win-win" situation, it did not turn out that way. The knowledge and methods of craftspeople put them in the vanguard of the Scientific Revolution in its initial stages, but in the end it was not they who emerged as the masters or beneficiaries of the new science. In the second half of the seventeenth century, a new scientific elite, consisting of gentlemen who looked to Bacon, Boyle, and Galileo as their ideal and inspiration, came to the fore on the basis of the artisanal knowledge they had exploited. Outposts of this new elite appeared from Naples to Stockholm and from London to St. Petersburg, and the Enlightenment secured its dominance throughout Europe.

There were individual exceptions, of course, but for the most part the artisans and tradesmen who revolutionized science found themselves worse off than before. As their trade secrets entered the public domain and their crafts were subsumed into the factory system, they lost their economic independence. Those fortunate enough to find employment in the new economy became wageworkers and were "deskilled" by interchangeable assembly-line jobs, while the others were left behind as human detritus.

The big winners were the captains of commerce and industry who applied the mechanical philosophy to the "rationalization" of production processes, ushering in the Industrial Revolution and its mechanized factory system. Other prime beneficiaries were the self-styled *virtuosi*—the gentlemen natural philosophers whom the Scientific Revolution raised to the status of a new scientific elite. The ascendance of that elite—the subject of this chapter—constitutes the downside of a people's history of science, but it is essential to understanding how science came to be the way it is today.

THE SCIENTIFIC REGIME CHANGE

THE SCIENTIFIC REVOLUTION resulted in a "regime change." Whereas official learned science had previously been under the hegemony of Catholic university scholars, the new science came to be dominated by a secular commercial elite. The Protestant Reformation and the discovery of the Americas both had a great deal to do with eroding the scholastics' intellectual authority, but the dramatic changing of the guard of elite science can best be understood as a subordinate feature—and a consequence—of the transformation of European society from aristocratic to capitalistic.

"This new understanding of nature," Pamela Smith explained, was "part of the much larger economic and social transformation of Europe by the growth of an exchange economy."[1] The proposition that the rise of capitalism was central to the development of modern science is known as the "Hessen thesis" after the author who formulated it, a Soviet physicist named

Boris Hessen. The Hessen thesis will be discussed below in the context of the career of Isaac Newton.

Although the rise of capitalism was a centuries-long process, the definitive triumph of a new scientific elite over the old one can be more precisely dated to the span of a single generation—the four decades from 1680 to 1720. It was then that science went from being "a body of knowledge once promoted by its select devotees in Florence, Paris, or London, to the cornerstone of progressive thought among the educated laity."[2]

The victory of the new elite was preceded by a century of intense propaganda on the part of those who have traditionally been hailed as the heroes of the new science, most notably Galileo and Bacon. One of their most powerful weapons was the vernacular press that had been developed by many generations of artisan-authors. Galileo and Bacon not only read vernacular books, they began to write them as well. In 1605 Bacon's first major appeal for a new science, *The Advancement of Learning,* appeared in English and was only subsequently translated into Latin. In 1610 Galileo published an account in Latin of what he had seen in the heavens with his improved telescope, but in 1613 he elaborated on the subject in Italian and thereafter did most of his writing in that language.[3]

By publishing their findings and proposals in the everyday languages of nonscholars, Galileo and Bacon broadened the social appeal of science and stimulated a substantial expansion of scientific discourse. The readership they were targeting was obviously not limited to the classically educated but included the burgeoning urban "middle classes" brought into existence by the growth of capitalism. The inevitable outcome was a partial democratization of science—of which a people's historian must approvingly take note—but another consequence was the establishment of a new scientific elite that would serve the interests of the new masters of society, the commercial and industrial classes. Although Bacon and Galileo did not live to see the completion of the regime change they had set in motion, by the end of the seventeenth century it was irreversible.

BRAHE, BOYLE, AND THE NEW SCIENTIFIC ELITE

TYCHO BRAHE'S OBSERVATORIES on the island of Hven had been "a training ground for scientific and scholarly elites."[4] The scholars and craftsmen he employed "scattered throughout Europe when they left Tycho Brahe's service, working in many fields," John Christianson wrote. "In the seventeenth century, they infused the scientific culture of Tycho's island into the mainstream of European life."[5]

The decentralization caused by the breakup of Brahe's scientific team in the wake of his death in 1601 had a significant sociological consequence: "Aristocratic big science gave way to a brilliant scattering of individualized science by middle-class scholars."[6] These were the harbingers of the new scientific elite that was to replace the Aristotelian schoolmen as the official arbiters of learned science throughout Europe. Armed with Baconian ideology, they perceived the mechanical and other manual arts as indispensable to scientific knowledge, but their attitude

> was conditioned by the concept of "virtuosity," an ideal that tended increasingly to fashion upper-class tastes in the early modern period. As long as the arts were pursued out of pleasure and curiosity, or used to fill idle moments, they could be considered the marks of a gentleman. Only when they were practiced to make a living did they become sordid. For the virtuosi, the arts were not so much useful as an avocation that set one apart from the crowd.[7]

This was the ideal that inspired gentleman scientists as they carried out their mission of gaining access to the natural knowledge of artisans and mechanics. The epitome of the *virtuoso* was Robert Boyle, who attributed his "disinterested mind" to the fact that he had been "through God's bounty furnished with a competent estate." His fortunate circumstances, he added, allowed him to seek enlightenment rather than money: "I had no need to pursue lucriferous experiments, to which I so much preferred luciferous ones."[8]

EXPERIMENTAL ACADEMIES

BUT BEFORE BOYLE, and even before the Brahe diaspora, associations of virtuosi devoted to natural philosophy had begun to form in Italy. Among the earliest of these "experimental academies" was an Accademia dei Secreti created in Naples in the 1560s by Giambattista Della Porta, an aristocrat who "wrote in Latin and not for the people." Della Porta "dominated the scene of Italian science" by the end of the century; "princes and prelates throughout Italy and Europe," including Holy Roman Emperor Rudolf II, the duke of Florence, and the duke of Mantua, "lavished their patronage upon him."[9]

Della Porta was yet another pre-Baconian advocate of importing craft knowledge into elite science. "His laboratory was the artisan's workshop, where he observed technical operations first hand, absorbed the lore of the crafts, and attempted to distinguish between folklore and empirically tested techniques." He "learned metallurgy, as he learned other crafts, by observing artisans at work and by experimenting on his own."[10] Due to Della Porta's preoccupation with secrecy, little information about how his Accademia dei Secreti operated has survived, but it is known that he hired at least two artisans, a distiller named Giovanni Battista Melfi and an herbalist named Favio Giordano, to assist with the academy's experiments.[11]

Although Della Porta did not reveal much about his Accademia dei Secreti, another author, the Venetian humanist Girolamo Ruscelli, has left us a "lengthy, detailed description of the organization, finances, and operation" of a remarkably similar organization (even down to its name, the Accademia Segreta) that he claimed to have founded in "a famous city" in the kingdom of Naples. Although Ruscelli's account lacks corroboration and may be partially or wholly fictional, it happens to be "the only contemporary description we have of a sixteenth-century Italian scientific society."[12]

Ruscelli's academy, which had twenty-four members, was supported by the patronage of a local nobleman. With this nobleman's assistance, the Accademia Segreta was able to construct a headquarters with facilities for meetings, a laboratory where experiments could be performed, and an herb garden:

> The society hired several artisans to assist them in doing their experiments, including two apothecaries, two goldsmiths, two perfumers, and four herbalists and gardeners. There were also domestic servants and a team of "choremen" assigned to the laboratory itself, where they attended the furnace, cleaned vessels, ground herbs and chemicals, and made lute for the distillation apparatus. Each of these workers was assigned to specific tasks according to his special skills, while the academicians acted as overseers, giving directions for conducting experiments: "All those who stood above these attendants gave orders to the apothecaries, goldsmiths, and perfumers."[13]

The hierarchical division of labor described by Ruscelli corresponds closely to that of Solomon's House, the ideal scientific research organization imagined by Bacon more than half a century later in *The New Atlantis*.[14] But regardless of the extent to which Bacon was directly inspired by the experimental academies on the Continent, he needed little encouragement to endow his utopian institution with an elitist structure.

Somewhat after the Accademia dei Secreti had come and gone, Della Porta became the guiding spirit of another Italian experimental society, the Accademia dei Lincei (Academy of Lynxes), so named because of the lynx's reputation for great sharpness of vision. The Lincei was founded in Rome in 1603 by a teenaged nobleman, Federico Cesi, marchese di Monticello, but Della Porta was its inspiration, and in 1610 he became a formal member. One year later, however, the fickle Cesi recruited the newly famous Galileo, and "almost immediately Galileo's influence on the society began to eclipse that of Della Porta." Subsequently, in 1616, "Cesi outlined a new vision for the society" in which he articulated the *virtuosi's* principle of disinterestedness. His academy would not be tainted by the base desire for "profit, honor, or reputation," he declared, but would pursue knowledge only "for its own sake and for the improvement of humanity's spiritual and material condition."[15]

The institutionalization of science took a significant leap forward in the second half of the seventeenth century. In Italy, the Lincei was succeeded by the Accademia del Cimento (Academy of Experiment) in 1657,[16] and in the 1660s, important new organizations appeared in London and Paris: the Royal Society and the Royal Academy of Sciences, respectively. The formation of these permanent scientific societies is often discussed in terms of the professionalization of modern science, but that is another way of describing its increasingly elitist character.

The Accademia del Cimento demonstrated the extent to which old prejudices that scorned hands-on experimental practice as the business of "low mechanicks" had diminished. Its Medici patrons, Grand Duke Ferdinand II of Tuscany and his younger brother Prince Leopold, not only provided the material backing for the organization but were themselves among its most active experimenters. Their largesse endowed the academy with impressive resources; a published account of its operations reports one experiment in which the researchers "piled fifty plates of gold one upon the other, [and] saw a needle placed on the top plate obey the motion of a magnet moved close to the under surface of the bottom one."[17]

The Accademia del Cimento proved to be short-lived—it lasted only ten years—but its counterparts in England and France sank deeper roots. Scientific organization took quite different forms in those two countries. Let us consider the British experience first.

THE ROYAL SOCIETY

The inception of the Royal Society, Derek de Solla Price suggested, may be directly attributable to the mathematical practitioners of London, who numbered in the hundreds by the middle of the seventeenth century. Before it was officially chartered, its adherents constituted an informal club that called itself the "Invisible College." Among its original meeting places were "the shops of the instrument-makers and the taverns (later coffeehouses) they   frequented." When it began to meet more regularly, "it seems to have been called together by Elias Allen, chief of the instrument-makers."[18]

But by the early 1660s, the club had been taken over by a group of *virtuosi* who, having gained at least the lukewarm support of the newly restored monarchy, named it The Royal Society of London for Promoting Natural Knowledge. The association's Baconian inclinations were expressed in the establishment of a Committee for Histories of Trades that was charged with investigating mining, refining, brewing, and other industrial processes. To enhance its social status, the Royal Society offered membership to "anyone of the rank of baron or over." As a result, "a number of courtiers and gentlemen were made Fellows, some of whom were genuinely interested in science, but many of whom were not."[19]

The Royal Society hired talented "demonstrators" such as Robert Hooke, Denis Papin, Francis Hauksbee (both the elder and the younger), Jean Theophilus Desaguliers, Daniel Fahrenheit, and Stephen Gray to do the actual work of performing experiments. It was these "paid servants"—"men whom the Society's Fellows did not value as experimental philosophers"—Stephen Pumphrey showed, who "gained the early Society its continuing reputation":

> Many of the famous experimental achievements of its "Golden Age," such as the development of pneumatics, the consolidation of Newton's mechanical and optical theory and the new science of electricity, depended wholly on their labour and in no small part on their genius. Although the rhetoric of experimentalism was created by gentleman philosophers, the practice was forged by their inferiors.[20]

Unlike the French Royal Academy of Sciences, the Royal Society was not an organ of the state; the British Crown's financial assistance to the new "royal" scientific institution was virtually nil. Robert Boyle's material support was crucial; the profits from his Irish lands "helped to maintain the Royal Society, which also benefited from the trade secrets that Boyle appropriated from the art and mystery of the Irish craftsman."[21] The organization's grateful members enthusiastically promoted Boyle's books and established his enduring image as the archetype of the experimental scientist. *Plus Ultra*, Joseph Glanvill's widely read paean to the Royal Society, singled out Boyle as the personification of its achievements.[22]

As for Bacon, the Royal Society hailed him as its inspirational beacon. Thomas Sprat, its first historian, identified "Lord Bacon" as the "one great Man, who had the true Imagination of the whole extent of this Enterprize, as it is now set on foot."[23] In his "Ode to the Royal Society," the poet Abraham Cowley gushed, "Bacon, like Moses, led us forth at last." It was Bacon, he added, who showed "the mechanic way" to knowledge.[24]

Bacon's own generation had paid little heed to his ideas, but by the mid-seventeenth century, Baconianism began to assert a strong claim as the ideology of modern science. His legacy was complex. Some of the radicals of the English Revolution simultaneously looked to Paracelsus and to Bacon for inspiration. Among the most notable was John Webster, "the celebrated enthusiast preacher physician who had led the radicals' attack on universities during the Revolution."[25] But although Bacon became a hero to some reformers who championed the people and attacked the elites, his political conservatism also affected the way later generations interpreted the Baconian program. The process of institutionalization by which science became a profession in the Western world ensured that it also became, as Bacon intended, an ideological adjunct of power politics and social stability.

THE ENGLISH REVOLUTION
AND "VALUE-FREE" SCIENCE

THE ROYAL SOCIETY was forged in an ideological struggle over the meaning of science that occurred during the Civil Wars of the 1640s and '50s, the great social revolution that "turned the world upside down."[26] The organization's founders were among the better-off people who supported the revolution against monarchical absolutism but who did not want it to go beyond that point. There were many other people, however, who wanted the revolution to go much further toward "leveling" the distinctions between the rich and the poor. In the "vacuum of control" in England of the 1650s

> radical sects (small groups of like-minded religious believers) flourished as never before or since. Their views often derived from the Radical Reformation of the previous century. The very names of these sects are evocative. The Levellers called for universal manhood suffrage and the elimination of the property qualification to vote. The Independents attacked the very idea of a state church and the taxes collected to support it. The Seekers asserted the capacity of every person, man or woman, to find his or her own way to God. The Ranters called for absolute freedom of expression. The Quakers exalted the divine light within each person that equalizes us all and used it to attack every kind of hierarchy. Perhaps most radical of all were the Diggers, whose leader, Gerrard Winstanley (ca. 1609–after 1660), asserted that God and nature are one and argued for, and took steps to effect, a redistribution of land so that it could be worked communally for the common good. These various sects preached and practiced what they preached.[27]

For these Seekers, Ranters, Diggers, and other radical sectaries, science did not exclude social concerns, and its essence was not neutrality. "The radicals of the English Revolution made a last attempt to see the universe as a whole, science and society as one," explains Christopher Hill. "Ending the distinction between specialists and laymen" was central to their program. "Winstanley wanted science, philosophy and politics to be taught in every parish by an elected non-specialist. . . . He and the radical scientists wanted science to be applied to the problems of human life."[28]

Winstanley was adamant that knowledge and education should not be monopolized by scholars in elite universities. He advocated "universal education, regardless of class or sex, to be combined with manual work so as to insure that no privileged class of idle scholars should arise."[29] In his own words,

> One sort of children shall not be trained up only to book learning and no other employment, called scholars, as they are in the government of monarchy; for then through idleness and exercised wit therein they spend their time to find out policies to advance themselves to be lords and masters above their labouring brethren.[30]

"It is sadly ironical," Hill commented,

> that the time when Winstanley was thus visualizing a democratization and widespread dissemination of all knowledge was almost precisely the time at which significant specialization began to set in. The last of the polymaths were dying out just as Winstanley hoped to establish a minor polymath in every parish. His scheme was not utterly utopian. . . . We can hardly say that Winstanley's vision was impossible; we can only say that it was never tried.[31]

Meanwhile, the Paracelsian movement had continued to grow in numbers and influence, peaking in the mid-seventeenth century. Its vitality was enhanced by its confluence with the revolutionary events in England, where the Paracelsians' vitalistic worldview and antagonism toward established authority appealed to the radical religious sects.[32] A focus of their ire was the Royal College of Physicians, a medical monopoly that "excluded surgeons and apothecaries, who were likely to be Paracelsians. In the heady atmosphere of the 1640s and 1650s, the champions of the excluded attacked the college as outdated and corrupt and called for a new, democratic system of public health."[33]

One radical opponent of the College of Physicians' monopolistic practices, an apothecary named Nicholas Culpeper, "translated into English the sacred text of the College, the *Pharmacoepia Londinensis*, so that medical prescriptions would be available to the poorest. He hoped it would make every man his own physician, as the translation of the Bible made every man his own theologian."[34] In general, the experience of the English Revolution reinforced in Bacon's heirs their mentor's fear of the political consequences of Paracelsian "enthusiasm" in science.

By 1660, the English Revolution was over. The hereditary monarchy had been restored, the radicals had been routed, and irreversible steps had been taken toward the establishment of a market-oriented economy. As for the fate of the people's science movement that the Paracelsians represented, the defeat of the radical sects in the English Revolution enabled "an increasingly conservative Baconianism" to succeed in "distancing itself from radical Paracelsian influences; it was the former that would go forward as the mainstream scientific movement, while the latter slipped into the shadows from the late 1650s on."[35]

This was a watershed event for a people's history of science. The radicals' downfall, Hill explained,

> also meant the end of dreams of an all-embracing *Weltanschauung* accessible to ordinary people. Newton was as incomprehensible to the average mechanic as Thomas Aquinas. Knowledge was no longer shut up in the Latin Bible, which priestly scholars had to interpret; it was increasingly shut up in the technical vocabulary of the sciences which the new specialists had to interpret.[36]

The restoration of the monarchy also ensured that the universities would survive

> almost untouched by the scientific ideas which had invaded them during the Revolution. . . . They also retained a classical emphasis when Latin had ceased to be either the main source of scientific information, or the language of international scholarship. . . . So Oxford and Cambridge became

isolated from the main stream of national and international intellectual life, a backwater.[37]

The social consequences, Hill concluded, were far-reaching:

Not only did England enter the epoch of the Industrial Revolution with a ruling élite ignorant of science; the scientists of the Royal Society themselves abandoned the radicals' "enthusiastic" schemes for equal educational opportunity. So the reservoir of scientific talent in the lower classes which the schemes had envisaged remained untapped, and "England advanced towards the technological age with a population ill-equipped to take the fullest advantage of its resources."[38]

Although the ruling class in Restoration England was at first unwilling fully to embrace the new science, a new intellectual elite had nonetheless begun its rise to prominence. A major milestone in that process was the creation of the Royal Society, which gave institutional form to a new scientific ideology. Anxious to put the turmoil and divisiveness of the revolutionary years behind them, the society's leaders explicitly ruled controversy over political, religious, and social problems outside the bounds of the new science. It was not the "business and design of the Royal Society," Robert Hooke declared, to be "meddling with Divinity, Metaphysics, Moralls, Politicks, Grammar, Rhetoric or Logick."[39]

"Enthusiasts" and "fanatics" would not be tolerated in their midst. Radical-minded people were thus excluded from the Royal Society, and the organization of science as a profession was consciously tailored to fit into a new kind of civil society that was rising in England. As the premier historian of the English Revolution explained, "The Society wanted science henceforth to be apolitical—which then as now meant conservative."[40]

The leaders of the Royal Society believed that by banning ideological discussion they had thereby exorcised ideology from science, but what they had actually accomplished was to assure the monopoly of their own elitist ideology. The neutrality they promoted as the ideal of scientific objectivity was a fine-sounding abstraction, but in practice some people and some viewpoints were always "more equal" than others. Because the social function of the Royal Society—empowering a new scientific elite— was originally underwritten by Baconianism, let us examine the biases inherent in that ideology.

BACON AND "THE PEOPLE"

"I do not like the word People," a biographer quoted Bacon by way of illustrating that he was not a champion of "the commonality" or what he would have called "the meaner sort."[41] Although as a philosopher he habitually wrote in universal terms, justifying his proposals by appeal to the welfare of all humanity, as a government official his social concerns were much more narrowly focused: to serve the interests of the privileged classes—"the greater sort." This was not necessarily a matter of hypocrisy on his part; as far as he was concerned, aristocrats and gentlefolk *were* all humanity—or all of it that mattered. The social majority was perceived mainly as a source of civil disorder—"the Material of seditions."[42]

Bacon's attitude toward the explosive growth of natural knowledge he was witnessing at the beginning of the seventeenth century is usually depicted as one of unalloyed jubilation, but that tells only one side of the story. He was also deeply concerned—even alarmed—by the social implications that accompanied the discrediting of traditional learning:

> A huge potential existed for social instability and political rebellion, as centralizing rulers often learned to their cost, and their anxieties in this regard extended to the production and, especially, to the dissemination of knowledge. Men publicly claiming to possess "true knowledge" were potentially dangerous to the authority of the ruler and the precarious legitimacy of his regime; as far as possible, such persons were to be harnessed and their products policed.[43]

Bacon's political career took him to the highest levels of influence in the state.[44] By 1612, he had become one of King James's most important advisors. He became attorney-general in 1613, privy councilor in 1616, lord keeper of the great seal (head of the Court of Chancery) in 1617, and lord chancellor in 1618. He was not born into the aristocracy, but in 1618, he was elevated to the peerage as Baron Verulam and in 1621 was created Viscount St. Albans. As "a member of the English governing elite," Bacon's "overriding ambition was the augmentation of the powers of the Crown in the state, and he believed his refashioned natural philosophy was [an] instrument by which to achieve this political aim." His proposed reform of science was essentially "a Royalist reaction" against efforts to establish independent forms of science that were not subject to royal control.[45]

Bacon has traditionally been portrayed by historians as a precursor of the Enlightenment whose philosophy was profoundly modern and humanistic.[46] Loren Eisley saw him as "a civilized man out of his time and place, dealing with barbarians," and to Benjamin Farrington he was the benign "philosopher of industrial science."[47] In his official capacities, however, he did not shrink from employing torture as a means of repression in dealing with rebellious commoners. After the Enslow Hill Rebellion of 1596, Bacon, equating attacks on land enclosures with high treason, subjected Bartholomew Steere, a carpenter, to "two months of examination and torture in London's Bridewell Prison." On another occasion, in an investigation of suspected heretical Anabaptists, "he stretched a schoolmaster, Samuel Peacock, on the rack until he fainted."[48]

We need not concern ourselves with how typical or aberrant Bacon's willingness to utilize torture was in the context of his time; it is sufficient to note that he viewed "mankind" through the lens of the dominant social class. His claim that the new science he advocated would benefit "the whole race of man" must be evaluated in that light.[49] The Baconian call for scholars to learn from craftsmen can thus be seen not as a benevolent program of knowledge-sharing, but rather as an appropriation of working people's knowledge in the interest of the ruling class.

RAPING AND TORTURING MOTHER NATURE

THE PATRIARCHAL IMAGERY in Bacon's writings reflected the social position of women at the beginning of the seventeenth century in England. Bacon invariably portrayed Nature as a female who was hiding her secrets. He wrote of the secrets "locked in nature's bosom" or "laid up in the womb of nature," and said she would have to be forcibly penetrated in order to make her give them up.[50]

"I am come in very truth," Bacon declared, "leading to you nature with all her children to bind her to your service and make her your slave." We cannot "expect nature to come to us," he said. "Nature must be taken by the forelock" (grabbed by the hair). It is necessary to "lay hold of her and capture her," to "conquer and subdue her, to shake her to her foundations." He cited the way women suspected of witchcraft were tortured by mechanical devices to extract confessions as a metaphor to indicate the methods of inquisition by which he thought Nature's secrets should be extracted from her:

> howsoever the use and practice of such arts is to be condemned . . . for the further disclosing of the secrets of nature . . . a man [ought not] make scruple of entering and penetrating into these holes and corners, when the inquisition of truth is his whole object.

Nature, he said, "exhibits herself more clearly under the trials and vexations of [mechanical devices] than when left to herself."

The sexual imagery of penetrating, torturing, and enslaving Mother Nature should not be dismissed as harmless figures of speech unrelated to the way seventeenth-century English gentleman scientists perceived the world. The subordination of women was an essential component of their worldview, which was entirely committed to maintaining male dominance in a patriarchal society. To believe that the early scientists' pronouncements were "value-free" with regard to women or any other social matters would be extremely naïve.

THE GREAT EUROPEAN WITCH CRAZE

THE STATUS OF women in the era of the Scientific Revolution is most starkly revealed in the witch-hunting mania that engulfed Europe at the very same time. Both the new science and the witch craze were reactions to a generalized social crisis that had been deepening throughout Europe during the preceding century: the breakdown of the traditional feudal society.

"In the sixteenth and seventeenth centuries," Hugh Trevor-Roper observed, belief in witches "was not, as the prophets of progress might suppose, a lingering ancient superstition, only waiting to dissolve. It was a new explosive force, constantly and fearfully expanding with the passage of time."[51] It may seem paradoxical that such an apparently irrational mass psychosis could have coincided in time and place with the birth of modern science, but it did. "In the 1630s, the decade in which Galileo published his *Dialogue Concerning the Two Chief World Systems*, more witches were being burnt throughout Europe than at any time previously."[52]

> Credulity in high places increased, its engines of expression were made more terrible, more victims were sacrificed to it. The years 1550–1600 were worse than the years 1500–1550, and the years 1600–1650 were worse still. . . . If those two centuries were an age of light, we have to admit that, in one respect at least, the Dark Age was more civilized. For in the Dark Age there was at least no witch-craze.[53]

The suppression of witchcraft mirrored Baconian efforts to stamp out independent claims to knowledge of nature. Witchcraft, some scholars have argued, was in part "a survival of pre-Christian nature worship," an echo of a pagan culture of the hinterlands that was alien to mainstream European culture and perceived as a threat by the social elite.[54] Secular and clerical authorities, however, did not attempt to justify their horrible, massive persecutions on political grounds but by charges that the old women had gained dangerous supernatural powers from an alliance with Satan.

This accusation, which in retrospect seems so patently absurd, cannot be attributed to the illiterate, superstitious masses; it was "the product of a cold-blooded campaign launched by self-interested clerics and inquisitors. It had no genuine social roots, but was imposed from above."[55] An elaborate ideology—a new "science" of demonology—was provided by elite intellectual leaders: "the scholars, lawyers and churchmen of the age of Scaliger and Lipsius, Bacon and Grotius, Bérulle and Pascal." It was the creation of "some of the most original and cultivated men of the time," who "devoted their genius to its propagation." Among the most important was Jean Bodin, who published (several decades after Copernicus's death, it should be noted) a key demonological text, *De la démonomanie des sorciers:*

> Bodin the Aristotle, the Montesquieu of the sixteenth century, the prophet of comparative history, of political theory, of the philosophy of law, of the quantitative theory of money, and of so much else, who yet, in 1580, wrote the book which, more than any other, reanimated the witch-fires throughout Europe . . . to see this great man, the undisputed intellectual master of the later sixteenth century, demanding death at the stake not only for witches, but for all who do not believe every grotesque detail of the new demonology, is a sobering experience.[56]

Among the leading theorists of the demonological science was King James VI of Scotland, who later, as James I of England, would become Francis Bacon's royal patron.[57] The king fancied himself a scholar, and his learned treatise *Daemonologie* was one of many such "encyclopaedias of witchcraft" that

> insist that every grotesque detail of demonology is true, that scepticism must be stifled, that sceptics and lawyers who defend witches are themselves witches, that all witches, "good" or "bad," must be burnt, that no excuse, no extenu-

ation is allowable, that mere denunciation by one witch is sufficient evidence to burn another.[58]

Some of the victims of the witch mania were men, but it was nonetheless a profoundly antifemale phenomenon. Statistical studies suggest that of a total of about 100,000 people put on trial for witchcraft in several European countries, about 83 percent were women.[59] The manifesto of the movement, *Malleus maleficarum* (The Hammer of the Witches), was "a misogynist's textbook."[60] *Malleus maleficarum* was published in 1486 by two Dominican priests, Henry Kramer and James Sprenger, who were appointed by Pope Innocent III to head up a special campaign against witchcraft. According to Kramer and Sprenger, one reason most witches are female is because women are mentally weaker than men, but the most important reason is the voracious sex drive of women: "All witchcraft comes from carnal lust, which is in women insatiable," they declared; "Wherefore for the sake of fulfilling their lusts they consort even with devils."[61]

By the mid-seventeenth century, the witch-hunts had become so frenetic that fanatical prosecutors were leveling accusations not only against defenseless rural women but against members of clerical and secular elites as well. Anyone, regardless of social position, who expressed tolerance or skepticism toward witchcraft might find themselves targeted for repression. As people with social weight began to feel seriously threatened, the witch craze subsided. One German town's experience admirably exemplifies how the change came about:

> Even Würzburg, where trials had raged until 1630, produced a solemn reconsideration when Bishop Philip Adolf von Ahrenberg found both his chancellor and himself accused of witchcraft. In an about-face that symbolizes the whole process throughout southwestern Germany, the bishop prohibited further trials and established regular memorial services for the innocent victims of justice.[62]

The leading theorists of demonology were Aristotelian scholars, but Bacon, Descartes, and the other familiar Heroes of Modern Science failed to offer an effective challenge; they were for the most part noncommittal on the issue of witchcraft. In the 1660s, when the witch craze had run its course as a major social phenomenon in Europe and was rapidly disappearing, influential leaders of the new scientific elite attempted to breathe new life into it. In England some of the most prominent spokesmen of the Royal Society, including Robert Boyle, Joseph Glanvill, and Henry More, joined the rearguard action in defense of witchcraft belief.

"It is a tenacious tradition," Charles Webster said, "that the new science and its agents, the distinguished Fellows of the Royal Society, played a major part in the decline of witchcraft beliefs in Britain." But despite its tenacity, it is untrue. In fact, "a substantial portion of the literature favourable to witchcraft belief was contributed by Glanvill, Aubrey and other writers having firm associations with the Royal Society."[63]

> The wave of interest in witchcraft and sorcery among the experimental philosophers began with the publication at Robert Boyle's behest of *The Devil of Mascon* (1658). . . . The devil of Mascon became the anchor man of subsequent compilations of authentic accounts of demons. Boyle continued to believe that the verification of supernatural phenomena represented the best means of invalidating the arguments of atheists. The "sceptical chymist" and his friends expressed no obvious reservations concerning witchcraft.[64]

"Boyle's cause was particularly energetically pursued by Joseph Glanvill, who together with Thomas Sprat, was the major apologist for the Royal Society."[65] In 1666, when the frenzy over witches had all but completely subsided, Glanvill published *A Philosophical Endeavor towards the Defence of the Being of Witches and Apparitions*. In that treatise, Glanvill argued Boyle's case that denial of the reality of witchcraft constituted cover for atheism: "those that dare not bluntly say, *There is no*

God, content themselves, for a fair step and Introduction, to deny there are *Spirits,* or *Witches.*" In 1677, when "a retired revolutionary and enthusiast who was a devotee of Paracelsus" attacked Glanvill's belief in witches, Robert Boyle and Henry More came to Glanvill's defense.[66]

Boyle and like-minded Fellows of the Royal Society were deeply concerned, in the wake of the disruptive mid-century Civil Wars, to maintain the social respectability of the new mechanical philosophy and to distance it and themselves from any taint of "atheistical" materialism. The issue of witchcraft "provided the Royal Society with an opportunity of demonstrating its religious and social conformity and freedom from suspected materialistic leanings."[67]

THE WITCH-HUNTS AND MEDICAL SCIENCE

A SUBORDINATE BUT significant effect of the witch-hunts was the support they provided to elite doctors in their bitter rivalry with female folk healers. "In all times," Francis Bacon noted, "witches and old women and impostors have had a competition with physicians." Although Bacon's sympathies were clearly with the latter, his admission that "empirics and old women are more happy many times in their cures than learned physicians" is revealing.[68]

The theorists of demonology were adamant about the need to suppress not only the black magic of evil witches but the *beneficial* magic of "white witches" as well. Village women reputedly possessing curative powers found themselves in the latter category:

> Of course, since it was all but impossible for female healers to enter university (reading for a degree being associated with training for the priesthood), there was virtually no way in which women could gain professional respectability. Consequently, as physicians demanded the professionalization of medicine and petitioned kings and parliaments to make the unqualified practice of medicine illegal, women

healers, especially the more successful of them, necessarily found themselves on the wrong side of the law.[69]

Guy de Chauliac, the most eminent surgeon of the fourteenth century, made it clear that the primary function of the university medical curriculum was not to teach students how to heal, but how to distinguish themselves from the hoi polloi: "If the doctors have not learned geometry, astronomy, dialectics, nor any other good discipline, soon the leather workers, carpenters, and furriers will quit their own occupations and become doctors."[70]

Physicians in England petitioned Parliament in 1421 to bar all women from medical practice, and to admit only those men who had attended the physicians' own "scholes of Fisyk."[71] In 1511 Parliament passed an act to prevent the "great multitude of ignorant persons" from practicing the healing arts. Among those so proscribed were "common artificers, as smythes, weauers, and women," who were said to have performed "greate cures and thinges of greate difficultie: in the whiche they partly use sorcerye, and witchcrafte."[72]

How else could the therapeutic superiority of illiterate women and other "common artificers" be explained? Educated medical men apparently felt they were facing unfair competition from supernatural forces and demanded that the civil authorities intercede on their behalf by enforcing the statutes against witches.

The persecution of witches can be further correlated with specific medical developments that negatively affected women:

> It was also the period in which men began to wrest control of reproduction from women (male midwives appeared in 1625 and the forceps soon thereafter); previously, "Childbirth and the lying-in period were a kind of ritual collectively staged and controlled by women, from which men were usually excluded." Since the ruling class had begun to recognize its interest in increased fecundity, "attention was focused on the 'population' as a fundamental category for economic and

political analysis." The simultaneous births of modern obstetrics and modern demography were responses to this crisis.[73]

It is evident that even demography, the quantitative social science whose objectivity would seem least open to question, was by no means "value free."

In conclusion, the rise of the new science was not a significant factor contributing to the decline of the witch craze. "To the degree that attitudes in general were changing about such questions as witchcraft persecution, scientists were dragged along with the tide," Charles Webster explained. "We must look in places other than science for the explanation of these changes."[74]

One place to look is in the socioeconomic transformation that Europe was experiencing in the seventeenth century—the transition from an aristocratic to a market-oriented society. The careers of Johann Rudolph Glauber and Joseph von Fraunhofer, separated by more than a century, illustrate the impact of the momentous social change on the practice of science.

JOHANN RUDOLPH GLAUBER: A NEW SCIENTIFIC TYPE?

IN 1658, A German admirer of Paracelsus named Johann Rudolph Glauber announced the discovery of a wondrous substance he had isolated from the waters of a spring reputed to have healing powers. He called it *sal mirabile*, "miraculous salt," and described its medicinal virtues at length.[75] He credited both Paracelsus and Agricola with having known of this salt, but claimed for himself the discovery of its composition, which gave him the ability to produce it on an industrial scale. It was a commercial success; Pamela Smith noted that the salt (hydrated sodium sulphate) is still known in Germany today as "Glaubersalz."[76]

Glauber's *sal mirabile* was but one of the chemical products that were the source of his livelihood. In 1640 when he moved to Amsterdam, "he entered a world in which it was possible to make one's fortune by selling goods on the market."[77] But the prestige he felt he was due was not automatically forthcoming; he had to demand it, and in doing so, he insisted on the superiority of practical workers to scholars as producers of knowledge of nature:

> Despite his lowly origins as a barber's son who had apprenticed as an apothecary and never been to university, he was proud of his capabilities as an artisan: "why should an experienced *Artist* not come closer [to the truth about the salt] when he industriously considers the matter? A scholar is good for preaching but not much else."[78]

Over the last quarter-century of his life, Glauber published some thirty books on chemistry and alchemy, "all containing in different measure practical processes, advertisement of secret preparations, and a theoretical structure for his inventions and recipes." Only one was dedicated to an individual; the rest were "addressed to the 'common man' or 'the peasants, vine-growers, and gardners,' or to his 'German Fatherland.'"[79] But in spite of his close connections with the artisanal class, he was also deeply affected by the elitist ethos of the *virtuosi* and sought to identify with them as well:

> As a figure liminal between artisans and scholars with no established qualifications in the republic of letters and no clear estate in society, Glauber was always in danger of slipping down to a purely mechanical or mercantile level. He was anxious to make clear that his activities were not just carried out for the pursuit of wealth. He was particularly sensitive to being called a gold-making alchemist, as he says his neighbors in Amsterdam called him.[80]

The examples of Glauber and Leeuwenhoek demonstrate that tradesmen and craftsmen of modest social origins continued to make major contributions to science throughout the second half of the seventeenth century and beyond. Nevertheless, the value system of the *virtuosi* had become dominant. Glauber's writings reveal that he had thoroughly embraced their notion of gentlemanly "disinterested" science and echoed it in his desire to achieve social respectability. Although he made his living by manufacturing and selling chemical products developed in his laboratory, he adamantly avowed that his scientific pursuits were driven by altruistic rather than pecuniary motives.

JOSEPH VON FRAUNHOFER

THE CONTRADICTIONS THAT characterized Glauber's career persisted through the eighteenth century and even into the nineteenth, as the case of Joseph von Fraunhofer illustrates. Fraunhofer "was a rather remarkable figure in the history of science and technology. He was a working-class optician whose work on physical optics revolutionized the production of achromatic glass, telescopes, heliometers, and ordnance surveying instruments."[81]

> Although he undoubtedly belonged to the artisan population, he strove for scientific recognition. He clearly contributed to the corpus of scientific knowledge. His work on the dark lines of the solar spectrum, which now bear his name, as well as his work on diffraction gratings, which was to support the nascent undulatory theory of light so eloquently proposed by Thomas Young and Augustin Fresnel, formed the cornerstone of an impressive spectrum of disciplinary research during the nineteenth century, including spectroscopy, photochemistry, and of course stellar and planetary astronomy.[82]

In spite of all he had achieved, Fraunhofer's artisanal status stood as a major obstacle to his acceptance into the German scientific elite. For one thing, he could not fully disclose his discoveries because he was legally bound to protect the commercial secrets of his employer, Joseph von Utzschneider of the Optical Institute. Although "secrecy was necessary to ensure the Optical Institute's monopoly of the optical-glass market," it "proved deleterious to Fraunhofer, as it thwarted his attempts to be recognized as a *Naturwissenschaftler*." Furthermore, "Utzschneider felt that he, as owner of the Optical Institute, was also the owner of the practical knowledge of optical glass production," and that "he owned any practical, artisanal knowledge produced by those employed by him."[83]

Fraunhofer was on the fringes of the scientific elite but was not allowed to penetrate its inner circles. The Royal Academy of Sciences in Munich accepted him as a corresponding member in 1817, but a ferocious debate erupted when he was proposed for full membership in 1820. The official nomination declared that his accomplishments "have secured an everlasting name for him in the history of science," but a number of influential members of the academy vehemently objected. Joseph von Baader appealed to the academy's constitution, which said that "members may be accepted only if the world of scholars has been convinced of the merit of the potential member's published works":

> Baader continued by emphasizing that Fraunhofer was not university educated, and indeed never even attended *Gymnasium* (high school). Although Fraunhofer was admittedly well versed in the [art] of practical optics as a result of his training in glassmaking, this knowledge was insufficient for Fraunhofer to be called a mathematician or a physicist.

Baader's polemic culminated in a warning that admitting the likes of Fraunhofer would put the academy in danger of becoming a "corporation of artists, factory owners, and artisans."[84]

Baader's objections were seconded by Julius Konrad Ritter von Yelin, a physicist and chemist:

> Yelin echoed Baader's concern that Fraunhofer was self-educated. Such a lack of formal education, Yelin argued, would result in an inability to follow the complex lectures that peri-

odically took place in the mathematics and physics section of the academy. Yelin's anger was most evident in his concluding remark: he found it personally insulting that Fraunhofer would join the same section, at the same rank, as he.[85]

Fraunhofer's candidacy was not without supporters. Bavarian Court astronomer Johann Georg von Soldner cited Fraunhofer's exact measurement of the dark lines of the solar spectrum and proclaimed, "I consider this discovery of Fraunhofer's to be the most important one in the area of light and colors since Newton."[86] Unfortunately, von Soldner's valuation of Fraunhofer's accomplishment was not widely shared. The result of ignoring the "Fraunhofer lines" was to delay for half a century the recognition that they provided a means of determining the chemical composition of the sun and the stars. Meanwhile, Fraunhofer continued to be denied full membership in the Academy, but as a consolation prize he was upgraded in 1821 from corresponding member to "extraordinary visiting member."

Fraunhofer was only the most visible example of a large class of artisans who received insufficient credit for their contributions to science. Although no one denied that instrument-makers and practitioners of related crafts were "crucial to the scientific enterprise," they were excluded from the halls of elite science on the grounds that their knowledge was a product of manual rather than intellectual processes—that is to say, not a result of creative genius but derived from "shamefully following the rules or techniques required of the craftsman."[87] But the prejudices of privileged German academicians jealously guarding their prestige should not be allowed to obscure the role of other participants in the construction of scientific knowledge.

Glauber and Fraunhofer exemplify a transitional type midway between the pure artisan and the professional scientist. Freed of guild restraints and dependence on patronage but bound instead by commercial imperatives, they reflected different stages of the larger social change occurring in Europe: the transition to capitalism. The implications of that momentous social transformation for the history of science were first explored by Boris Hessen in an analysis of Isaac Newton's career.[88]

ISAAC NEWTON AND THE HESSEN THESIS

"HAVE YOU EVER stopped to count the number of books that either begin or end with Isaac Newton?" asked B. J. T. Dobbs, illustrating her observation that the canonical account of the Scientific Revolution treats him as its "final cause and first mover."[89] Newton has barely been mentioned so far in this discussion because the focus here is not on the Great Minds. There is no denying that he was an extraordinary individual, but recapitulating his innovative ideas is more the business of the biographer than the historian; it does not add much to understanding the root causes of the rise of modern science.

The frequent occurrence of simultaneous discoveries is evidence that scientific ideas are not autonomous historical agents. There are many examples, but the two best known are Newton's and Leibniz's independent invention of the calculus and Darwin's and Wallace's independent formulation of biological evolution by natural selection. In both cases, the Great Idea was "in the air" with all of its antecedents in place awaiting inevitable recognition. If one extraordinary individual had not found the final piece of the puzzle, another would have soon enough.

In general, then, a glorification of Newton's ideas does not enhance historical understanding; it is the appearance of those ideas at a particular time and in a particular place that requires explanation. How did they come to be "in the air"? Why did Newton's gravitational theory surface in England in the last half of the seventeenth century rather than, say, in China in the fourteenth century? That is the question Boris Hessen attempted to answer in a seminal paper entitled "The Social and Economic Roots of Newton's *Principia*."[90]

In the *Principia*, or *Mathematical Principles of Natural Philosophy*, Newton demonstrated that the same force that causes objects to fall to earth also controls the orbits of the moon and the planets. Although he is credited with a number of other major innovations in mathematics and optics, the "Newtonian synthesis" of terrestrial gravity and celestial motion was certainly his most celebrated accomplishment.

Traditional history of science treated the *Principia* as a work of pure intellectual activity—the logical outcome of a series of ideas put forward especially by Galileo and Kepler. Hessen challenged that notion by showing that the necessary preconditions of Newton's theory were not "in the empyrean of abstract thought" but in his social environment. On the whole, Hessen contended, "the brilliant successes of natural science during the sixteenth and seventeenth centuries were conditioned by the disintegration of the feudal economy, the development of merchant capital, of international maritime relationships and of heavy (mining) industry."[91]

Hessen focused on "the historical demands imposed by the emergence of merchant capital and of its development," which characterized the era of Newton's activities, and the "technical problems the newly developing economy raised for solution." That led in turn to consideration of the "grouping of physical problems and of science necessary to the solution of those technical problems."[92]

A vital problem for a commercial nation dependent on oceanic transport was to find a means of measuring longitude at sea. Long before Newton was born,

> in order to determine the longitude recourse was made to measurement of the distance between the moon and the fixed stars. This method, put forward in 1498 by Amerigo Vespucci, demands an exact knowledge of the anomalies in the moon's movement and constitutes one of the most complicated tasks of the mechanics of the heavens.[93]

"The study of the laws of the moon's movement was of fundamental importance," Hessen continued, "in compiling exact tables for determining longitude, and the English 'Council of

Longitude' instituted a high reward for work on the moon's movement." Parliament passed bills to stimulate that research. There was a perceived need for "the creation of a harmonious structure of theoretical mechanics which would supply general methods of resolving the tasks of the mechanics of earth and sky. The explanation of this work fell to Newton to supply."[94] The third section of the *Principia*

> is devoted to the problems of the movements of planets, the movement of the moon and the anomalies of that movement, the acceleration of the force of gravity and its variations, in connection with the problem of the inequality of movement of chronometers in sea-voyages and the problem of tides.[95]

As was noted in chapter 4, it was not Newton's theory but a timepiece produced by a skilled watchmaker that finally solved the longitude problem. But in studying the moon's trajectory, Newton came to realize that it could be explained as the combination of two mechanical influences: the momentum of an object traveling in a straight line and the force of terrestrial gravitation pulling that object toward the earth. That serendipitous discovery was widely perceived as the crowning achievement of the mechanical philosophy, thus assuring the final triumph of the new science. As industrial capital surpassed mercantile capital in importance, the use of machinery—and consequently the prestige of mechanical philosophy—increased exponentially. The *Principia* was put on a pedestal as the shining ideal to which scientists in all fields were expected to aspire for centuries to come. Newton's work has therefore traditionally been hailed as the culmination of the Scientific Revolution.

Hessen provided a lengthy list of other technical problems that the nascent capitalist economy posed—problems in mechanics, hydrodynamics, aerodynamics, ballistics, optics, and metallurgy, to name a few—with which Newton was also concerned. Although the *Principia* treats them only on the theoretical level, Newton's interest in their practical implications —and therefore his motives for choosing to study them in the first place—is clearly revealed in his correspondence. The most thorough statement of that interest is in a letter of May 18, 1669, that Newton wrote to a Cambridge colleague named Francis Aston, who had asked his advice about what to observe while traveling abroad.[96]

Newton began by instructing Aston in the art of dissimulation as a means of worming information out of foreigners concerning "their trades and arts" and other matters. In order to make them "more ready to communicate what they know to you," Newton advised him to "insinuate [yourself] into men's favour" by "seeming to approve and commend what they like." Among the areas of knowledge into which Aston was urged to inquire were "such fortifications as you shall meet with, their fashion, strength, and advantages for defence, and other such military affairs," and "the mechanism and manner of guiding ships."[97]

"As for particulars," Newton continued,

> these that follow are all that I can think of, viz. whether at Chemnitium in Hungary (where there are mines of gold, copper, iron, vitriol, antimony, &c.) they change iron into copper by dissolving it in a vitriolate water, which they find in cavities of rocks in the mines, and then melting the slimy solution in a strong fire, which, in the cooling proves copper. The like is said to be done in other places which I cannot now remember. Perhaps too it may be done in Italy.[98]

He also asked Aston to try to find out

> Whether in Hungary, Sclavonia, Bohemia, near the town Eila, or at the mountains of Bohemia near Silesia, there be rivers whose waters are impregnated with gold; perhaps the gold being dissolved by some corrosive waters like aqua regis, and the solution carried along with the stream that runs through the mines. And whether the practice of laying mercury in the rivers till it be tinged with gold, and then straining the mercury through leather, that the gold may stay behind, be a secret yet, or openly practised.[99]

He suggested that Aston visit "a mill to grind glasses" that he had heard of in Holland, and while in that country to look up a man named Bory, who possessed "some secrets (as I am told) of great worth, both as to medicine and profit." It would also be worthwhile, he said, to "inform yourself whether the Dutch have any tricks to keep their ships from being all worm-eaten in their voyages to the Indies—whether pendulum clocks be of any service in finding out the longitude, &c."[100]

Hessen's critics protest that this letter is not typical of Newton's correspondence and therefore does not constitute a balanced view of his interests and concerns.[101] But at the very least, it demonstrates an *awareness* on Newton's part of a range of specific technical problems that were stimulating scientific investigations, including his own. And as Hessen pointed out, they happened to be the same technical problems for which contemporary capitalism was demanding solutions.

In summary, Boris Hessen's contribution to a people's history of science was a compelling argument that the capitalist socioeconomic context of late-seventeenth-century England was crucial to the emergence of the theory of universal gravitation. Giving credit to Newton for formulating that theory is certainly justified, but the collective activity of the many, many people who constituted the social soil from which it sprouted should be recognized as a more fundamental causal factor than Newton's individual genius.

NEWTONIANISM: "SUCKING SPIRIT OUT OF NATURE"

DUE TO THE great prestige he had attained, Newton's personal philosophy of nature became enormously influential among his contemporaries. The Newtonian version of mechanical philosophy provided an unemotional rationalism that aspired to explain all phenomena as resulting from impersonal, lawful interactions of lifeless particles (Newton's providentialism notwithstanding). This outlook resonated with the scientific elite's desire for a dispassionate science intent on minding its own business and not threatening the social status quo. Margaret Jacob has argued persuasively that Newtonianism became the dominant scientific paradigm in England in the late seventeenth century not primarily as a result of its truth content but because it was an integral element of the ideology of the social class that came out on top in the long revolutionary struggle that erupted in the 1640s and culminated in the Glorious Revolution of 1688.[102]

Newtonianism struck the final blow against the "vogue of Paracelsus"[103] that had swept England during the revolutionary years. The Paracelsians promoted vitalist notions of nature wherein all creatures democratically share in the life-spirit. From that premise followed the essential equality and kinship of all humans, which posed a challenge to traditional defenses of social inequality.

By "sucking spirit out of nature" Newtonianism countered the socially disruptive enthusiasm of the radical sects.[104] To the popular passions it opposed logic and reason; to knowledge born of inspiration it opposed knowledge produced by patient and disciplined experimentation; to knowledge accessible to all individuals it opposed knowledge controlled by a scientific elite. The Newtonian worldview thus rose to dominance in scientific thought on the coattails of a new ruling elite after a revolution. As James Jacob explained, "modern science developed in part in response to popular disorder and elite perceptions of the threat from below."[105]

And the Newtonian outlook did indeed become a *world* view. Although Newtonianism—the physics and the ideology alike—was exported through many channels,[106] it was the ardent publicity Voltaire gave it that was undoubtedly most responsible for making it a pillar of Enlightenment thought. By the end of the eighteenth century, Pierre Simon Laplace had gained recognition as the "successor to Newton," and Paris had long since supplanted London as the international center of elite science.

SCIENCE AND THE FRENCH REVOLUTION

IN THE MEANTIME, France had experienced a social revolution of its own. As in the earlier English Revolution, discord between the scientific elite and plebeian knowledge-seekers was exacerbated to the point of open conflict. The principal institution of elite science in France, the Royal Academy of Sciences, was obliterated, but when the dust settled, the practice of science was dominated more thoroughly than ever by a professional elite. The French Revolution—"the great turning point of modern civilization"—was thus a pivotal event for science also.[107]

Despite a great deal of conservative hand-wringing over the destruction wrought by Jacobinism, the traditional assessment of the Revolution's overall impact on science holds that the great social transformation was, in the final analysis, beneficial.[108] The destruction of the royal guilds, corporations, and monopolies (including the Academy of Sciences) succeeded in "opening careers to talent," allowing the sciences to draw their human resources from a considerably larger social pool than was possible under the old regime. Particularly notable in this regard was the breaking of the old academic stranglehold over the medical sciences. Elimination of the caste system that had separated physicians, surgeons, and pharmacists opened the way for a new and better kind of medical profession.[109]

The establishment of formal equal access to careers in science was important, but the Revolutionary governments went further by creating educational institutions to guarantee the state an expanding supply of scientific talent, a commitment that was reinforced and expanded by the Napoleonic regime. Counterbalancing the democratizing effects of these changes, however, was a rapid increase in specialization of the sciences; "the age of the generalist was over in the eighteenth century."[110]

With scientific practice ever more restricted to specialists, opportunities for artisans and other outsiders to make significant contributions sharply diminished, and the hegemony of professional scientists that characterizes science today was solidified.

Specialization was an inevitable consequence of the great expansion of scientific practices, but the elitist form it took was not. Although often said to have been a consequence of the Revolution, this transformation can with more justice be attributed to the Thermidorian reaction, the wave of conservatism following Robespierre's downfall on July 27, 1794, that eventually engulfed the Revolution and brought it to a close.[111]

Understanding how these changes came to pass requires an examination of the origins of elite science in France. The institutionalization of science in that country proceeded in a different manner from the way it occurred in Italy and England.

THE PARISIAN ACADEMY OF SCIENCES

NEITHER THE ACCADEMIA del Cimento (despite the active participation of the Grand Duke of Tuscany) nor the Royal Society (despite its name) were official state institutions. By contrast, the Parisian Royal Academy of Sciences was from its inception "a new organ of government" that provided "the means by which the French virtuosi were metamorphosed into professional scientists."[112] It was founded in 1666 by Jean Baptiste Colbert, Louis XIV's Minister of Finance, and its initial meeting was held in the monarch's private library.

"Elitism was the very essence of the institution." The Academy of Sciences "did not follow egalitarian principles within its own ranks, and it assumed its members were by definition superior to the common citizen."[113] Its members were handpicked and paid a salary from the royal treasury. Like the Royal Society in England, it was expected to remain apolitical and aloof from social issues; one of its creators decreed, "In the meetings, there will never be a discussion of the mysteries of religion or the affairs of state."[114] Although the selection of academicians was influenced by prior reputations, the new scientific elite in France was nonetheless essentially created from above.

The Academy of Sciences was "the central theater in which science's intricate plot was unraveled during the Enlightenment."[115] Its class composition was unambiguous: "Artisans were clearly excluded." In Old Regime France, "avenues of learning were open principally to the wealthy and the titled," and because membership in the academy "presupposed a good education, men of lower social rank for whom the possibilities of formal schooling were almost nonexistent were ruled out."[116]

In 1699, the Abbé Bignon explicitly introduced "notions of etiquette, status, rank, and propriety so prevalent in Old Regime society" into the Academy. "A more elitist or authoritarian system compatible with scientific expertise could hardly have been conceived." Academicians came to be characterized by "a sense of benevolent superiority that often drifted into arrogance."[117]

> Election to the Academy quickly became the most coveted prize for the ambitious savant who, if he acquired the title *membre de l'Académie,* cherished it as proudly as if it were a title of nobility. To be an academician, or to have one's views discussed favorably by the Academy, was in fact to receive the supreme accolade of that loosely knit community of scientists known as the "Republic of Science."[118]

Although artisans were personally unwelcome in the Republic of Science, their knowledge was nonetheless coveted as indispensable to its progress. D'Alembert, in his famous introduction to the *Encyclopédie,* hailed "the sagacity, the patience, and the resources of the mind" of craftsmen.[119] As Roger Hahn, author of the definitive history of the Academy, observed:

> The new seeker of knowledge and the artisan ought to have become brothers-in-arms. Yet the typical intellectual . . . was unable to turn his back upon the weight of tradition that relegated the manual artisan to a lower echelon of society. . . . The age-old disdain for *techne* was especially persistent in France.[120]

"Elitism," Hahn continued, "found a strong ally in the centralizing and bureaucratizing tendencies of the absolutist state."[121] That alliance allowed the academicians to gain control over the knowledge of artisans. In 1675, Colbert

> instructed the Academy to begin a description of the mechanical arts. Throughout the eighteenth century, the members of the Academy continued to collect and publish information about the crafts, eventually forming an impressive collection of twenty-seven folio volumes, the *Description des Arts et Métiers*.[122]

As official arbiter of French science, the Academy was given the power to rule on what was scientific and what was not. A proposal written in 1720 by a highly influential academician recommended that the Academy "be given technocratic direction of all enterprises remotely concerned with science."[123] In the course of the eighteenth century, "practices for administering the [patent] laws evolved in such fashion that the Academy or its individual members were placed in a central position of control." By gaining the right to exercise judgment over the creative efforts of artisans, technicians, and engineers, "academicians came to dominate technological activities in France."[124]

The idea of the patent had arisen several centuries earlier as a bridge between the artisan's desire to protect hard-earned knowledge and the social benefits of sharing knowledge. If governmental power could guarantee inventors a legal monopoly over the profits generated by their new ideas, inventors would be encouraged to go public with their innovations. The first documented patent in European history was granted to Filippo Brunelleschi in 1421 by the Florentine city council. The practice spread to other Italian cities; in 1474, Venice enacted the first general patent law. During the following century, the issuance of patents became commonplace in cities throughout much of Europe.

The struggle over patents in pre–Revolutionary France centered on who would control the process. The academicians' legal right to rule on patent applications allowed them to transfer the artisans' knowledge into the public domain. That was justified as being in the best interests of society as a whole, but the inventors themselves often perceived it as a violation of their rights, leading to considerable friction between artisans and the Academy. The artisan was understandably

> intent upon maintaining the ownership of his invention in order to capitalize on it. . . . The disclosure of technical details was considered a baring of secrets, to be sold for cash or offered only in return for capital investment. It was therefore difficult for an inventor to bring himself to release his secret to a select group of academicians without the assurance of a favorable report in advance. . . . by discussing his invention before a public institution, he was in effect relinquishing ownership.[125]

This is not to suggest that the academicians were simply elite bureaucrats who collected and compiled artisanal knowledge. The illustrious members of the Academy certainly made many worthwhile scientific contributions, but the basic stimulus of the eighteenth century's scientific progress was technical innovation, for which primary credit is due to now-forgotten artisans, mechanics, technicians, and engineers.

The tensions between the creators of natural knowledge and those who were striving to control it are not accorded prominence in traditional histories of science because for the most part their struggle remained beneath the surface—until the French Revolution, when it exploded into the open. Then "a revolt of technology" erupted, "one among the many rebellions swelling into the great Revolution." The intensity of the conflict between artisans and academicians indicates that artisanal science had been a potent force in France all along. "To follow its course," however,

> is to move down into obscure places among people whose history is hard to come by: craftsmen, engineers, inventors, minor manufacturers—small but solid people. . . . Nevertheless, fragmentary traces do remain of the societies which

they formed to attain their interests, popular societies of inventors and artisans.[126]

THE SHOWDOWN BETWEEN ARTISANS AND ACADEMICIANS

"IT IS NOT surprising," Charles Gillispie observed, "that the Academy's role of referee earned it deep hatred among the artisans whose work it judged." Imagine, he said,

> the worthy mechanic, the salt of France, the hero of the *Encyclopédie*, all unversed in polite ways, standing before a committee of scientists with his new machine, on which he has lavished years of labor and pinned his hopes, twisting his hat and trying to answer incomprehensible questions about the laws of statics and dynamics. The artisans insist with very deep feeling on their rights: not only on their right to private property in the fruits of their ideas, but with even greater feeling of their right to be judged by their peers, who look at mechanical problems from their own point of view and not from some high theoretical plane.[127]

The artisans were not averse to the "high theoretical plane" because they lacked sufficient intelligence to understand it but because they deemed it irrelevant to their endeavors. More often than not, that perception was accurate. Technology had a great deal to offer theory in the eighteenth century, but theory had little to offer in return. Even the staunchest defenders of heroic history of science acknowledge that the celebrated theoretical innovations of the Scientific Revolution contributed only minimally to technological progress leading toward the Industrial Revolution.[128]

The outbreak of the Revolution in 1789 initiated a profound change in the balance of power between artisans and academicians. The collapse of censorship gave rise to freedom of the press, of which "a legion of less educated artisans who had long felt oppressed by officialdom" took full advantage:

The broadsides, pamphlets, and newspapers of the Revolution opened new channels of expression and advertisement, and the emergence of voluntary associations gave the artisan greater hopes of finding a secure political platform from which to demand an end to injustice. . . . The inhibiting effect of academic domination and arrogance was now being replaced with a self-confidence that produced the most detailed and telling criticism against the learned societies.[129]

Freedom of expression and freedom of association were mutually reinforcing. "A bewildering chorus of voices called for the creation of voluntary associations for the artisans, free from the clutches of academic domination."[130] Independent scientific societies *(sociétés libres)* came forth and multiplied. They were the organized expression of "the revolutionaries' profound desire to democratize science by making its fruits understandable and applicable to all members of society. . . . They came to represent the antithesis of the elitist and professional traditions forged for over a century by the Academy."[131]

One of the new organizations, the Société des Inventions et Découvertes, spearheaded a drive for a new patent law that would take the power to rule on new inventions out of the academicians' hands. The law they sought was passed in 1791, and its principles continue to govern French patent practices today. Artisan lobbying efforts also led to the creation of two new governmental agencies, the Bureau de Consultation des Arts et Métiers and the Bureau des Brevets et Inventions, "both of which were designed to protect the interests of artisans by taking over some of the functions previously exercised by the Academy."[132]

As the Revolution radicalized, more militant artisans' organizations came to the fore. The Société du Point Central des Arts et Métiers, which proudly claimed to represent "all truly *sans-culottes* artisans,"[133] led an impassioned campaign against a national education plan submitted to the National Convention by the Academy's principal leader, Condorcet. The plan called for the creation of a new "superacademy" with "even greater control of the scientific and technological life of France than the old one had in the Old Regime." But "when Condorcet unfurled his thoroughly elitist project in April 1792, the artisans were prepared to oppose it with an effective barrage of counter-arguments."[134]

Condorcet's plan had the effect of driving public opinion about the academies into two opposing camps. "On one side were the proponents of the academic system. . . . On the other side were their antagonists, who rejected the bureaucratization of cultural elites. . . . Knowledge of nature and discovery of truth were meant to be a public affair and not the exclusive province of an elite."[135] The Société du Point Central published an appeal to all artisans that

> condemned the Condorcet project as a most pernicious effort to repress artisans' rights and to circumvent the role of voluntary associations in determining their own future. The tract was a call to arms asking the scientific proletariat to rise up against oppression and to overturn, once and for all, the arrogant pretensions of academicians by defeating the Condorcet proposal.

This was "the artisans' declaration of war against any form of academic domination."[136]

But the artisans' struggle did not unfold in isolation from the Revolution as a whole. It ebbed and flowed with the tide of the general movement for egalitarianism. On August 8, 1793, the artisans must have thought total victory was theirs when the National Convention decreed an end to the academic system in its entirety. But two years later, following the fall of the Jacobin Republic and the onset of the Thermidorian reaction, a new "superacademy," the Institut de France, came into being after all. The Academy of Sciences was reconstituted as the Institut's First Class, representing—at least symbolically—the final triumph of scientific elitism.

Although its founders expected the new institution to play the same role as its predecessor, it could not. Even long before the Revolution, the former Academy of Sciences, with its "vested interest in the permanence of its environment," had become "an essentially conservative institution," increasingly incapable of guiding scientific progress. On top of that, times had changed, making the new Institut all the more "superfluous for the pursuit of science." The day of the specialist had dawned, and the Institut was unable to overcome its generalist orientation. The effects of the Revolution in France radiated throughout the Continent. "Everywhere in Europe, the age of professionalized science cultivated in institutions of higher learning and perfected in specialized laboratories was replacing the age of academies that had dominated the scene since the middle of the seventeenth century."[137]

The new research facilities and technical schools were the embodiment of a reconstituted scientific elite in France at the beginning of the nineteenth century. But the Institut still had a significant part to play. It served "to effect a firm alliance between the governing and intellectual elites of France," symbolized in the person of Napoleon Bonaparte, who was inducted into its First Class in 1797 "at a time when he had no scientific credentials." The reborn Academy of Sciences "became an obsolete institution. . . . Today it is a glorious relic of the past, more akin to a Hall of Fame than an Olympic stadium."[138]

The meaning for modern science of all that had happened in France was that a qualitative, irreversible change had occurred. "With the founding of a new set of institutions of science during the Revolution and the Empire, French science became establishment science."[139] The rest of the world would follow suit—nineteenth-century science would everywhere be even more elitist in character than eighteenth-century science had been.

GRUB STREET SCIENCE: THE "SPIRIT OF SYSTEM" VERSUS THE "SYSTEMATIC SPIRIT"

THE INVENTOR-ARTISANS' grievances against the Academy were serious, but a more thorough ideological challenge was mounted by a diverse band of popular-science writers and practitioners. These alienated natural philosophers constituted a scientific underground of the Old Regime analogous to the literary underground described by Robert Darnton.[140] There was, in fact, an overlap between the two groups, because some of those identified by Darnton as "Grub Street writers" could equally well be categorized as "Grub Street scientists."[141]

Grub Street science represented a remarkable surge of natural-philosophy speculation that occurred in France in the decade prior to the Revolution and during the early years of the Revolution itself. It was evidenced, among other things, by a massive literary production of "systems of the world," not in the narrow Laplacean sense of technical astronomy but in the earlier sense of works that promised integrated scientific explanations of all natural phenomena.[142] As early as 1781, the most important scientific journal at the time commented, "Never have so many systems, so many theories of the universe, appeared as during the last few years."[143] Authors of voluminous works of natural philosophy believed they were making significant contributions to the scientific quest for knowledge about nature, and the large readership they attracted apparently concurred. But this "spirit of system" was anathema to the ideologues of the Academy.

One of the Academy's foremost spokesmen, Jean d'Alembert, had earlier contrasted what he called the *esprit de système*, the spirit of system, with the *esprit systèmatique*, the systematic spirit.[144] To d'Alembert, the systematic spirit was the mark of a true scientist; it signified devotion to patient, rigorous, rational, analytical, quantitative methods of inquiry into the workings of nature. The spirit of system, however, signified the impatient habit of leaping to conclusions on the basis of insufficient evidence. Those who were infected with the spirit of system were wont to create grandiose, hypothetical explanations of everything in the universe. The practice of system-building, as d'Alembert saw it, was not good science.

In the years leading up to the Revolution, d'Alembert's point of view gained strength among the scientific elite. In 1769 the abbé Nollet wrote to a collaborator who had asked his opinion of a speculative paper, "I ought not hide from you that the Academy is getting more and more difficult about this way of philosophizing."[145] The defining characteristic of scientific orthodoxy in France became opposition to system-building.

The Grub Street scientists were repelled by the Academy's narrow orthodoxy. They stood for a fundamentally different approach to seeking knowledge. They saw themselves as synthesizers rather than analyzers and bold hypothesizers rather than fact-gatherers. They were stalking the answers to the "big questions," disdaining the narrower pursuits that formed the orthodox ideal. It was the right and duty of science, they insisted, to go beyond known facts and to speculate on possible and probable causes.

The respective systems of the Grub Street scientists embodied a broad range of varied and conflicting notions, but they exhibited a number of shared characteristics. Many of them reflected an organic conception of nature as a unified whole and of the fabric of reality as continuous rather than atomistic. They generally resisted the reduction of natural phenomena to mathematical abstractions. Most significantly, as heirs of Rousseau, most of them insisted on the *moral* component of science that transcends purely physical questions and demands a true science of humanity with the goal of improving society. What would it profit us, they might argue, to gain perfect knowledge of the entire "objective" world yet understand nothing of our own moral, mental, political, and social universe?

Charles Gillispie identified the element in the worldview of the Grub Street scientists that most fundamentally connected it with radical social thought. They were less concerned with what *is* than with what *ought to be;* their focus was on "metamorphosis" and "*becoming* rather than *being.*"[146] Their elite opponents, by contrast, treated nature as essentially static, or dynamic only in the mechanical sense wherein all change is reduced to displacement—that is, to the motion of material atoms through empty space.

The ideological utility of this latter outlook for defenders of the social status quo is obvious. To make "what is" the only legitimate focus of science implies that no reality other than the presently existing reality is possible. From this starting point, a conservative ideologue can readily demonstrate that revolutionary social aspirations are foolish, irrational, and futile. Limiting science to "being" and excluding "becoming" not only excludes the possibility of a meaningful science of society, however; it also blocks the progress of the natural sciences. To take only the most salient example, ruling out "becoming" is to rule out the evolution of species.

The ultimate victory of the Academy over its Grub Street critics contributed to the consolidation of the conservative ideology that has characterized modern science ever since: the narrow outlook in which physics and mathematics reign supreme over all other scientific disciplines. What was at stake in the conflict between the system-builders and the guardians of scientific orthodoxy is evident in the thoughts and deeds of some of the leading participants. The nonelite practitioners are represented here by Jacque-Henri Bernardin de Saint-Pierre and Nicolas Bergasse, and their elite adversaries by Georges Cuvier and Napoleon Bonaparte.

BERNARDIN DE SAINT-PIERRE AND THE INTERCONNECTEDNESS OF NATURE

Jacques-Henri Bernardin de Saint-Pierre was an articulate eighteenth-century proponent of the main theme of this book: that knowledge of nature originates not with scholars but with ordinary working people in the course of their everyday productive activities. "Let the academies accumulate machines, systems, books, and eulogies," he declared. "The principal credit for them is due to ignorant men who provided them with raw materials." He illustrated that contention with an anecdote about the horse chestnut.[147] After a great deal of deliberation, several learned academies had determined that the horse chestnut served no alimentary purposes in nature and relegated it to the status of a substance useful only for making candles or preparing face powder. But Saint-Pierre, without the erudition of academicians, learned from a simple child goatherd that the horse chestnut served as a stimulant for the production of milk in goats.[148]

Bernardin de Saint-Pierre is today regarded as a major figure of French literature; his novel *Paul et Virginie* is a classic familiar to French schoolchildren.[149] But he was known to his contemporaries as a man of science as well as a man of letters. His multivolume natural philosophy *Etudes de la nature* provides a prime example of a body of nonelite scientific work that was neglected, in spite of its undeniable value, as a result of post-Thermidorian social attitudes. Bernardin's approach to understanding nature was out of step with the current orthodoxy in that it was antireductionist,[150] synthetic as opposed to analytic, and imbued with a strong Rousseauian moral component.

If a Rousseauian approach to science can be said to have existed at the time of the Revolution, it is most likely to be found in its purest form in Bernardin's works. Bernardin was Jean-Jacques Rousseau's closest spiritual heir. The two men first met in 1772, when Jean-Jacques was sixty years old. They botanized together on long walks on the outskirts of Paris, and their close relationship lasted until Rousseau's death in 1778. It has been suggested that a study of Bernardin's words and deeds during the Revolution might provide insight into the perennial controversy over how Rousseau would have reacted had he lived to participate in the tumultuous events of 1789 and 1793.[151]

Bernardin's *Etudes de la nature* received widespread public acclaim when it appeared in 1784. In 1792, at the height of the Revolution, he was appointed Intendant of the Jardin des Plantes, an important administrative position that was one of the most prestigious in the world of Parisian science. Although Bernardin's tenure was brief and unsuccessful, his statue sits today in the Jardin des Plantes. It shows him seated, deep in contemplation, above a depiction of the arcadian innocents of his creation, Paul and Virginia.

Bernardin's scientific credentials, however, were not accepted by everyone. "There was a striking unanimity as to the value of his scientific or *soi-disant* scientific pronouncements and deductions," one commentator claimed. "The universal opinion was that with scarcely a single exception they are grotesquely wrong."[152] This judgment, however, is extremely misleading. The allegation that Bernardin's science was completely rejected by his contemporaries derived from the common error of identifying "universal opinion" with that of the scientific elite. It was only the latter that with "striking unanimity" deemed Bernardin's scientific work unworthy of consideration.[153] *Etudes de la nature* was translated into several languages and attained considerable international popularity. An author writing in English in 1893 explained that although Bernardin's volumes had been for the most part forgotten by that time, "our grandparents read them greedily enough either in the original or in the excellent translation of Dr. Henry Hunter."[154]

Bernardin's relations with the French scientific elite were particularly bitter. Members of the Academy of Sciences were alienated by the anthropocentrism of his *Etudes de la nature* and gave it the cold shoulder. Bernardin in turn wielded his mighty pen against the academicians; his were "the most devastating of public utterances belittling the academic system."[155] The Academy of Sciences, he charged, was a closed corporation, typical of the Old Regime, dedicated not to science but to preserving its members' privileges.

A famous anecdote has it that Bernardin, in an interview with the Emperor Napoleon (a great admirer of *Paul et Virginie*), demanded to know why the scientific establishment refused to take him seriously. Bonaparte responded by asking him if he was proficient in the differential calculus. When Bernardin acknowledged that he was not, the emperor replied, "Then go learn it and your question will answer itself."[156]

This tale is usually told to illustrate Bernardin's scientific irrelevance, but it illustrates equally well the role of the

Napoleonic state's influence in establishing the hegemony of a narrow type of orthodox science. The refusal to take Bernardin seriously meant, among other things, rejecting his approach of studying natural phenomena in environmental context, a narrowness of scope that was to characterize French science (and, by extension, modern science in general) for a long time to come. Napoleon's put-down of Bernardin suggested that as far as the pinnacle of state power was concerned, the only science that mattered was mathematically oriented science.

As noted earlier, traditional history of science has for the most part uncritically adopted the viewpoint of the academicians. Like other works of natural philosophy and nonelite science of the same period, *Etudes de la nature* has been treated harshly by most modern commentators. Arthur Lovejoy, for example, cited it as a masterpiece of the teleological and anthropocentric genre that stands as "one of the most curious monuments of human imbecility."[157] Gaston Bachelard pointed to Bernardin de Saint-Pierre as an example of the "nonscientific" or "prescientific" mind.[158] Nonetheless, the dismissal of Bernardin's work as scientifically worthless is unjustified.

Bernardin's ecological approach to natural history led him to conclude that large-scale agricultural practices and industrial activity were detrimental to the natural environment that sustains human life.[159] "What crimes pride and avarice have committed against nature!" he lamented. "Our treatises on agriculture show us nothing in the fields of Ceres but sacks of grain; in the meadows of the nymphs, nothing but haystacks; and in the majestic forests, nothing but log piles and bundles of kindling wood."[160]

This was not a scientific ideology that the industry-boosting Napoleonic state and its elite scientific institutions were likely to find congenial. Bernardin's science appeared incompatible with economic progress. The orthodox scientists did not mount an attack on Bernardin's environmental ethic, however; they ignored it and instead focused on some of his more vulnerable positions.[161]

Clarence Glacken, in a seminal study of the history of environmental thought, noted an important aspect of European science that historians have frequently left out of account: "It is wrong to say that the Western tradition has emphasized the contrast between man and nature without adding that it has also emphasized the union of the two."[162] Glacken listed Bernardin de Saint-Pierre with Linnaeus and Buffon as noteworthy eighteenth-century natural historians in whose hands organic themes flourished.[163]

Glacken's assessment is based not only on Bernardin's carefulness and thoroughness as an observer but on the fact that he was among the first to insist that the proper focus for the study of nature is on the interconnections between organisms and their environment. Bernardin began *Etudes de la nature* by explaining that he had started out several years earlier to write a "general history of nature, in the manner of Aristotle, Pliny, or Bacon."[164] He quickly discovered, however, that "not only the general history of nature, but even that of the smallest plant, is well beyond my ability."[165]

He had reached that conclusion by observing a strawberry plant on his windowsill. To each of the enormous variety of tiny flies that it attracted, it presented a different aspect. How could he ever hope to comprehend the plant from their many perspectives or from those of the microscopic animals to which the strawberry plant was a world in itself?

> But even if, like them, I could have acquired an intimate knowledge of this new world, I still could not hope to know its history. It would have been necessary to study its relationships with the rest of nature, with the sun that makes it flourish, with the winds that spread its seeds, and the streams whose banks it fortifies and embellishes.[166]

Bernardin also recognized that no science of the strawberry plant could be complete without comprehending the environmental factors accounting for its remarkable geographic distribution:

It would have been necessary to discover ... how a fragile creeping plant could ... spread over the globe from north to south, from mountain to mountain ... how it could have extended from the mountains of Kashmir to Archangel, and from the peaks of Norway to Kamchatka; and how, finally, it should be found in both Americas, even though an infinite number of animals make war on it, and no gardener bothers to sow its seeds.[167]

"Everything in nature is linked in a single chain," he declared.[168] One of the primary goals of botany should be to discover the factors that prompted nature to vary each plant's form "and to create so many species of the same genus, and so many varieties of the same species, by coming to understand the wonderful interrelationships they have, in every latitude, with the sun, the winds, the waters, and the soil."[169]

A most important part of an organism's environment, of course, consists of the other living organisms surrounding it:

If it now be considered that all these species, varieties, analogs, and affinities, in every latitude, have necessary relationships with a multitude of animals, and that we are completely ignorant of these relationships, it becomes obvious that the complete history of the strawberry plant would be enough to keep all the naturalists in the world busy.[170]

Even if we were acquainted in infinite detail with the part of which plants are composed, he said, it would be a "useless science"; it is "the harmonies that they form" that are of real interest.[171] Bernardin qualified this harsh assessment of analytic/reductionist science by saying he "did not claim that no useful discovery has ever resulted from these partial methods," but he believed it very wrong to wrap up all of science in them "and to reject in bad faith all that cannot be comprehended by them."[172]

As for the program of investigating nature's interrelationships that he had proposed, "I do not know of anything of this sort that has ever been tried before."[173] He knew such a program could never be completed, but he thought it important at least to make a start. His *Etudes de la nature* represents an admirable beginning. Bernardin's impressive erudition and his insights into environmental interconnections equipped him to make an innovative and potentially important contribution to natural science. The contemporary scientific establishment, however, failed to recognize its value and chose instead to dismiss it with ridicule.

The ridicule was directed first of all against the anthropocentric framework that gave *Etudes de la nature* its structure. The supreme author of nature, Bernardin believed, had made nothing in vain, and nothing in nature exists other than that which is useful to humankind.

Providentialism was not uncommon in the late eighteenth century, even in the writings of "respectable" orthodox naturalists, but most merely tipped their hats to the concept by attributing to divine intervention such major benefits to humankind as wheat or sunlight or air. Bernardin's providentialism was distinctive for its merciless consistency; he insisted that literally everything in the world could be explained as gifts of a beneficent creator. Volcanoes, he asserted, serve to purify the world's waters; earthquakes purify the atmosphere. Scorpions are useful because they frighten us away from damp and unhealthy places. The death of pets helps children learn to deal with grief. Fleas are useful because they oblige the rich to give employment to the poor as domestics to keep their homes and clothes clean. There are no useless pests or harmful weeds in Bernardin's nature.

In the era of Voltaire's Dr. Pangloss this could be expected to invite derision. Yet, as Glacken pointed out, in the eighteenth century naturalists could operate productively "within a teleological framework or outside of it."[174] In his efforts to reveal the details of the workings of providence, Bernardin explored such interesting biological phenomena as mimicry, symbiosis, and protective resemblances, which later theories, following Darwin, would explain in terms of adaptation and survival value.[175] It is also worth noting that although Bernardin's scientific writings have frequently been scorned for their teleological character, Cuvier, the pillar of orthodoxy, likewise built his work in comparative anatomy on a thoroughly teleological foundation.

Bernardin's *Etudes de la nature* included some erroneous geophysical conclusions, but neither they nor his providentialism were sufficient reason to dismiss the work as a whole. The scientific elite's failure to recognize its potential value was a far more grievous error than any Bernardin had committed. By ignoring Bernardin's concept of nature as a delicately balanced, interactive system, the defenders of scientific orthodoxy stifled innovative ecological and environmentalist lines of inquiry in the life sciences that would only much later independently resurface and be accepted into the realm of legitimate science.

The fate of Bernardin's science was largely decided by conservative social pressures rather than on the basis of scientific criteria. It was neglected after Thermidor not because it was worthless but because it was unorthodox. Though he was not at all politically associated with the Jacobin movement, Bernardin shared Rousseau's critique of sterile Enlightenment rationalism, his contempt for academic science, and his insistence that science's ultimate mission is social reform. Bernardin de Saint-Pierre thus fit the profile of the "Jacobin scientist," and his scientific reputation suffered accordingly.

NICOLAS BERGASSE
AND THE MESMERIST MOVEMENT

NICOLAS BERGASSE, AN important political figure during the French Revolution, was also the principal expositor of Mesmerist natural philosophy. His grand synthesis of the laws of nature was perhaps the most influential of the Grub Street systems because it had the force of organization behind it. Mesmerism was a remarkable "people's science" movement that took Paris by storm in the early 1780s and presented a formidable challenge to the scientific orthodoxy of the Academy.[176]

The initiator of the theory and practice of animal magnetism, Franz Anton Mesmer, maintained a certain distance between himself and his public; the appearance of being above worldly concerns was an important element of his personal mystique. Most of the literature of Mesmerism was produced by his followers, of whom Bergasse was the most prominent at a crucial time. Bergasse's writings constitute the most authoritative statement of the Mesmerist natural philosophy of the 1780s, and he should be accorded a large share of the credit for creating rather than merely propagating it. It was he who took Mesmer's principle of a universal vital energy and "made of it a system."[177]

Bergasse was not Mesmer's only spokesman, but in the critical formative years of the movement he held first place in the official ranking.[178] In 1783, he and his close friend Guillaume Kornmann, a wealthy banker, founded the Société de l'Harmonie Universelle (Universal Harmony Society); Bergasse became its main theoretician and organizer. The Universal Harmony Society proved financially and organizationally successful. Within a few years, the Parisian base had several hundred members, and sister sections had formed in Strasbourg, Lyons, Bordeaux, Montpellier, Nantes, Bayonne, Grenoble, Dijon, Marseilles, Castres, Douai, Nîmes, and a number of other towns.[179]

By 1785, however, a split had occurred, and Mesmer had expelled Bergasse, Kornmann, and other prominent members of the organization. Although Bergasse was no longer willing to serve as Mesmer's lieutenant, his dedication to promoting animal magnetism was undiminished. He and Kornmann rallied their supporters and quickly set up a rival organization. Bergasse was its principal leader, but because it was headquartered at Kornmann's home it became known as the "Kornmann group."

The split between Mesmer and the Kornmann group reflected the political tensions of pre-Revolutionary French society. Bergasse and Kornmann wanted to "extend the movement's original fight against the 'despotism of the academies' . . . into the larger battle against political despotism."[180] Mesmer, however, had no interest in alienating Parisian high society; his insistence that the movement remain apolitical made a parting of the ways inevitable.

Although the new group was unable to mount much of an organizational challenge to the mainstream Mesmerist movement, it attracted the most politicized elements away from the Universal Harmony Society and soon emerged as an important focus of revolutionary activity. "The Kornmann group," Darnton wrote, "represents the culmination of mesmerism's movement into politics." Its members numbered among the main leaders of the opposition. "The resistance to the government was led in the Parlement by d'Eprémesnil and Duport, in the Notables by Lafayette, at the Bourse by Clavière, and among the reading public by Brissot, Carra, Gorsas, and Bergasse."[181]

"By the time of its greatest activity, 1787–1789," Darnton explained, the Kornmann group "had neglected mesmerism in order to devote itself fully to the political crisis."[182] Its members successfully "rallied popular support for the Parlement's opposition to the programs of the Calonne and Brienne ministries and the Parlement's call for the convocation of the Estates General."[183]

Jacques-Pierre Brissot, who had joined the Kornmann group in the summer of 1785, later advanced the claim that Bergasse's advocacy of animal magnetism was *primarily* a cover for radical politics: "Bergasse did not hide from me the fact that in raising an altar to mesmerism, he intended only to raise one to liberty." Brissot quoted Bergasse as saying, "It is necessary to unite men under the pretext of experiments in physics, but, in reality, for the overthrow of despotism."[184]

Brissot's testimony would suggest that Bergasse's adherence to Mesmerism was merely tactical. It is more likely that Bergasse perceived it as intertwined with revolutionary organizing as dual means to a single end. Mesmerism was the expression of a Rousseau-inspired natural philosophy that provided Bergasse with an ideological basis and a social theory for revolutionary action.

Bergasse rose to national political prominence in 1787 as a lawyer in a scandalous high-society divorce case involving his friend Kornmann. In the course of the highly publicized proceedings, Bergasse leveled a charge of moral terpitude against the royal government and, echoing Rousseau, against Old Regime society as a whole. Although he eventually lost the divorce case, Bergasse's great talents as an orator and a pamphleteer had made him a hero to opponents of the monarchy. He had championed the Parisian Parlement's battle against the king's ministers; when that struggle resulted in the calling of the Estates General, Bergasse emerged as one of its leaders.

The case of Nicolas Bergasse illustrates perhaps best of all that the radical social and political implications of Grub Street science cannot be linked to any specific brand of radicalism such as Jacobinism. Bergasse represented the "aristocratic rebellion" and turned against the Revolution at the first hints of its democratic character.[185] He withdrew from the National Assembly before the end of 1789 and was soon in the direct service of the Crown. He was arrested during the period of Jacobin rule but by chance survived. Bergasse's transformation from revolutionary to counterrevolutionary was rapid, and his reactionary tendencies continued to deepen with age.[186]

But what makes Bergasse relevant to a people's history of science was his defense of Mesmerism. Before the Revolution, he and his associates devoted a great deal of energy to a public-relations campaign aimed at winning scientific acceptance for animal magnetism. The interest Mesmerism had generated among the Parisian upper crust obliged the Academy of Sciences and the Royal Society of Medicine, after some hesitation, to give it a respectful hearing. The royal government appointed a commission to evaluate it, but the result was not what Bergasse had hoped for. The commission ruled that animal magnetism was nonexistent and that Mesmerism therefore had no scientific basis.[187]

It is evident that some of the royal commissioners had approached the task with less than open minds: "The investigation of animal magnetism thus started with two of its chief members, [Benjamin] Franklin and [Antoine-Laurent] Lavoisier, having to some extent already passed judgement.

The results obtained by the commissioners along these lines were exactly what Lavoisier had foreseen."[188] Indeed, the commission's final report acknowledged that before undertaking their investigation, its members had "armed themselves with that philosophic doubt which ought always to accompany enquiry." Thus armed, they "felt none of those sensations, which were experienced by the three patients of the lower class." Their conclusion that those patients' sensations were "the fruit of anticipated persuasion" might apply equally to their own lack of perceived sensations.[189]

The commission's report was not, however, a purely subjective rejection of animal magnetism. Its success in countering the Mesmerists' influence derived from experiments with blindfolded patients who behaved as if magnetized when they had not been and behaved as if not magnetized when they had been. The claimed effects of animal magnetism were therefore attributed to the power of suggestion on the patients' imaginations.

Bergasse adamantly challenged the academicians' claims that Mesmerism was unscientific. He explicitly denied any imputation of mystical or occult knowledge and insisted on the scientific truth and demonstrability of animal magnetism, which "has wrought a revolution in our entire system of physical knowledge."[190] He presented the system as a whole as deducible from fundamental axioms of Newtonian physics:

> If there is one truth that is no longer challenged in physics, it is that all objects moving through space at some distance exert a mutual force on each other, that this force is stronger or weaker according to how close or far they are apart and to how more or less considerable their mass is. . . . Of all of nature's forces, this one is the most profound, invariable, and universal.[191]

The mutual influences that all beings have on each other were said to be effected by means of a subtle fluid transmitting universal magnetism. Mesmerist therapy consisted in channeling the magnetic fluid through a patient as a means of restoring whatever internal imbalance had resulted in discomfort, pain, and illness.

Often, though not always, the magnetic power was directed into a patient by a laying on of hands. The patient would often report feeling an intense heat in an afflicted part of the body; in many cases patients trembled violently, went into trances, suffered seizures, or fainted dead away. These dramatic "crises" signaled the power of the magnetic force fulfilling its curative function.

The royal commission's report alerted the king to the possibility that Mesmerism could entail dangerous social consequences. Noting that magnetizers were proficient at manipulating the imaginations of excitable subjects and stirring up their passions, it warned, "The same cause is deeply concerned in rebellions; the multitude are governed by the imagination; the individuals in a numerous assembly are more subjected to their senses and less capable of submitting to the dictates of reason."[192] The commissioners also worried about the moral implications of male practitioners laying their hands on excitable females and inducing what can only be described as orgasmic responses. This concern was omitted from the public report for reasons of propriety but was communicated to the king's ministers in a secret memorandum.

Bergasse was especially concerned to refute the royal commission's conclusion that because the effects attributed to animal magnetism could be explained by suggestion, there is no evidence for the existence of animal magnetism. First of all, he said, there is no denying the reality of the power of suggestion, but the commissioners had got things backward. "Suggestion" is not an explanation; it is precisely that which must be explained. Why is it, he asked, that when one is in the midst of an emotional crowd it is difficult to resist sharing the crowd's emotion? This is a case of suggestion, to be sure, but is it not obviously a product of the overwhelming impact of the crowd's collective animal magnetism?

Along the same lines, Bergasse asked why it is "that people who live in the same society are disposed toward receiving, as if involuntarily, the same opinions, the same prejudices, the same habits?"[193] The consideration of "involuntary opinions" could certainly have opened up a new and fertile field of research had it been pursued further. In any case, Bergasse's epistemological point was well taken: attributing physical effects to suggestion raises more questions than it answers.

The physical proof of Mesmer's doctrines, Bergasse contended, lies in the many, many irrefutable cures that his methods had accomplished, not all of which could be attributed to suggestion and imagination. "Teach me, if you will, how the imagination can cure blindness, deafness, wounds, or local paralysis."[194] Finally, he added, it is obvious that cures of infants and animals cannot be explained by imagination.

Bergasse's arguments are not those of a naïve and credulous person but rather of someone unwilling to accept reductionist explanations of complex phenomena. The royal commission's experiments with blindfolds were very clever, but the only explanation the commissioners could offer for cures that Bergasse had witnessed was to declare them fraudulent. From his point of view, to forego the immense benefits of this science on the grounds of a sterile skepticism would be inexcusable.

Although widespread interest in animal magnetism made it a focus of a socially significant movement well into the nineteenth century, after the Revolution it was treated only with derision in official scientific institutions and dismissed out of hand as pseudoscience. The words "Mesmerism" and "animal magnetism" today still evoke smiles and expressions of wonderment at the widespread gullibility they represented. When this condescending reflex reaction is set aside, however, a certain historical injustice is recognizable.

The movement's initial impact and its staying power were based on the ability of Mesmerist practitioners to induce seizures in their patients or to put them in trances. The challenge for science was to explain those powerful effects. When the orthodox scientists determined to their satisfaction that no *external physical force* was responsible for them, they concluded that the phenomena of animal magnetism were unreal and Mesmerism was fraudulent.

Leading historians of science continued to echo that judgment almost two centuries later. Gillispie faulted "the scientific community" for "its maladresse" in answering Mesmer: "At first, in refusing to examine his claims, it took the line that what everyone was tremendously interested in was not worth the attention of serious people, *which was true* but not tactful."[195] Is it true that the dramatic effects demonstrated by the Mesmerists were unworthy of serious consideration? To the contrary, the academicians' myopia in this affair represented a monumental missed opportunity for the extension and advance of science.

As the partisans of animal magnetism correctly argued, the inability to detect a cause cannot be grounds for ruling out the reality of a perceived effect. In the case of Mesmerism, the absence of an external physical cause should not have ended the investigation of effects occurring in human subjects. The fact that Mesmer and Bergasse were mistaken in believing animal magnetism to be an external physical force did not absolve their critics of a responsibility to look further.

By dismissing the effects of animal magnetism as imaginary, the academicians were declaring them unreal, nonexisting, or at least beyond the province of scientific investigation. The effects produced by the Mesmerists on their patients' imaginations, however, were certainly real. "Justly it can be said that Mesmer's discovery of animal magnetism was a pivotal moment in the evolution of modern psychology and psychotherapy. It led to Puysegur's investigation of the consciousness manifested in magnetic sleep [hypnotism] and the eventual discovery of a subconscious realm of mental activity."[196] Historian of medicine Erwin Ackerknecht adds that "psychoanalysts themselves trace modern psychotherapy back to F. Anton Mesmer."[197]

It would be anachronistic in the extreme to fault the royal commissioners for failing to perceive the problem in modern psychological terms, but the point is that their epistemological blinders prevented them from even acknowledging the existence of an important class of phenomena. The Mesmerist natural philosophy stimulated the development of a science of the mind; the skepticism and narrow "physics only" bias of the orthodox academicians was once again a retarding influence.

THE IMPACT OF THERMIDOR
ON FRENCH SCIENCE

THE DISMISSIVE TREATMENT Bergasse and Bernardin de Saint-Pierre received at the hands of the scientific elite was symptomatic of a narrow-mindedness that would continue to afflict the practice of science in France throughout the nineteenth century. Given the high degree of centralization and state control over French science, no conspiracy theory is necessary to make the case that conservative social pressures shaped its course after the Revolution.

The social conservatism of Thermidor engendered scientific conservatism. As the revolutionary tide subsided and radical politics became strongly out of favor in influential circles, the spirit of system likewise became ever more unfashionable in official scientific milieus. Speculation and system-building were associated with socially dangerous habits of thought and were eliminated from official French science, a tendency reinforced by the scientific policies of Napoleon I's empire. The downfall of the Grub Street scientists did not result from their failure to contribute to the growth of knowledge or their inability to survive the rigors of scientific combat; their fate was sealed by their perceived association with political radicalism.

Although the suppression of system-building has traditionally been hailed as indispensable for the progress of science, an opposing argument can be made for its having retarded its growth. These conflicting judgments are not irreconcilable; the demise of unorthodox natural philosophy had genuinely contradictory consequences. The predominance of analytical and quantitative methods did indeed expand the boundaries of some areas of natural knowledge while simultaneously restricting development in others.

The acceleration of specialization was another reflection of the defeat of the Grub Street scientists, who stood for the integration of knowledge as opposed to chopping it up along disciplinary lines. That their holistic perspective was not at all unreasonable serves as a reminder that specialization was a mixed blessing: its benefits have to be weighed against the losses that resulted from shattering knowledge into fenced-off fields.

The complete victory of Lavoisier's heirs in French science after the Revolution can be attributed in part to the social usefulness of their ideology as a set of legitimating ideas supporting Thermidorian and Napoleonic society.[198] The Newtonian credo *Hypotheses non fingo* ("I frame no hypotheses")—by discouraging speculative thinking, by restricting the scope of scientific investigation, and by abstracting phenomena from their context and studying them in isolation—was an ideal basis for a science that would not disrupt existing social arrangements.

Orthodox science was above all insistent on separating science from political, social, and moral issues. Rousseau's program of integrating moral and social values into science was utterly rejected. Morally and socially neutral science was a corollary of the laissez-faire ideology justifying Thermidorian legal equality and social inequality. The campaign against nonelite science was conducted by the leaders of the reconstituted Academy of Sciences, especially Georges Cuvier, with the full support of its most influential member, Napoleon Bonaparte.

GEORGES CUVIER, "LEGISLATOR OF SCIENCE"

IN 1803, GEORGES Cuvier was named permanent secretary of the Academy of Sciences.[199] "This position, perhaps the most coveted in the sciences, gave Cuvier unparalleled opportunity to pass judgment on his predecessors and contemporaries and to legislate the progress of natural history."[200] Fully aware of the importance of the social function he was expected to perform, in 1808 he enumerated the social uses of science in a major report to the emperor:

> To spread healthy ideas among even the lowest classes of people, to remove men from the influence of prejudice and passion, to make reason the arbiter and supreme guide of public opinion: that is the essential goal of the sciences; that is how science will contribute to the advancement of

civilization, and that is what deserves the protection of governments who want to insure the stability of their power.[201]

Cuvier's formulation summarizes the classic themes of the social uses of science; the operative phrases are "healthy ideas," "lowest classes of people," "reason as arbiter of public opinion," and "passion." It is to be inferred that, with regard to the lower classes of people, unhealthy ideas are revolutionary ones. Ideas that challenge the social status quo must be shown to be unscientific, and the dangerous classes must be educated to accept that kind of science. To be dissatisfied with the state of society, to desire social change, or even to think that social change is possible must be demonstrated to be contrary to nature and contrary to reason.

Most of all, "passion" in science or politics must be understood as unhealthy, unnatural, unreasonable, and unscientific. Passion leads to enthusiasm and from there to social unrest. Passion overstimulates the imagination and prompts unreasonable conjectures about what is possible. The intrusion of passion into science stems from Rousseau's admonition that science should above all focus on the most deeply felt human concerns. Passion, in sum, is a source of social evils that must be definitively exorcised from responsible science.

It is noteworthy that Cuvier stressed the social uses of science not as a secondary concern but as *the essential goal* of the sciences." That goal was not knowledge for its own sake or for the satisfaction of curiosity, nor technological advance for the sake of material betterment, but the *ideological conquest of public opinion in the interests of stabilizing the state power.*

Cuvier's report emphasized the social dangers of the speculative spirit in science. He had been "sufficiently chastened by the experience of the French Revolution to begin identifying unbridled scientific theorizing with social and moral chaos":

A political elitist who advocated government by expert administrators, Cuvier feared above all the return of revolution and rule by the uncontrolled mob. Scientific ideas, if unrestrained, might be exploited by unscrupulous men to weaken the social order. Thus, it was partly for political reasons that Cuvier insisted so strongly on limiting science to "positive facts."[202]

Cuvier decreed what would and would not thenceforth qualify as science. "All the hypotheses, all the more or less ingenious suppositions," he wrote, "are today rejected by real scientists." On the other hand, "experiment alone, precise experiments with scales, measurements, calculations and comparisons of all the substances used and obtained: that is today the only legitimate path of reasoning and demonstration."[203] There was more than a touch of the double standard evident in this proclamation. His own scientific work "was far from empirical," but "Cuvier adopted and promulgated an empiricist ideology in order to combat opposing points of view."[204]

In 1797 Napoleon Bonaparte was himself elected to membership in the reorganized Academy of Sciences, in the section of Mechanical Arts. In spite of Bonaparte's evident interest in science, his election to its leading institution was obviously not based on considerations of scientific merit. It stemmed from a recognition by the Academy that he was in a position to advance its interests. His joining the Academy did not initiate the merger of political power with institutional science but certainly extended and consolidated it.

The Napoleonic intervention in post-Revolutionary scientific politics need not be interpreted according to the Great Man Theory of History. He did not create either the scientific elite or its conservative scientific ideology, nor did he bring about its dominance by imperial fiat. By showering his favor on it, however, he did contribute greatly to augmenting its social weight, thereby ensuring its continued hegemony.

In summary, the French Revolution qualitatively transformed all aspects of human culture, including science, for better and for worse. The institutional and ideological changes wrought in French science by the Revolution and its aftermath shaped the subsequent course of modern science everywhere. The essential underlying factor, as the Hessen thesis maintains, was the victory of capitalism, which the Revolution consolidated. The new social order spread to Europe and the rest of the world, everywhere subordinating the further development of science to capitalist interests.

NOTES

1. Pamela H. Smith, "Vital Spirits," p. 135.
2. Margaret C. Jacob, *The Cultural Meaning of the Scientific Revolution*, p. 105.
3. Galileo Galilei, *Sidereus nuncius;* Galileo, *Istoria e dimostrationi intorno alle macchie solari.*
4. John Robert Christianson, *On Tycho's Island*, p. 168.
5. Ibid., pp. 5–6.
6. Ibid., p. 237.
7. William Eamon, *Science and the Secrets of Nature*, p. 146 (see also pp. 224–225).
8. Robert Boyle, *The Works*, vol. I, pp. cxxx–cxxxi.
9. Eamon, *Science and the Secrets of Nature*, pp.137, 139, 222.
10. Ibid., pp. 219–220.
11. Ibid., pp. 201, 402, n. 38.
12. Ibid., p. 148. Ruscelli's description of the Accademia Segreta appeared in his *Secreti nuovi di maravigliosa virtù*, published posthumously in 1567.
13. Eamon, *Science and the Secrets of Nature*, p. 149. Eamon's source is Pompeo Sarnelli's biography of Della Porta.
14. See chapter 5.
15. Eamon, *Science and the Secrets of Nature*, p. 231.
16. The Lincei went out of existence in 1629, but it was nonetheless "the true precursor of the Cimento." W.E. Knowles Middleton, *The Experimenters*, p. 7.
17. Quoted and translated by Middleton, *Experimenters*, p. 53. The original source is the Accademia del Cimento's sole publication, *Saggi di naturali esperienze* (1667).
18. Derek J. de Solla Price, *Science Since Babylon*, pp. 54–55.
19. Christopher Hill, "Newton and His Society," p. 30.
20. Stephen Pumphrey, "Who Did the Work?" pp. 139, 152.
21. Peter Linebaugh and Marcus Rediker, *The Many-Headed Hydra*, p. 123.
22. Joseph Glanvill, *Plus Ultra*, pp. 92–109.
23. Thomas Sprat, *The History of the Royal-Society of London*, p. 35.

24. Cowley's ode was published as a preface to Sprat, *History of the Royal-Society.*

25. Charles Webster, *From Paracelsus to Newton*, p. 96.

26. As Christopher Hill has shown in *The World Turned Upside Down*, this phrase was commonly used in the seventeenth century to denote what we would call a social revolution.

27. James Jacob, *The Scientific Revolution*, pp. 95–96.

28. Hill, *World Turned Upside Down*, p. 296.

29. Ibid., p. 301.

30. Gerrard Winstanley, *The Law of Freedom* (1652), quoted in Hill, *World Turned Upside Down*, p. 287.

31. Hill, *World Turned Upside Down*, p. 304.

32. See P. M. Rattansi, "Paracelsus and the Puritan Revolution."

33. Jacob, *Scientific Revolution*, p. 102.

34. Hill, *World Turned Upside Down*, p. 298. For more on Culpeper, see F. N. L. Poynter, "Nicholas Culpeper and His Books."

35. Jacob, *Scientific Revolution*, pp. 102–103.

36. Hill, *World Turned Upside Down*, p. 296.

37. Ibid., pp. 304–305.

38. Ibid., p. 305. The quotation within the quotation is from Charles Webster, "Science and the Challenge to the Scholastic Curriculum, 1640–1660," in *The Changing Curriculum* (History of Education Society, 1971), pp. 32–34.

39. Quoted in Henry G. Lyons, *The Royal Society, 1660–1940*, p. 41.

40. Christopher Hill, "Science and Magic," p. 287.

41. Catherine Drinker Bowen, *Francis Bacon*, p. 26.

42. Francis Bacon, "Of Seditions and Troubles," p. 32.

43. Julian Martin, "Natural Philosophy and Its Public Concerns," p. 101.

44. He did not remain at the peak of political power, as is well known, but there is no need to go into that here.

45. Martin, "Natural Philosophy and Its Public Concerns," pp. 105, 108.

46. The responsibility for creating that image lies ultimately with the leading figures of the Enlightenment themselves—Voltaire, Diderot, d'Alembert, Rousseau, Condorcet—who hailed Bacon as the greatest of their philosophical predecessors. See, e.g., d'Alembert's introduction to the *Encyclopédie.*

47. Loren Eisley, *The Man Who Saw Through Time*, p. 22; Benjamin Farrington, *Francis Bacon.*

48. Linebaugh and Rediker, *Many-Headed Hydra*, pp. 19, 37, 65.

49. See the epigraph at the beginning of this chapter.

50. The quotations from Bacon in this and the following paragraph are from Carolyn Merchant, *The Death of Nature*, pp. 168–172. The particular works of Bacon that she cites are "The Masculine Birth of Time," "De Dignitate et Augmentis Scientiarum," and "Thoughts and Conclusions on the Interpretation of Nature or A Science of Productive Works."

51. H. R. Trevor-Roper, "The European Witch-Craze of the Sixteenth and Seventeenth Centuries," pp. 90–91.
52. Brian Easlea, *Witch Hunting, Magic and the New Philosophy*, p. 33.
53. Trevor-Roper, "European Witch-Craze," p. 91.
54. A. L. Morton, *A People's History of England*, p. 146. The qualifying words "in part" in this sentence are important. Morton overstated the extent to which the witch craze was a reaction to a vast organized secretive social movement. That thesis, propounded most thoroughly by M. A. Murray in *The Witch-Cult in Western Europe* and *The God of the Witches*, seemed tenable at the time Morton wrote, but it has since been refuted. See Keith Thomas, *Religion and the Decline of Magic*, pp. 514–516.
55. Thomas, *Religion and the Decline of Magic*, p. 456, paraphrasing the views of R. H. Robbins, *Encyclopedia of Demonology and Witchcraft* (1959).
56. Trevor-Roper, "European Witch–Craze," p. 91, 122.
57. James published his *Daemonologie* in 1597. His passionate belief in witches cooled considerably and even evolved into skepticism in later years.
58. Trevor-Roper, "European Witch-Craze," p. 151. Suspected witches were executed by burning in Continental Europe and by hanging in England.
59. Merchant, *The Death of Nature*, p. 138.
60. Easlea, *Witch Hunting, Magic and the New Philosophy*, p. 8.
61. Kramer and Sprenger, *Malleus maleficarum*; quoted in Easlea, *Witch Hunting, Magic and the New Philosophy*, p. 8.
62. H. C. E. Midelfort, *Witch Hunting in Southwestern Germany 1562–1684*, p. 192.
63. Webster, *From Paracelsus to Newton*, p. 99.
64. Ibid., pp. 92–93.
65. Ibid., p. 93.
66. The retired revolutionary was John Webster, who had himself become a Royal Society member. He took Glanvill to task in *Displaying of Supposed Witchcraft* (1677). See Webster, *From Paracelsus to Newton*, pp. 96–100.
67. Webster, *From Paracelsus to Newton*, p. 100.
68. Francis Bacon, *Advancement of Learning* (1605), quoted in Easlea, *Witch Hunting, Magic and the New Philosophy*, p. 38.
69. Easlea, *Witch Hunting, Magic and the New Philosophy*, p. 38.
70. Quoted in Vern L. Bullough, *The Development of Medicine as a Profession*, p. 95.
71. M. J. Hughes, *Women Healers in Medieval Life and Literature*, p. 85.
72. "An Acte Concernyng the Approbation of Phisicions and Surgions" (3 Hen. VII, cap. XI); quoted in Sidney Young, *Annals of the Barber-Surgeons of London*, pp. 72–73.
73. Linebaugh and Rediker, *Many-Headed Hydra*, p. 92. The source of the quotations within the quotation is Silvia Federici, "The Great Witch Hunt," *The Maine Scholar* (1988).
74. Webster, *From Paracelsus to Newton*, p. 100.

75. The contents of Glauber's *Tractatus de natura salium* (1658) are described in Smith, "Vital Spirits."

76. Smith, "Vital Spirits," p. 120.

77. Ibid., p. 121.

78. Ibid., p. 120. The source of the quotation from Glauber is *Tractatus de natura salium* (Smith's translation; emphasis in original).

79. Smith, "Vital Spirits," pp. 124, 135.

80. Ibid., p. 130.

81. Myles W. Jackson, "Can Artisans Be Scientific Authors?" p. 113.

82. Ibid., p. 114.

83. Ibid., pp. 118–119.

84. Ibid., pp. 123–124.

85. Ibid., p. 125.

86. Quoted in Jackson, "Can Artisans Be Scientific Authors?" p. 126.

87. Jackson, "Can Artisans Be Scientific Authors?" pp. 120–121.

88. Max Weber preceded Hessen in connecting the rise of science with the rise of capitalism, but Weber's primary focus was on Protestantism, whereas Hessen placed capitalism at center stage.

89. B. J. T. Dobbs, "Newton as Final Cause and First Mover," p. 29.

90. Boris Hessen, *The Social and Economic Roots of Newton's "Principia."* Originally published as a contribution to a collection of papers, *Science at the Crossroads* (1931).

91. Hessen, *Social and Economic Roots*, pp. 5, 24.

92. Ibid., p. 6.

93. Ibid., p. 9.

94. Ibid., p. 21.

95. Ibid., p. 26.

96. Newton to Aston, May 18, 1669, in Newton, *Correspondence*, vol. 1, pp. 9–11. I have modernized some of the spelling.

97. Ibid., pp. 9–10.

98. Ibid., p. 11.

99. Ibid., p. 11.

100. Ibid., p. 11.

101. On the uniqueness of the letter, see Richard S. Westfall, *Never at Rest*, p. 193. Westfall suggested that Newton may never have actually sent this letter to Aston.

102. Margaret C. Jacob, *The Newtonians and the English Revolution*; Jacob, *The Cultural Meaning of the Scientific Revolution*.

103. Rattansi, "Paracelsus and the Puritan Revolution," p. 24.

104. On the other side, Newtonian ideology was useful in combating materialists like Hobbes who sucked too much spirit out of nature, thereby calling the existence of God into question. See James R. Jacob and Margaret C. Jacob, "Anglican Origins of Modern Science," p. 257.

105. James R. Jacob, "'By an Orphean Charm,'" p. 242.

[106.] See esp. Henry Guerlac, *Newton on the Continent*.

[107.] The quoted phrase, from R. R. Palmer's preface to his translation of Georges Lefebvre's classic *Quatre-vingt-neuf*, represented the general assessment of historians until challenged by a "revisionist" school of thought initiated by Alfred Cobban's "Myth of the French Revolution" lecture in May 1954; see Cobban, *The Social Interpretation of the French Revolution*. I find the revisionist challenge thoroughly unconvincing and continue to consider the Revolution a decisive historical turning point.

[108.] This line of historical interpretation was solidly established by 1803, as evidenced by the physicist Jean Baptiste Biot's *Essai sur l'histoire générale des sciences pendant la Révolution française*.

[109.] Biot, *Essai sur l'histoire générale des sciences pendant la Révolution française*, pp. 74ff.; Donald M. Vess, *Medical Revolution in France, 1789–1795*.

[110.] James McClellan, *Science Reorganized*, p. 257. See also Roger Hahn, *The Anatomy of a Scientific Institution*, pp. 264, 285.

[111.] This assertion is defended in "The Impact of Thermidor on French Science" in this chapter. The Thermidorian reaction takes its name from Thermidor, the month in the French Revolutionary calendar in which Robespierre and the Jacobin Republic fell.

[112.] Hahn, *Anatomy of a Scientific Institution*, p. 4.

[113.] Ibid., pp. 224–225.

[114.] Christiaan Huygens, *Oeuvres complètes*, vol. IV, p. 328. Quoted in Hahn, *Anatomy of a Scientific Institution*, p. 12.

[115.] Hahn, *Anatomy of a Scientific Institution*, p. ix.

[116.] Ibid., p. 15.

[117.] Ibid., pp. 41, 76, 80.

[118.] Ibid., p. 35.

[119.] d'Alembert, *Preliminary Discourse*, p. 42.

[120.] Hahn, *Anatomy of a Scientific Institution*, p. 41.

[121.] Ibid., p. 45.

[122.] Ibid., p. 68.

[123.] Ibid., p. 68. The author of the proposal was probably René-Antoine Ferchault de Réamur.

[124.] Ibid., pp. 66, 68. The word "patent" is used here to indicate what contemporaries would have called a royal *privilège*.

[125.] Ibid., p. 67.

[126.] C. C. Gillispie, "The *Encyclopédie* and the Jacobin Philosophy of Science," pp. 270–271.

[127.] Ibid., pp. 271–272.

[128.] See, e.g., A. Rupert Hall, "Epilogue" in *From Galileo to Newton*. A more thorough exposition (although I certainly do not place these authors in the category of "defenders of heroic history of science") is in James McClellan and Harold Dorn, *Science and Technology in World History*.

[129.] Hahn, *Anatomy of a Scientific Institution*, p. 185.

130. Ibid., p. 185.

131. Ibid., pp. 176–177.

132. Ibid., p. 186.

133. The Société described itself as *"composé de tous artistes vrai sans-culottes."* Quoted in Gillispie, "The *Encyclopédie* and the Jacobin Philosophy of Science," p. 274.

134. Hahn, *Anatomy of a Scientific Institution*, p. 192.

135. Ibid., p. 240.

136. Ibid., pp. 215–216. The Point Central's pamphlet was *A tous les artistes et autres citoyens.* See Maurice Tourneux, *Bibliographie de l'histoire de Paris pendant la Révolution français* (Paris, 1890–1913), vol. III, p. 661.

137. Hahn, *Anatomy of a Scientific Institution*, pp. 83, 275, 288.

138. Ibid., pp. 303, 318.

139. Toby Appel, *The Cuvier-Geoffroy Debate*, p. 8.

140. Robert Darnton, *The Literary Underground of the Old Regime*. See also Clifford Conner, "Jean Paul Marat and the Scientific Underground of the Old Regime."

141. I will use the term "Grub Street scientists" in spite of its somewhat condescending overtones because of its usefulness for collectively denoting the "outsiders," the nonelite natural philosophers. Among the Grub Street literati who might also be considered Grub Street scientists were Jean Paul Marat, Jacques-Pierre Brissot, Jean-Louis Carra, Louis Sébastien Mercier, Nicolas Edmé Restif de la Bretonne, Jean Jacques Fillasier, and Louis Jacques Goussier.

142. For example: "The totality and consequences of necessary laws constitute mechanics. We call the totality of the other laws the 'System of the World,' which we cannot know entirely without knowing all phenomena." Condorcet, *Lettre sur le Système du Monde et sur le Calcul Intégral* (1768), quoted in Roger Hahn, *Laplace as a Newtonian Scientist*, p. 16.

143. *Observations sur la physique*, December 1781, p. 503. This journal is better known by its later name, the *Journal de physique*. Ironically, it was soon to come under the editorship of Jean-Claude Delamétherie, who would himself prove to be one of the most prolific system-builders of the 1780s.

144. d'Alembert, *Preliminary Discourse*, p. 23. The opposition to system-building did not originate with d'Alembert; he was restating the views of Fontenelle, Condillac, and others. "Esprit" can be translated either "spirit" or "mind"; the ambiguity of the French word served d'Alembert's purpose.

145. Letter of March 13, 1769, from Nollet to E. F. Dutour, quoted in J. L. Heilbron, *Elements of Early Modern Physics*, p. 69.

146. C. C. Gillispie, *The Edge of Objectivity*, p. 199; also Gillispie, "*Encyclopédie* and Jacobin Philosophy," p. 282 (emphasis added).

147. Jacques Henri Bernardin de Saint-Pierre, *Etudes de la nature*, vol. I, pp. 39–40, 47.

148. Hahn, *Anatomy of a Scientific Institution*, p. 156.

149. *Paul et Virginie* first appeared in volume IV of the third edition of *Etudes de la nature,* published in 1788. The first separate edition of the novel was published in 1789. It is available in English translation as *Paul and Virginia.*

150. The word "reductionist" means many things to many people. It has in recent years been misused to characterize proponents of social history of science. As Bruno Latour noted, "to those who reject 'social' explanations of science . . . all social studies of science are thought to be reductionist and are held to ignore the most important features of science." Latour, *The Pasteurization of France,* p. 153. In this study, however, "reductionist" is used to denote the tendency to reduce all of the laws of nature to the laws of physics.

151. Maurice Souriau, *Bernardin de Saint-Pierre d'après ses manuscrits,* p. 255.

152. John Nottingham Ware, "The Vocabulary of Bernardin de Saint-Pierre and Its Relation to the French Romantic School," p. 2.

153. After the Revolution, Bernardin continued to gather material for another multivolume work of natural philosophy, which would be published after his death as *Harmonies de la nature.* His posthumous reputation was considerably distorted by an adoring disciple, Aimé Martin, who sought to defend his master in conservative times by editing all of his writings to cleanse them of their anticlericalism and other radical implications. Martin was a legitimist before the Restoration; he integrated his own political and religious convictions into Bernardin's works. Only the earlier, unbowdlerized editions of *Etudes de la nature,* then, can be considered trustworthy representatives of Bernardin's thought. Maurice Souriau carefully documented and analyzed what he called "les méfaits de Aimé Martin." Souriau, *Bernardin de Saint-Pierre d'après ses manuscrits.*

154. Augustine Dirrell, "Preface" to Arvède Barine, *Bernardin de Saint-Pierre.* Arvède Barine was the pseudonym of Cécile Vincens.

155. Hahn, *Anatomy of a Scientific Institution,* 183.

156. E. de las Casas, *Le mémorial de Sainte-Hélène;* quoted in Nicole Dhombres and Jean Dhombres, *Naissance d'un nouveau pouvoir,* p. 685.

157. Arthur O. Lovejoy, *The Great Chain of Being,* 186.

158. Gaston Bachelard, *The Psychoanalysis of Fire,* pp. 28–30. Bachelard accorded the late-eighteenth-century system-builders serious attention and respect but insisted that the significance of their writings is as poetry rather than science; as a source of psychoanalytic insights rather than as material useful to the historian of science. He characterized them as "muddled thinkers" and examples of their work as "nonsense," "stupid observations," and "ridiculous statements."

159. It may seem anachronistic to use the word "ecology" with regard to Bernardin, but, as Donald Worster noted, "the idea of ecology is much older than its name. Its modern history begins in the eighteenth century, when it emerged as a more comprehensive way of looking at the earth's fabric of life: a point of view that sought to describe all of the living organ-

isms of the earth as an interacting whole." Worster, *Nature's Economy*, p. xiv.

160. Bernardin de Saint-Pierre, *Etudes de la nature*, vol. 1, p. 36.

161. Bernardin's critics targeted two prominent geophysical errors in *Etudes de la nature* in order to discredit the entire work. The first error was his claim that oceanic tides were caused not by the moon's gravitational influence but by the partial melting and refreezing of the ice at the North and South Poles. The second error was his insistence that the earth was elongated rather than flattened at the poles.

162. Clarence J. Glacken, *Traces on the Rhodian Shore*, p. 550.

163. Ibid., p. 501.

164. Bernardin de Saint-Pierre, *Etudes de la nature*, vol. 1, p.1.

165 Ibid., p. 2.

166. Ibid., pp. 10–11.

167. Ibid., p. 11.

168. Ibid., p. 60. It should be noted that Bernardin explicitly disavowed the doctrine of the Great Chain of Being. His use of the chain metaphor simply called attention to the bonds of dependency that link all life forms.

169. Ibid., p. 53.

170. Ibid., p. 13.

171. Ibid., p. 18.

172. Ibid., p. 37.

173. Ibid., p. 18.

174. Glacken, *Traces on the Rhodian Shore*, p. 548.

175. See Louis Roule, *Bernardin de Saint-Pierre et l'harmonie de la nature*.

176. For a social history of Mesmerism as a people's science movement, see Robert Darnton, *Mesmerism and the End of the Enlightenment in France*.

177. Etienne Lamy, "Introduction," in Aristide G. H. N. Bergasse du Petit-Thouars, *Nicolas Bergasse*, p. x.

178. Mesmer presented diplomas to each of his lieutenants that "certified his place in the hierarchy of disciples: Bergasse was first." Darnton, *Mesmerism*, p. 75.

179. Ibid., p. 51.

180. Ibid., p. 78.

181. Ibid., p. 163.

182. Ibid., p. 79.

183. Ibid., p. 88.

184. J. P. Brissot, *Mémoires*, vol. II, pp. 53–56; quoted in Darnton, *Mesmerism*, p. 79. Darnton warned that these recollections of Brissot's should not be accepted uncritically because Brissot, writing while facing charges of counterrevolutionary conspiracy, had good reason to portray his past as completely devoted to the Revolution.

185. The French monarchy, in the 1780s, was in the grip of a deepening fiscal crisis. The Crown's frantic search for new sources of revenue forced it to turn to the aristocracy, which had traditionally enjoyed extensive exemp-

tion from taxation. In spite of their privileged social position, the aristocrats had long been excluded as a class from a direct political role in governing the country. The threat to their economic interests prompted them to begin agitating for political rights as a means of self-defense. The "aristocratic rebellion" succeeded in shaking the monarchy, but in doing so, it opened the door for other social layers to begin raising demands as well. The great irony of the French Revolution is that the social class that initiated it was the one that was ultimately destroyed by it.

186. After marching into Paris at the head of his triumphant armies in 1815, Tsar Alexander I of Russia visited Bergasse and recognized in him an ideological soul mate. They shared an obsessive fear (by no means purely delusional) of the return of revolution throughout Europe. Alexander, a mystic with messianic ambitions, spearheaded the formation of a Holy Alliance of the thrones of Russia, Prussia, and Austria as a "Christian answer" to the French Revolution. Bergasse, as the Tsar's confidant and spiritual advisor, contributed to shaping the ideology and policies of that counterrevolutionary crusade.

187. There were in fact two such commissions; all references here refer to the more important one, which issued a report in August 1784 over the signatures of Franklin, Lavoisier, Bailly, Le Roy, Guillotin, D'Arcet, Sallin, De Bory, and Majault. Quotations from that report are from the English translation published in London in 1785 as *Report of Dr. Benjamin Franklin and Other Commissioners Charged by the King of France, with the Examination of the animal Magnetism as Now Practised at Paris.*

188. Denis I. Duveen and Herbert S. Klickstein, *Bibliography of the Works of Antoine Laurent Lavoisier, 1743–1794*, pp. 249–261.

189. Anonymous, *Report of Dr. Benjamin Franklin and Other Commissioners*, p. 54.

190. Nicolas Bergasse, *Considérations sur le Magnétisme animal*, p. 60.

191. Ibid., p. 38.

192. Franklin report. See note 187.

193. Nicolas Bergasse, *Considérations sur le Magnétisme animal*, p. 54.

194. Ibid., p. 131.

195. Gillispie, "*Encyclopédie* and Jacobin Philosophy," p. 267 (emphasis added).

196. Adam Crabtree, *Animal Magnetism, Early Hypnotism, and Psychical Research 1766–1925*, p. xi.

197. Erwin H. Ackerknecht, *A Short History of Medicine*, p. 207.

198. Keith Baker has argued that science offered the postrevolutionary statesmen "more than a repertory of technology derived from scientific knowledge. It also held out the potential for a new source of legitimacy, a system of authority resting on principles of reason and nature." Baker, *Inventing the French Revolution*, p. 159.

199. There were at that time two *sécretaires perpétuels* of the Academy; the other was J. B. Delambre.

200. Appel, *Cuvier-Geoffroy Debate*, p. 42.

201. "répandre des idées saines jusque dans les classes les moins élevées du peuple; soustraire les hommes à l'empire des préjugés et des passions; faire de la raison l'arbitre et le guide suprême de l'opinion publique, voilà l'objet essentiel des sciences; voilà comment elles concourent à avancer la civilisation, et ce qui doit leur mériter la protection des Gouvernemens qui veulent rendre leur puissance inébranlable." Georges Cuvier, *Rapport historique sur les progrès des sciences naturelles*, p. 387.

202. Appel, *Cuvier-Geoffroy Debate*, pp. 9–12.

203. Cuvier, *Rapport historique sur les progrès des sciences naturelles*, p. 389.

204. Appel, *Cuvier-Geoffroy Debate*, p. 41.

7

THE "UNION OF CAPITAL AND SCIENCE"

THE NINETEENTH CENTURY

THE LAWS OF commerce are the laws of Nature, and therefore the laws of God.
—EDMUND BURKE, *Thoughts and Details on Scarcity* (1800)

THE GROWTH OF a large business is merely the survival of the fittest.
—JOHN D. ROCKEFELLER (c. 1900)

BY THE INFIRMITY of nature it happens, that the more skilful the workman, the more self-willed and intractable he is apt to become, and, of course, the less fit a component of a mechanical system. . . . The grand object therefore of the modern manufacturer is, through the union of capital and science to reduce the task of his work-people to the exercise of vigilance and dexterity.

. . . how surely science, at the call of capital, will defeat every unjustifiable union which the labourers may form.

. . . when capital enlists science in her service, the refractory hand of labour will always be taught docility.
—ANDREW URE, *The Philosophy of Manufactures* (1835)

T HE "UNION OF capital and science" was not and is not an alliance of equals; it has always been a master–servant relationship, with capital as the dominant partner. The dazzling successes of modern science have created the illusion that it is an autonomous factor driving the process of historical change, but as Andrew Ure correctly observed, science has long been "at the call of capital" and "in her service."[1] Today, the production of knowledge occurs on an industrial scale in science factories, also known as research laboratories. Almost all scientific research is the work of professional scientists either directly employed or indirectly funded by capitalist corporations and governments.[2]

As a consequence, knowledge and even nature itself have become increasingly "commodified"—converted into things that can be bought and sold. Scientific knowledge production in the nineteenth and twentieth centuries was shaped not by considerations of human needs but by the profit motive. This tends to be masked by the dogma of scientific neutrality—that the impartiality of scientists shields their findings from external influences and guarantees that the content of modern science is an ever-expanding ocean of objective truth. Obvious counterexamples, however, are plentiful. Is it not notorious that the medical research performed by pharmaceutical companies routinely subordinates human well-being to narrow proprietary interests? Are the tobacco industry's "scientific" studies suggesting smoking is neither cancer-causing nor addictive not laughable? As we approach the present, then, the attention of a people's history of science must increasingly be directed toward popular efforts to rein in a socially blind Big Science.[3]

But the dominance of Big Science developed over more than two hundred years. In the nineteenth century, it was still possible for people who were shunned by the scientific establishment to make momentous contributions to knowledge of nature. In contrast to their earlier counterparts, these were not, for the most part, independent artisans but wageworkers in the employ of capitalists. In England, a self-taught surveyor working for coal and canal companies, William Smith, revolutionized knowledge about the geological processes that formed the earth, and nonelite medical practitioners—"medical democrats"—laid the groundwork for evolutionary biology well in advance of Darwin.

Before those stories are recounted, however, plebeian groups of even greater historical significance deserve the spotlight: the miners, metalworkers, and mechanics who created the knowledge that gave rise to the Industrial Revolution. They merit a prominent place in the history of science for many reasons, not least of which is that the Industrial Revolution was the takeoff point for the "miracles" of modern technology customarily misattributed to theoretical science.

TWO PROBLEMS

WHAT, EXACTLY, WAS the crucial knowledge without which the Industrial Revolution could not have been launched in England in the decades just before and after the turn of the nineteenth century? In retrospect, two interrelated problems stand out. First of all, industrialization required a rapidly increasing supply of iron, but the availability of that all-important material was sharply limited by a shortage of the energy source needed to produce it: charcoal.

Even a century earlier, fast-rising demand for wood to build ships and to make charcoal for fuel had caused a rapid depletion of England's forests, creating a serious energy crisis. Some two hundred acres of trees had to be cut down yearly to produce the charcoal necessary to keep a single smelting furnace in operation. An alternative fuel was known—coal—but in its natural state it was unsuitable for smelting iron. The first crucial problem, then, exposed a deficiency in knowledge of the properties of coal and iron: Could coal be altered to make it into a usable fuel for the production of iron—and if so, how could it be done?

Once that knowledge was gained, it led to a surge in demand for coal, which posed another problem: as more coal was needed, miners had to dig ever deeper into the earth to get it, but the deeper they went, the more likely their mine shafts were to be flooded. The prevailing knowledge of how to raise large quantities of water was inadequate; coal production would have come to a standstill without the creation of a new kind of pump designed according to new principles of physics.

The solution to the second problem—pumps powered by steam and atmospheric pressure—proved to be of far greater consequence than its creators could have envisioned. It led to the invention of the "workhorse of the Industrial Revolution"— the steam engine—which provided a source of virtually unlimited power for machinery capable of vastly expanding industrial output.

The knowledge of natural processes that facilitated the Industrial Revolution opened the way to the high-tech world of today. Who was responsible for creating it? "All of the technical innovations that formed the basis of the Industrial Revolution," James McClellan and Harold Dorn explained, "were made by men who can best be described as craftsmen, artisans, or engineers. Few of them were university educated, and all of them achieved their results without the benefit of scientific theory."[4] The engineers alluded to here were not members of an elite profession. In eighteenth-century Britain, "engineers for the most part began as simple workmen, skilful and ambitious but usually illiterate or self-taught. They were either millwrights like [Joseph] Bramah, mechanics like [William] Murdoch and George Stephenson, or smiths like [Thomas] Newcomen and [Henry] Maudslay."[5]

That those craftsmen and engineers approached and solved problems empirically rather than theoretically does not mean their accomplishments are less worthy of being deemed scientific. The theories that eventually explained the behavior of gases and the principles of atmospheric pressure and thermodynamics were generated by thinkers trying to understand— after the fact—what artisans had created by experimental means. The laws of conservation and transformation of energy, for example, were not discovered by disinterested theoreticians motivated by sheer intellectual curiosity; they were the fruit of deliberate attempts to increase the efficiency (and hence the profitability) of steam engines. Once again, the history of science is founded on empiricism, not theory; *in the beginning was the deed, not the word.*

FROM COAL TO COKE

DISCOVERY OF AN alternative fuel for smelting iron is usually attributed to Abraham Darby, and not without justification. It was Darby, the proprietor of a small ironworks, who in 1709 produced the first iron of good quality using coke rather than charcoal. Coke is to coal as charcoal is to wood; it is coal that has been partially burned in a controlled manner in order to drive out impurities and make it a purer fuel.

"Neither scientific theory nor organized or institutional science played any role" in Darby's achievement, said McClellan and Dorn. "Applicable theoretical principles of metallurgy had not yet come into being, and even 'carbon' and 'oxygen' were entities yet to be defined. As a typical artisan-engineer, Darby left no record of his experiments or, rather, tinkering."[6] It was, in other words, the experimenting of "tinkerers" rather than the theorizing of established scientists that produced this immeasurably important body of knowledge about the properties of iron and coal under varying conditions of heat. Would carbon have ever been "defined" if artisans had not first learned how to isolate it in the form of charcoal and coke? As for oxygen, the role of another category of artisans in the discovery of gases will be discussed later in this chapter.

It does not diminish Darby's accomplishment to acknowledge that he was influenced and aided by many predecessors, collaborators, and successors. Brewers constituted an occupational group who made a particularly significant contribution. Like other tradespeople, beer producers had long felt economic pressure to find an alternative to charcoal to fuel their malt-drying ovens. Coal was not the answer, because the sulfur it contained imparted a disagreeable taste to the beer. But as early as 1603, Hugh Plat (the compiler of artisanal science introduced in a previous chapter) proposed subjecting coal to a charring process analogous to that used in making charcoal from wood. That was, however, easier said than done; it took another four decades before a workable method of charring coal was first put into practice by the brewers of Derbyshire. Although coke then became the fuel of choice for brewers, its use was for the most part restricted to that industry.

And then along came Abraham Darby. As a young man Darby was apprenticed to a company in Birmingham that manufactured malt-grinding mills for making beer. In addition to learning metal-casting techniques, he also familiarized himself with the brewers' craft. It was Darby's experience with the brewers' alternative fuel that led him to experiment with its use for smelting iron.

Darby's successful 1709 demonstration was but the starting point of a "cumulative accomplishment" that took his family firm, the Coalbrookdale Company, sixty years to complete. One historian of technology has credited the Darby "multigenerational family partnership" with providing

> the only way in which, in the iron industry at that time, a stable cadre of mechanicians and managers could be assembled and perpetuated over several generations, passing on the paradigm from one cohort to the next. . . . The significant thing about the Darbys and coke-iron is not that the first Abraham Darby "invented" a new process but that five generations of the Darby connection were able to perfect it and develop most of its applications.[7]

"RAISING WATER BY FIRE"

MECHANICS AND THEORETICIANS alike devoted a great deal of attention to the problem of raising water to expel it from deep mine shafts. The first working steam-powered pump—Thomas Savery's in 1698—was designed on the basis of theoretical principles. Although Savery's device represented a significant attempt at a solution, it was a practical failure in the mines.

More than a decade later, a blacksmith, Thomas Newcomen, in collaboration with a plumber, John Calley, produced the first commercially successful machine for "raising water by fire."[8] Newcomen's "contemporaries were as much surprised as we," said Lynn White, "that a provincial craftsman who lacked all contact with, or knowledge of, the Savery engine or the scientific researches on which Savery's work was based, could have solved such a problem as harnessing steam."[9] Their surprise was a product of social prejudice: "the scientists of the day looked upon the practical man, who lived by the skill of hand and eye, with a certain arrogance and contempt that is typified in their attitude towards Thomas Newcomen."[10]

"There is no machine or mechanism," an early-nineteenth-century commentator said of the steam engine, "in which the little the theorists have done is more useless. It arose, was improved and perfected by working mechanics—and by them only."[11] Newcomen could not have based his design on prevailing scientific theory, White argued, because his engine relied on the dissolution of air in steam, and "scientists in his day were not aware that air dissolves in water." Evidently "the mastery of steam power" was a product of empirical science and was "not influenced by Gallilean science."[12]

The specific innovations that made the steam engine an efficient source of power for driving machinery were the work of a craftsman of modest means and little formal education, James Watt. An instrument-maker whose shop was located in the vicinity of Glasgow University, Watt was asked by a professor to repair a model of a Newcomen engine, which stimulated him to improve on its design. He located its inefficiency in the massive heat loss that occurred as steam was condensed in the engine's cylinder. His solution was to add a separate vessel where the steam could be condensed with cold water without drastically lowering the temperature of the main cylinder.

Although Watt was an artisan, his academic surroundings no doubt influenced him, but the extent of that influence has frequently been exaggerated. Watt was friendly with Joseph Black, who at that time taught chemistry at Glasgow University.[13] The traditional tale is that Dr. Black's theory of latent heat provided Watt with the theoretical key to understanding the Newcomen engine's limitations. But as McClellan and Dorn point out, the claim "that Watt had applied Joseph Black's theory of latent heat in arriving at the separate condenser" has "been discredited by historical research."[14]

Furthermore, Watt's crucial mechanical improvements to the steam engine "could not even begin to be studied scientifically until kinematic synthesis developed the appropriate analytic techniques in the last quarter of the nineteenth century." Watt's career thus epitomizes the general relation between theory and practice in that era: That "thermodynamics owes much more to the steam engine than ever the steam engine owed to thermodynamics" has become a cliché.[15]

BREWERS, DISTILLERS, AND THE MULTIPLICITY OF AIRS

IRON AND STEAM were at the heart of the Industrial Revolution, but that was far from all there was to it. Neither were mechanics and metalworkers the only artisans of that era who augmented the knowledge of nature. Brewers were mentioned earlier in connection with the discovery of coke as a substitute fuel. Also, it was "the experience of the distillers and salt-makers, who were accustomed to boiling and condensing liquids on a large scale," that gave rise to a quantitative science of heat.[16] Joseph Black's theory of latent heat emerged from trying to explain certain elements of knowledge common among distillers: that much more heat is needed to boil water away than to raise it to the boiling point, and that the heat absorbed in boiling reappears during the process of condensing the steam.

Brewers also played a noteworthy part in understanding the nature of gases. Cornelius Drebbel's ability to produce and bottle oxygen early in the seventeenth century, and his recognition that not all "airs" are the same, were mentioned in chapter 5, but little more was known about gases until the final decades of the eighteenth century. In the 1770s, Joseph Priestley, a preacher with no scientific education, began systematically to examine the properties of various gases. Priestley later explained that he was "led into" his experiments "in consequence of inhabiting a house adjoining to a public brewery, where I first amused myself with making experiments on the fixed air [carbon dioxide] which I found ready made in the process of fermentation."[17]

Priestley's experiments with gases led him to the same one Cornelius Drebbel had used a century and a half earlier to keep his submarine passengers alive. Priestley called it "dephlogisticated air," in keeping with his theoretical understanding of combustion, and noted that it was, for purposes of animal respiration, a purer version of the ordinary air of the atmosphere. He shared his research with the famous French chemist Lavoisier, who eventually named the gas oxygen and provided a new theoretical framework for understanding it. Although Lavoisier's "new chemistry" took knowledge about gases to a higher level, the debt this field owed to the empirical knowledge of brewers and distillers should not be forgotten.

That was not the last of the brewers' contributions to chemistry. At the end of the nineteenth century, an Australian brewer, Charles Potter, observed that the bubbles rising in fermenting beer carried impurities to the surface. Potter recognized in this phenomenon a possible answer to a long-standing mining problem. Although not a miner himself, he lived in a mining district, interacted with miners, and was familiar with their procedures. He knew, in particular, that a large proportion of the metal that was being taken out of the Melbourne mines was routinely discarded as waste because it could not be efficiently separated from the ores. More than half of the silver, a third of the lead, and almost all of the zinc was considered unrecoverable.

In 1901, after more than ten years of experimentation, Potter patented a new method of separating minerals from their ores. His process called for grinding the ore into small particles, mixing it with a liquid, and blowing a strong current of air into the mixture to whip up a mass of small bubbles. To this slurry he added chemicals that would coat the metals and adhere to the bubbles, producing a froth rich in the desired minerals. The froth could then be skimmed from the surface and the metals extracted from it. Potter's flotation method represented a major advance in metallurgical knowledge and practice. We should not be completely surprised by this brewer's accomplishment: "It was a likely accident to happen in Australia, where beer-drinking and mining had become inseparable cultures."[18]

THE INDUSTRIAL BIOGRAPHER

IF THE ARTISANS and engineers of the Industrial Revolution seem less shrouded in anonymity than their earlier counterparts, it is because many of their accomplishments were indeed recorded and publicized, thus bringing them to the attention of an admiring public. The social context of industrializing Britain no doubt made a change in attitude toward the value of manual labor inevitable: "The period from the 1850s saw a great burst of celebration of engineers through heroic biography, individual and collective. The writings of Samuel Smiles are only the best known examples of this genre."[19]

Smiles dedicated himself to documenting the lives and deeds of the people he deemed primarily responsible for creating the modern world. The names of his most important books testify to his intentions: *Lives of the Engineers* (in five volumes), *Men of Invention and Industry*, and *Industrial Biography*. In addition to collections of biographical sketches, he wrote full biographies of George Stephenson, James Watt, Josiah Wedgwood, and numerous not-so-well-known practitioners of the empirical arts. His works were put to political use as ammunition in a struggle similar to the one in pre-Revolutionary France: defending "the patent system and Britain's legion of inventors from those who would have deprived them of their legal monopoly."[20]

It was a "constant succession of noble workers—the artisans of civilisation," Smiles declared, that "has served to create order out of chaos in industry, science, and art":

> Patient and persevering labourers in all ranks and conditions of life, cultivators of the soil and explorers of the mine, inventors and discoverers, manufacturers, mechanics and artisans, poets,

philosophers, and politicians, all have contributed towards the grand result, one generation building upon another's labours, and carrying them forward to still higher stages.[21]

Smiles did not deny the Great Thinkers their due but insisted that their place in history had been exaggerated and was long overdue for correction:

> Rising above the heads of the mass, there were always to be found a series of individuals distinguished beyond others, who commanded the public homage. But our progress has also been owing to multitudes of smaller and less known men. Though only the generals' names may be remembered in the history of any great campaign, it has been in a great measure through the individual valour and heroism of the privates that victories have been won. And life, too, is "a soldiers' battle"—men in the ranks having in all times been amongst the greatest of workers. Many are the lives of men unwritten, which have nevertheless as powerfully influenced civilisation and progress as the more fortunate Great whose names are recorded in biography.[22]

Smiles's panegyrics were successful in helping Stephenson, Watt, and others gain entry into the pantheon of Great Men. More generally, his biographies facilitated and reflected the rise of engineers from artisanal status to that of a professional elite. In spite of his effusive style, Smiles was an honest and competent scholar who deserves credit for his seminal contributions as a pioneering people's historian.

MINES, CANALS, AND
THE SCIENCES OF THE EARTH

METALLURGY WAS NOT the only science that benefited from the knowledge and activities of miners.[23] The origins of geology are intimately connected with mining, and paleontology began with miners who routinely unearthed interesting seashells, bones, and petrified plants and brought them to the surface for further study.[24]

Miners had for centuries studied the sequences of rocks in the earth's crust in an effort to correlate their patterns with the presence of the minerals they were seeking.[25] "Mining," E. P. Hamm contended,

> is crucial for understanding the ways in which the sciences of the earth were constituted. Mines provided much more than economic grounds for studying the earth; they were the place, both intellectual and social, for the *making of knowledge about the earth*, especially in German-speaking lands.[26]

But because miners rarely committed their knowledge to paper,[27] it is useful to consult the writings of one of the canonical Great Minds to shed some light on the miners' place in the history of geology. Gottfried Wilhelm Leibniz's *Protogaea* (c. 1690), an essay on how the earth was formed, demonstrated that the earliest theoretical treatments of the subject were based on information provided by working miners.

Leibniz did not deny the name of science to the empirical knowledge of practical people: "Although the German term for mine surveying, *Markscheidekunst,* points to its origin in the crafts, Leibniz called it a science."[28] *Protogaea* was an early effort by a scholar to assemble what miners had learned about the earth, reflect on it, and import that knowledge into the world of scholarship.

Mining was not a matter of abstract intellectual interest to Leibniz; he "was a practically-minded philosopher fascinated with scientific and technical matters, not least those having to do with mines, and his study of the earth's history needs to be seen in the context of a culture where mines mattered a great deal."[29] From Leibniz's desire to create a new science of "natural geography" in order to systematize the discovery of valuable ores to today's geology-major university graduates commanding high salaries from oil companies who covet their expertise in locating new petroleum deposits, the influence of capitalism's "hidden hand" is nowhere more evident than in the field of geology.

The new science Leibniz called for "would both draw from and contribute to the sciences of the mines. Who could care more about the location of minerals than miners?" He made thirty-one trips to the mines of the Harz mountains and worked there for more than three years.[30]

Leibniz recognized the centrality of fossils to gaining an understanding of the earth's physical history. A process used by goldsmiths to cast golden replicas of insects or small animals provided him with an analogy to explain how fossils came into existence. "The world of the chemists, assayer, artisan and miner was also important for the discussion of fossils and their organic origin, which makes up by far the greatest part of the *Protogaea*."[31]

About seventy years later, Johann Gottlob Lehmann, a chemistry professor at St. Petersburg, also began studying the Harz. Lehmann believed that exploring nature's "subterranean workshops" was "essential for knowledge of the earth."[32] He heaped scorn on those he called "Messrs Pedants" for their reluctance "to sully themselves, to don such strange raiments, to crawl about with such a wretched and coarse people—as one occasionally finds among one or another miner."[33] In the following generation, the official confluence of mining operations and academia underscored the importance to the state of the geological and metallurgical knowledge produced in mines. In 1765, the first mining academy was founded in Freiberg, renowned for its silver mines, and it rapidly became a leading international center of elite science, attracting such luminaries as Abraham Gottlob Werner and Alexander von Humboldt.

Meanwhile, in England, the expansion of coal mining and the need to transport it led to an increase in canal-building; both coal mining and canal-building brought previously unexposed layers of the earth's surface to the view of more and more astute observers. As T. S. Ashton explained, "The discoveries that made [James Hutton] the most famous geologist of his day owed something to the navvies who were cutting the clays and blasting the rock to provide England with canals."[34] In the generation after Hutton, a self-taught surveyor named William Smith was engaged in a mine survey when he became aware of regularities in the layers of sedimentary rocks he was uncovering and in the fossils that appeared in them. In 1815 Smith published a detailed map of the rock strata of a large part of England. By demonstrating that the layers were always in the same order and could be identified by their fossil contents, he managed "to open up a science that was saturated with theory but starved of data."[35]

The story of William "Strata" Smith need not be recounted in detail here because it has been admirably told by Simon Winchester, among others.[36] For our purposes, it is sufficient to note that Smith's humble social origins caused a significant delay in the scientific elite's recognition of the value of his work. Furthermore, his case offers a particularly clear example of knowledge robbery perpetrated by gentleman scientists. George Bellas Greenough and the other patrician officers of the Geological Society of London—a "self-validating knowledge elite"[37]—looked down their noses at the "low-born" surveyor. But when Smith published his gigantic stratigraphical map of England, they not only plagiarized it but published their derivative version at a lower price, dealing a severe financial blow to Smith and contributing to his being thrown into debtor's prison. Fortunately justice would prevail, albeit belatedly, and before his death in 1839, "Strata" Smith had gained recognition as the Father of English Geology.

MECHANICS' INSTITUTES: INSTITUTIONALIZING PEOPLE'S SCIENCE?

AT THE TIME Smith was carrying out his stratigraphical studies, interest in science was growing in Britain, affecting all social classes. A people's science movement of sorts appeared, but it did not spontaneously arise from the working people themselves. It took the form of mechanics' institutes initiated by middle-class reformers to bring enlightenment to the "lower orders." The first mechanics' institute grew out of free science lectures provided for people of modest means by a natural-philosophy professor at the University of Glasgow, George Birkbeck. His initiative met with an enthusiastic response; the fourth lecture attracted some five hundred people.

Birkbeck later moved to London and joined forces with Henry Brougham and other members of the reform movement inspired by Jeremy Bentham. In 1823 they founded the London Mechanics' Institute; two thousand people converged on the Crown and Anchor tavern for the founding meeting. Three years later there were a hundred mechanics' institutes throughout England, Scotland, and Wales, and their numbers continued to grow rapidly.[38] In 1827, Brougham also established the Society for the Diffusion of Useful Knowledge, which published material on scientific subjects at prices working people could afford.

The mechanics' institutes arose in a context of sharp class polarization and social turmoil characterized by heated, often violent, agitation for the extension of political rights to working people. The social tensions were reflected in scientific discourse. Britain's best-known scientist, Sir Humphry Davy, was among the era's most prominent ideologues. In his inaugural address at the Royal Institution in 1802, Davy underscored the social role of science, which, he said, "will uniformly appear as the friend of tranquility and order." He left no doubt as to his belief in its elite character. Science, he declared, is "in a high degree cultivated . . . by the rich and privileged orders," who then extend its benefits to "the labouring part of the community." Class distinctions, he added, were not to be deplored or struggled against: "The unequal division of property and of labour, the difference of rank and condition amongst mankind, are the sources of power in civilized life, its moving causes, and even its very soul."[39]

Scientific ideas were anything but neutral in early-nineteenth-century Britain. The mechanics' institutes were created "to provide acceptable science for the querulous working classes."[40] Their principal political patron, Lord Brougham, was a progressive Whig who, as a lawyer, had defended trade unionists when trade unions were illegal and as a parliamentarian had vigorously opposed slavery and supported women's rights. Nonetheless, the founding of the mechanics' institutes "was mainly informed by an interest in social control." Brougham and the movement's other leaders made no attempt to conceal their belief "that a regimen of scientific education for certain members of the working class would render them, and their class as a whole, more docile, less troublesome, and more accepting of the emerging structure of industrial society."[41]

The mechanics' institutes were financed by wealthy industrialists who saw them as a means of keeping "dangerous" ideas in check. The most dangerous of scientific ideas, from the standpoint of the ruling circles, was the ungodly notion of evolution.

"ARTISAN EVOLUTIONISTS" BEFORE DARWIN

THE THEORY OF the evolution of species has become so closely linked to Charles Darwin's name that it is easy to forget that evolutionary biology did not begin with him. Darwin's contribution was to provide a plausible explanation of *how* species evolved—that is, by natural selection—and to support it with mountains of evidence. That another naturalist, Alfred Russel Wallace, simultaneously offered the same explanation indicates the existence of a significant current of evolutionary thought well before Darwin published his famous *Origin of Species* in 1859.

Fortunately, Adrian Desmond has made the pre-Darwinian evolutionists the focus of a superlative work of people's history. Desmond's book *The Politics of Evolution* "is not about polite or 'responsible' science—the sort promoted at Oxford or Cambridge—but about angry, dissident views. It is about science to change society." Desmond argued that "a history of biology 'from below' is long overdue. If we are to cease being 'dazzled by the great,' then we need to pry into those social worlds where the mass of people lived."[42]

It was in the generation before Darwin came on the scene that "the real fight to establish a lawful, evolutionary worldview among the 'people'" took place.[43] Prior to Darwin and Wallace, scientific opponents of the biblical doctrine of "special creation" relied on a theory put forward by the French naturalist Jean-Baptiste Lamarck, which held that species evolved through the transmission of acquired characteristics from parents to their offspring. Lamarck's radical departure from orthodoxy emerged from the ferment of the French Revolution and came under heavy attack during the Thermidorian reaction. The retarding effect on French science of Georges Cuvier's heavy hand was nowhere more evident; as American paleontologist O. C. Marsh observed in 1879, Cuvier's influence "delayed the progress of evolution for half a century."[44]

Radical Lamarckism was thriving, however, in England during the tumultuous Reform Bill era of the late 1820s and early 1830s. Its banner was held aloft by nonelite medical practitioners and anatomy teachers in secular private schools. Desmond has identified them as proponents of a "republican science," a "social Lamarckian science of progressivism, materialism, and environmental determinism to underwrite the change to a democratic, cooperative society."[45]

For the "artisan evolutionists," Lamarck and his sympathizers "became symbols of resistance; they were the tricolor banners waved by the medical democrats massing outside the corporation porticos."[46] On the other side of the class divide were "Oxbridge dons and wealthy London allies, gentlemen specialists who channeled funds to their friends to ensure an acceptable Anglican science."[47] These "squires of science . . . made establishment science a recipe for social stability and Anglican supremacy."[48] They perceived evolutionary biology as a French import tainted with the radicalism of the French Revolution and a threat to core doctrines of Church and state.

The early proponents of evolution have usually been overlooked, Desmond contended, "because historians have been looking too closely at the gentlemen of science and their Anglican ministers."[49] A few of the radicals' more noteworthy spokesmen were Thomas Wakley, Robert Grant, Robert Knox, George Dermott, Patrick Matthew, Marshall Hall, and Hewett Watson.[50] Wakley, who founded the combative medical journal *The Lancet* in 1823, expressed the desire to "bring science within the reach of the sons of the humbler classes"; in 1831, he chaired the militant National Union of the Working Classes. Dermott, whose private school of anatomy charged tuition low enough for poor pupils to afford, was an "uncompromising physical-force radical"; that is, he believed that social progress could be accomplished only by violent revolution.[51]

The spread of evolutionary thought owed a great deal to a popular-science writer named Robert Chambers. Desirous of creating a "people's science," Chambers synthesized the artisan evolutionists' ideas (though giving them "middle-class cachet" by downplaying their radical social implications) in an 1844 book entitled *Vestiges of the Natural History of Creation*. The anonymously published book "caused a furor among the old university elite, but it did notch up unprecedentedly high street sales, especially in London." Compounding the book's offense in the eyes of the conservatives, it "was giving indelicate topics such as pregnancy, abortion, and monstrosities (previously the province of medical men) a plebeian scientific importance. The trouble was that women—and not only emancipated socialist women—were actually enjoying the book; it was reaching a huge audience." The medical elite was distraught at seeing evolutionary ideas "leaving the shabby medical schools and entering the middle-class parlors."[52]

When not ignored altogether, the plebeian English Lamarckians have traditionally been remembered only as losers in the ideological battle with Darwinism, but they should rather be hailed as winners in the context of the larger issue. By defending and developing the idea of species evolution, they did the early spadework that led to the eventual acceptance of Darwin's and Wallace's theory.[53] Remembering their contribution to evolutionary biology is important, Desmond insisted, because "if we are not to see science as a monolithic creation of the conservative elite, then we must get this dissident dimension back in the picture."[54]

"DARWIN'S BULLDOG" AND
THE WORKING PEOPLE

DARWIN PUBLISHED *ORIGIN of Species* in 1859, but he lacked the stomach to confront the firestorm of controversy that greeted it. Into the breach leapt T. H. Huxley, who earned himself the sobriquet "Darwin's bulldog" for acting as natural selection's leading public defender. But to Huxley, the defense of Darwinism was part of a broader campaign in the science wars of the era.

According to Desmond, it was the self-styled "proletarian" science of Huxley and his circle that directly prepared the way for Darwin.[55] In the decade preceding the appearance of Darwin's momentous work, Huxley and many other young would-be scientists were painfully aware that the English scientific establishment was in need of reform. Huxley himself "had struggled against odds to secure a job teaching science, only to find it as poorly paid as it was lacking in prestige." Before the middle of the nineteenth century, to belong to the elite of science in England "one generally had to be wealthy. The reformers now wanted it out of private hands and placed in the public sector. They wanted, in short, a nationally-organised professional community, equipped with modern research facilities, paid for and supported by the Exchequer."[56]

Huxley and his friends constituted "the vanguard of the new middle-class movement in science." Their aim was "to professionalise science and put it at the disposal of the mercantile middle classes," in opposition to "the Oxbridge dons, in the service, so to speak, of the traditionally antagonistic landed party." To that end, they "deliberately wooed the wage-earners." Building on the tradition of the mechanics' institutes, in 1855 Huxley began delivering "working men's lectures" and crusading for technical education in order "to win working class support for the new movement." He "portrayed the 'scientist' in his working men's lectures as a proletarian" and depicted artisans and anatomists as "labouring brothers." Desmond commented that Huxley and his fellow science reformers were "outdoing even the pamphleteering Methodists in their drive to reach Everyman."[57]

Huxley had much in common with the earlier founders of the mechanics' institutes. The sincerity of his "plebeian commitment" is not in question, but his lectures and essays were also "designed to appeal to bosses, and to persuade them to initiate workers into scientific modes of thought *for the stabilization of capitalist society.* In short, he was wooing both sides from a middle position." In an age of rising social discontent, "Huxley was presenting science as an alternative to more drastic socialist remedies."[58]

The reformers were largely successful in achieving their goals. Between 1868 and 1874, "the new middle-class professionals had begun to seize the organs of scientific power" in London.[59] The movement's attempt to bridge irreconcilable class interests, however, produced contradictory results. On the one hand, a significant democratization of science occurred, as scientific careers were opened to many more people. On the other hand, science was to become ever more dominated by a state-supported professional elite.

DARWINIST IDEOLOGY: SOCIALISM OR BARBARISM?

THE RELATION BETWEEN science and people cannot be understood apart from the ideological uses to which science has been put. Darwinism is usually assumed in retrospect to have triumphed simply because it states an objective truth about how nature operates, but the favorable reception of scientific theories is a far more complicated matter.

Darwin was able to get a hearing when he did because his theory lacked the most radical of the social implications attached to Lamarckism and thus offered a concept of evolution ideologically tolerable to important segments of the scientific elite. Whereas Lamarck's ideas were believed to imply the malleability of human societies, Darwin's could be interpreted as nonthreatening to social hierarchies. "Darwinism," one of its leading proponents crowed, "is thoroughly aristocratic; it is based upon the survival of the best."[60]

Darwin was directly inspired by the writings of the Reverend Thomas Malthus, who believed that poverty and starvation should not be combated because they were "nature's way" of controlling population. A veneer of mathematical hocus-pocus lent the apparent authority of science to Malthus's pronouncements. In his autobiography, Darwin wrote:

> I happened [in 1838] to read for amusement Malthus on population, and being well prepared to appreciate the struggle for existence which everywhere goes on from long continuous observation of the habits of animals and plants, it at once struck me that under these circumstances favorable variations would tend to be preserved and unfavorable ones to be destroyed. The result of this would be the formation of new species. Here, then I had at last got a theory by which to work.[61]

The profoundly reactionary social consequences of Malthus's "scientific" assertions had already been made clear; in 1834, the English government had appealed to them to justify ending relief payments to its poorest citizens and forcing them into brutal workhouses. Darwin was a political supporter of that Whig government at a time when working-class radicals were campaigning against the pitiless legislation inspired by the Reverend Malthus.

The arbitrariness of reading social meaning into biological theories is evidenced by the sharply conflicting ways Darwinism was interpreted. Very soon after *Origin of Species* was published, Karl Marx exuberantly wrote to Frederick Engels, "This is the book which contains the basis in natural history for our view."[62] Marx saw in Darwin's theory a confirmation of the dialectical-materialist philosophy that underpinned his own theory of social revolution.

In Germany, Rudolf Virchow, a leading scientist as well as a highly influential politician, agreed with Marx, linking Darwinism with socialism and the revolutionary uprising in Paris in 1871. "Be careful of this theory," the antisocialist Virchow warned, "for this theory is very nearly related to the theory that caused so much dread in our neighboring country."[63]

But Darwin's partisans vehemently denied Virchow's charge; Ernst Haeckel insisted that socialism and Darwinism "endure each other as fire and water."[64] Darwin himself exclaimed, "What a foolish idea seems to prevail in Germany on the connection between Socialism and Evolution through Natural Selection."[65]

Marx and Virchow notwithstanding, the primary ideological use to which Darwin's theory was put was decidedly antisocialistic. The social philosopher Herbert Spencer, a member of Huxley's circle, picked it up and transformed it into "social Darwinism" as the ultimate justification for the brutality and rapacity of unfettered capitalism. Spencer maintained that the Darwinian imperative of "the survival of the fittest" operates not only in biological evolution but also in human society.[66]

Social Darwinism promoted the view that the dog-eat-dog world of laissez-faire capitalism is the most "natural" of all possible economic systems: Rich people become rich because of their natural superiority, and poor people are poor because they are born deficient in the talents and abilities that lead to economic success. At its most virulent, social Darwinism suggested that the human race would be best served by allowing the poor to perish rather than helping them survive by means of private charity or state-sponsored welfare programs. Echoing Malthus, some capitalist apologists claimed that letting the poor die of starvation is "nature's way" of weeding out inferior individuals and is necessary to permit the human species to evolve and improve over time.

Because social Darwinism was little more than old Malthusian wine in a new bottle, it should not be blamed completely on

Darwin. Nevertheless, "his notebooks make plain that competition, free trade, imperialism, racial extermination, and sexual inequality were written into the equation from the start—'Darwinism' was always intended to explain human society."[67]

EUGENICS

A SLIGHTLY—but only slightly—less noxious variant of the notion that masses of poor people should be left to starve to death was the idea that the human race could hasten its evolution toward perfection by restricting the reproductive capacity of "inferior" types of people—which almost invariably meant the darker-skinned ones. That was the inspiration behind eugenics, a new science founded by a first cousin of Darwin, Sir Francis Galton. Galton urged the "gifted class" to produce more offspring of their own and to take measures to limit the procreation of "children inferior in moral, intellectual and physical qualities."[68]

Among other things, Galton attempted to provide a biological justification for scientific elitism. The primary thesis of his books *Hereditary Genius* and *English Men of Science* was "that great men, including creative scientists, tend to be related and that therefore a series of elite families contributed perhaps the majority of distinguished statesmen, scientists, poets, judges, and military commanders, of his day and of the past."[69]

Galton was a wealthy polymath who is traditionally awarded a place in the pantheon of Great Minds for pioneering methods of applying mathematics to the study of human behavior. He has been called the father of intelligence testing, he is frequently credited with the invention of fingerprinting as a means of identifying individuals, and modern statistical analysis owes an immense debt to his undeniably brilliant innovations.[70] But his eugenics—as well as his other attempts to apply scientific methodology to the solution of social problems—was constructed on the false foundation of social prejudice.

He held "axiomatically" that "certain marked types of character" can be "justly associated" with the "different races of men." Galton maintained, for example, that "the typical West African Negro" has "strong impulsive passions, and neither patience, reticence, nor dignity. . . . He is eminently gregarious, for he is always jabbering, quarrelling, tom-tom-ing, or dancing . . . and he is endowed with such constitutional vigour, and is so prolific, that his race is irrepressible."[71] Eugenics was, among other things, the supposedly scientific answer to the demographic threat posed by the sexually hyperactive black race and its infernal drumming.

Because it seemed to offer a scientific explanation of why privileged social groups deserved their privileges, Galton's eugenics gained and maintained an influence far exceeding what it would have achieved if science really were a disinterested, objective quest for truth. That influence was strong and growing stronger at the end of the nineteenth century and persisted well into the twentieth. Unfortunately, it has not entirely disappeared even in the twenty-first.

NOTES

1. Andrew Ure (1778–1857) was a chemist and chemistry professor who became an ardent proponent of free trade. His paeans to the factory system included apologetics for child labor. See esp. Ure, *The Philosophy of Manufactures.*

2. A significant exception is the important but relatively small proportion of scientific work carried out in countries such as Cuba, where economic planning trumps capitalist decision-making, but even there, the direction of science is heavily influenced by international financial markets and military competition with capitalist nations.

3. The capitalized phrase "Big Science" was coined by Alvin Weinberg, director of the Oak Ridge National Laboratory, in an article in *Science*, July 21, 1961; see also his *Reflections on Big Science*. The "citation classic" on the subject is Derek J. de Solla Price, *Little Science, Big Science*. A more recent study (valuable in spite of being limited to Big American Physics) is Peter Galison and Bruce Hevly, eds., *Big Science*.

4. James McClellan and Harold Dorn, *Science and Technology in World History*, p. 287.

5. J. D. Bernal, *Science in History*, vol. 2, p. 591.

6. McClellan and Dorn, *Science and Technology in World History*, p. 280.

7. Anthony F. C. Wallace, *The Social Context of Innovation*, pp. 91, 101.

8. Newcomen's pump, in contrast to Savery's, did not raise water by means of pressure created by expanding steam but by means of atmospheric pressure on a partial vacuum created by the condensation of steam. The subsequent development of the steam engine followed Newcomen's precedent.

9. Lynn White, Jr., "Pumps and Pendula," p. 107.

10. L. T. C. Rolt and J. S. Allen, *The Steam Engine of Thomas Newcomen*, p. 12.

11. R. S. Meikleham, *Descriptive History of the Steam Engine* (London, 1824); quoted in Bernal, *Science in History*, vol. 2, p. 580.

12. White, "Pumps and Pendula," pp. 107–108. White rigidly counterposes science and technology, but his narrative makes clear that the labors of Newcomen, "an empirical genius of awesome proportions," resulted in knowledge of nature that was previously unknown to the Gallilean scientists. No solid evidence exists to refute the claim that Newcomen's invention was independent of Savery's, but it has been challenged nonetheless; see, e.g., Wallace, *Social Context of Innovation*, pp. 55–57.

13. The extent and character of their relationship can be seen in their correspondence, which has been collected in a volume edited by Eric Robinson and Douglas McKie, *Partners in Science*.

14. McClellan and Dorn, *Science and Technology in World History*, p. 288. Watt expressed great appreciation for "the knowledge I acquired in conversation from . . . Dr. Black," but insisted that "it certainly did not directly point out the improvements I have made upon the Steam Engine." Robinson and McKie, eds., *Partners in Science*, p. 416. Bernal was among those misled by the claim of Watt's dependence on Black's theory; see Bernal, *Science in History*, vol. 2, p. 582.

15. Derek J. de Solla Price, *Little Science, Big Science . . . and Beyond*, p. 240.

16. Bernal, *Science in History*, vol. 2, p. 580.

17. Robert E. Schofield, ed., *A Scientific Autobiography of Joseph Priestley*, p. 51.

18. Robert Raymond, *Out of the Fiery Furnace*, p. 234.

19. David Philip Miller, "'Puffing Jamie,'" p. 9.

20. Ibid., p. 17.

21. Samuel Smiles, *Self-Help*. Page numbers are not given because the book is in digital form and the text is therefore searchable. Among other books by Smiles available in digital form are *Industrial Biography* and *Men of Invention and Industry*.

22. Ibid.

23. See chapter 5.

24. Are we justified in using the word "geology" when discussing a period before the term was coined? E. P. Hamm argued (and I fully concur):

"Those who are rigorously devoted to eighteenth-century terminology might write separate histories of cosmography, subterranean geography, oryctognosy, oryctology, physical descriptions of the earth, natural geography, sacred physics, mineralogical travelogues, the terraqueous globe, natural histories of the mineral kingdom and geognosy. Yet exchanging anachronism for fractiousness is neither desirable nor unavoidable." E. P. Hamm, "Knowledge from Underground," pp. 77–78.

25. Mott T. Greene, *Geology in the Nineteenth Century*, p. 39.

26. Hamm, "Knowledge from Underground," p. 79 (emphasis added).

27. A few notable exceptions were mentioned in chapter 5.

28. Hamm, "Knowledge from Underground," p. 82.

29. Ibid., p. 77.

30. Ibid., pp. 81–82.

31. Ibid., p. 81.

32. Ibid., p. 84.

33. Johann Gottlob Lehmann, *Abhandlung . . .* (1753); quoted in Hamm, "Knowledge from Underground," p. 86.

34. T. S. Ashton, *The Industrial Revolution, 1760–1830*, p. 16.

35. Cecil J. Schneer, "William Smith's Geological Map of England and Wales and Part of Scotland, 1815–1817."

36. Simon Winchester, *The Map That Changed the World*.

37. Roy Porter, "Gentlemen and Geology," p. 810.

38. See Thomas Kelly, *George Birbeck*.

39. Humphry Davy, "Discourse Introductory to a Course of Lectures on Chemistry," vol. 2, pp. 323, 326.

40. Adrian Desmond, *The Politics of Evolution*, p. 27.

41. Steven Shapin and Barry Barnes, "Science, Nature and Control," p. 32.

42. Desmond, *Politics of Evolution*, pp. 3, 20.

43. Ibid., p. 1.

44. O. C. Marsh, *History and Methods of Paleontological Discovery: An Address Delivered before the AAAS at Saratoga, N.Y., Aug. 28, 1879*; quoted in Adrian Desmond, *Archetypes and Ancestors*, p. 173.

45. Desmond, *Politics of Evolution*, p. 329.

46. Ibid., pp. ix, 24.

47. Ibid., p. 135.

48. Ibid., pp. 2, 20.

49. Ibid., p. 1.

50. See Appendices A and B to Desmond, *Politics of Evolution*, pp. 415–429.

51. Desmond, *Politics of Evolution*, pp. 121, 125, 166.

52. Ibid., pp. 177, 379.

53. The direct connection is most evident in Robert Grant's influence on Charles Darwin. See Desmond, *Politics of Evolution*, pp. 398–403.

54. Desmond, *Politics of Evolution*, p. 21.

55. Desmond, *Archetypes and Ancestors*, p. 13.
56. Ibid., pp. 109–110.
57. Ibid., pp. 13, 17, 40, 122, 139.
58. Ibid., pp. 40, 160, 162 (emphasis in original).
59. Ibid., p. 142.
60. Ernst Haeckel; quoted in Anton Pannekoek, *Marxism and Darwin*.
61. Charles Darwin, *Autobiography* (1876), p. 120.
62. Karl Marx to Friedrich Engels, December 19, 1860. *Selected Correspondence*, p. 126.
63. Quoted in Pannekoek, *Marxism and Darwin*.
64. Quoted in ibid.
65. Darwin to Karl von Scherzer, December 26, 1879. *The Life and Letters of Charles Darwin*, vol. 2, p. 413.
66. Of the originator of social Darwinism, Herbert Spencer, Darwin wrote, "I suspect that hereafter he will be looked at as by far the greatest living philosopher in England; perhaps equal to any that have lived." *Life and Letters of Charles Darwin*, vol. 2, p. 301.
67. Adrian Desmond and James Moore, *Darwin*, p. xxi.
68. Francis Galton, "Hereditary Improvement," p. 129.
69. Price, *Little Science, Big Science . . . and Beyond*, p. 31.
70. Galton can be better described as a popularizer rather than as the inventor of fingerprinting. See Martin Brookes, *Extreme Measures*, pp. 247–255.
71. Francis Galton, "Hereditary Talent and Character," pp. 320–321.

8

THE SCIENTIFIC-INDUSTRIAL COMPLEX

THE TWENTIETH CENTURY AND BEYOND

ADVANCES IN SCIENCE when put to practical use mean more jobs, higher wages, shorter hours, more abundant crops, more leisure for recreation, for study, for learning how to live without the deadening drudgery which has been the burden of the common man for ages past. Advances in science will also bring higher standards of living, will lead to the prevention or cure of diseases, will promote conservation of our limited national resources, and will assure means of defense against aggression.

—VANNEVAR BUSH, "Science: The Endless Frontier" (July 1945)

BETTER LIVING THROUGH Chemistry
—E. I. DU PONT DE NEMOURS AND COMPANY (1939–1980s)

A T THE TURN of the twentieth century, confidence in the beneficence of modern science was almost limitless. As late as July 1945, Vannevar Bush's triumphal promotion of science as "the endless frontier" did not seem quaint or naïve. The following month, however, perceptions began to shift when the obliteration of Hiroshima and Nagasaki revealed the awesome destructive power nuclear physicists had spawned. A growing uneasiness was temporarily masked by the stifling conformity of the Cold War era. Glorification of Big Science by DuPont, Union Carbide, and other mammoth corporations dominated the public discourse. But by the end of the century, the prestige of modern science had declined precipitously.

"Scientism"—blind faith in the ability of science to solve all problems—reigned supreme in the first decades of the twentieth century. Eugenics, for example, was not then an adjunct of right-wing ideology; it attracted avid supporters from the far right to the far left of the political spectrum. People of all political persuasions, it seemed, hoped that selective human breeding would provide an easy scientific answer for intractable social ills. Francis Galton's disciple and successor as chief eugenics theoretician, Karl Pearson, was an outspoken socialist who in his spare time gave lectures on Marxism. Margaret Sanger appealed to eugenic rationalizations in her pioneering campaign for birth control. Moderate socialists like George Bernard Shaw and the Fabians were enthusiastic eugenicists, as were more radical socialists like J. B. S. Haldane and Lancelot Hogben. One of the most influential promoters of eugenics among leading American scientists was Hermann J. Muller, whose socialist views prompted him to leave the United States in 1932 to work in the USSR.[1]

The horrific practical implications of eugenics emerged in political movements in the United States and Germany. The American eugenicists were sufficiently influential to win passage of the Johnson-Reed Act in 1924, which restricted immigration from southern and eastern Europe, the Balkans, and Russia. "America must be kept American," President Calvin Coolidge declared; "Biological laws show . . . that Nordics deteriorate when mixed with other races."[2]

Coolidge was echoing the scientific rationale provided by eugenics movement leader Harry H. Laughlin, who had determined "that eastern Europeans, Mediterraneans, and Russian Jews, among others, harbored a large number of defective genes in their populations." Not coincidentally, those immigrant populations also accounted for a great deal of the labor radicalism in the United States. "For the wealthy benefactors that supported eugenics, such as the Carnegie, Rockefeller, Harriman, and Kellogg philanthropies, eugenics provided a means of social control in a period of unprecedented upheaval and violence."[3]

Harry Laughlin and his movement

> also lobbied at the state level for the passage of eugenic sterilization laws, which would allow individuals in state institutions to be forcibly sterilized if they were judged to be genetically defective. Over 35 states passed, and used, such laws. By the 1960s, when most of these laws were beginning to be repealed, more than 60,000 people had been sterilized for eugenic purposes. In Germany, the National Socialists used Laughlin's model as one of the bases of their sweeping sterilization law of 1933, which ultimately led to the sterilization of over 400,000 people.[4]

It was the Holocaust that turned international public opinion decisively against eugenics. Eugenics was not discredited as a scientific doctrine by means of research generating new evidence but due to momentous events external to the world of science. In 1938, the Nazis' efforts at racial purification reached their logical, unspeakable culmination in a "euthanasia program" under the supervision of German psychiatrists that condemned tens of thousands of people deemed mentally ill, including children, to execution in gas chambers. Social Darwinism and eugenics were brought into disrepute as their implementation by the Nazis revealed them to be a slippery slope to genocide.[5]

HUMAN GUINEA PIGS

WORLD WAR II exposed the moral depths to which establishment science and scientists could descend when under the influence of antihuman ideologies. "The enthusiasm of German physicians in endorsing ideas of racial degeneracy and implementing race hygiene policies," Roy Porter pointed out, was an "expression of widely held biomedical and anthropological doctrines":

Physicians and scientists participated eagerly in the administration of key elements of Nazi policies such as the sterilization of the genetically unfit. Presiding at genetic health courts to adjudicate cases, physicians ordered sterilization of nearly 400,000 mentally handicapped and ill persons, epileptics and alcoholics even before the outbreak of war in September 1939. Thereafter, "mercy deaths," including "euthanasia by starvation," became routine at mental hospitals. Between January 1940 and September 1942, 70,723 mental patients were gassed, chosen from lists of those whose "lives were not worth living," drawn up by nine leading professors of psychiatry and thirty-nine top physicians.[6]

German scientists treated human beings as laboratory animals in research projects. Concentration-camp inmates were used "to study the effects of mustard gas, gangrene, freezing, and typhus and other fatal diseases. Children were injected with petrol, frozen to death, drowned or simply slain for dissection purposes." The wartime practices of Japanese doctors and scientists were equally inhumane:

> Hundreds of doctors, scientists and technicians led by Dr. Shiro Ishii were set up in the small town of Pingfan in northern Manchuria, then under Japanese occupation, to pioneer bacterial warfare research, producing enough lethal microbes—anthrax, dysentery, typhoid, cholera and, especially, bubonic plague—to wipe out the world several times over. Disease bombs were tested in raids on China.[7]

Dr. Ishii conducted experiments on some three thousand human subjects "to investigate infection patterns and to ascertain the quantity of lethal bacteria necessary to ensure epidemics. Other experimental victims were shot in ballistic tests, were frozen to death to investigate frostbite, were electrocuted, boiled alive, exposed to lethal radiation or vivisected." The cruelty involved in performing vivisections without anesthesia is almost beyond imagination. Most of Dr. Ishii's subjects were

Chinese, but some were American and British prisoners of war. Nonetheless, at the end of the war, "the American government chose to keep these atrocities secret" because "Dr. Ishii and his team did a deal with the American authorities, trading their research to avoid prosecution as war criminals."[8]

Although the transgressions of American scientists fell far short of those of their German and Japanese counterparts, during and after the war some also collaborated with morally indefensible military research in which, as part of the atomic research program, American GIs were deliberately exposed to radiation in secret tests.[9] Unfortunately, unconscionable abuses of science cannot be safely consigned to the distant past. In a 2004 issue of the *New England Journal of Medicine*, Robert Jay Lifton reported, "There is increasing evidence that U.S. doctors, nurses, and medics have been complicit in torture and other illegal procedures in Iraq, Afghanistan, and Guantánamo Bay."[10] A few months later, a report produced by the International Red Cross corroborated Lifton's charges. At the Guantánamo prison, it revealed, a group called the Behavioral Science Consultation Team, or BSCT (pronounced "biscuit"), composed of psychologists, psychiatrists, and doctors, had closely collaborated with military interrogators in constructing "an intentional system of cruel, unusual and degrading treatment and a form of torture."[11]

FROM SOCIAL DARWINISM TO SOCIOBIOLOGY

ALTHOUGH EUGENICS NO longer has many open supporters, other applications of Darwinian thinking in support of reactionary political agendas have arisen and flourished. Among the most prominent in recent decades are "sociobiology" and "evolutionary psychology," which promote the idea that human social behavior is essentially a product of inherited traits. If such attributes as maternal warmth, docility, aggressiveness, intelligence, or criminality are genetically determined, then individuals are for the most part locked into their social roles from birth, much like worker ants or queen bees. (It is not

irrelevant that sociobiology's leading theoretician is an ento-
mologist.)[12] "The hereditarian position," Stephen Rose observed,

> would have us believe that the working class, the Blacks, the
> Irish, are genetically stupider than the middle class, the
> Whites, the English; that women have genes for being sec-
> retaries and men for being executives—and therefore that the
> exploitation and justification of a class bound, racially and
> sexually divided society, lies *not* in social institutions and
> structures (which we can change) but in our genes (which we
> cannot).[13]

Hereditarian ideas have often been used to justify educational policies that condemn children of working-class families
to inferior schools on the grounds that they are biologically
incapable of serious intellectual achievement. Between 1943
and 1966, Sir Cyril Burt—a pupil of Karl Pearson who became
a highly respected experimental psychologist in his own right—
claimed that his experiments with identical twins separated at
birth and raised in different social-class settings proved that
intelligence was far more a matter of nature than nurture. The
implication for social policy was that providing higher education
for working-class youth was a waste of society's resources.

The British government heeded Burt's results and instituted an educational system that "tracked" many generations of
less affluent students into trade schools and other programs
deemed appropriate to their class origins. Only after Sir Cyril's
death—and after untold damage to the children of working-
class families—was it revealed that his studies were fraudulent.
A statistical analysis of his data showed that he had fabricated
them to support his social prejudices.[14]

Another implication of sociobiology is that "biology is
woman's destiny," a proposition that the feminist movement
has vigorously resisted as untrue and detrimental to the interests
of women. Yet another is the corollary of the doctrine of genetically determined intelligence that consigns people of African

descent to permanent intellectual inferiority—a pernicious falsehood that has been thoroughly refuted yet never seems to die. As recently as 1994, a widely read book, *The Bell Curve*, attempted to provide it with a semblance of academic respectability.[15]

Darwinism has also been called on to support the ideological proposition that social change must proceed slowly. Natural selection, in Darwin's treatment, exerts its influence in a very gradual manner over long, long stretches of time. What this supposedly teaches us is summed up in a famous Latin phrase: *Natura non facit saltum*—Nature does not make leaps. Extrapolating that idea into the social realm suggests that the "natural" path to social change is gradual reform rather than abrupt revolution.

Paleontologists Stephen Jay Gould and Niles Eldridge subjected Darwin's gradualism to close scrutiny. According to their reading of the fossil record (which is far more extensive now than it was in Darwin's day), speciation is a process characterized by very long periods of equilibrium that are occasionally punctuated by relatively sudden bursts of change. If they are right, it would seem that nature is more revolutionary than reformist after all. Gould's and Eldridge's "punctuated equilibrium" idea was at first a contentious issue among evolutionary biologists, but many, if not most, have since been won over to their point of view. The controversy over punctuated equilibrium, however, reveals that Darwinian gradualism is not a "fact" but an ideological construct.

The important lesson to be learned from all of this is that the social meanings attributed to biological theories are "not logically inherent in the theories themselves."[16] Whether evolution proceeds slowly or in bursts has nothing whatsoever to do with the way political struggles unfold. Nor is the red-in-tooth-and-claw competition of biological evolution a model human societies should aspire to emulate. In general, attempts to reduce the laws of the science of society to the laws of biology is bad science that encourages bad social policy.

FREDERICK TAYLOR AND
SCIENTIFIC MANAGEMENT

THE "UNION OF capital and science" found its most explicit ideological expression in social Darwinism, but its most direct manifestation in practice was the Scientific Management movement, known as Taylorism after its chief theoretician, Frederick W. Taylor. Its stated purpose was to apply the methodology of modern science to problems of capitalist enterprise. Although Taylor provided the inspiration, it was "the same economic elites and their business interests" bankrolling the eugenics movement who "introduced scientific management and organizational control into the industrial sector."[17]

The centerpiece of Taylor's application of scientific methods to manufacturing processes is time-and-motion study, in which the tasks of manual workers are subjected to close analysis in order to optimize their efficiency. The ultimate goal is to increase the productivity of labor, which Taylor insisted was in the common interests of employers and employees. Taylor adorned his proposals with paeans to "harmony between workmen and the management," and warned that the knowledge obtained by his methods should not be used "as a club to drive the workmen into doing a larger day's work for approximately the same pay that they received in the past."[18] But his moral strictures were unenforceable; they depended on the willingness of employers to heed them.

Meanwhile, the subjects of the time-and-motion studies understandably feared a number of unwelcome consequences: forcing them to speed up the rate at which they worked, intensifying the tyranny of the clock, increasing the mind-numbing repetitiveness of their tasks, decreasing their paychecks by reducing piece-rate wages, or even eliminating their jobs altogether. Small wonder that they tended to be hostile toward efficiency experts with stopwatches and clipboards who observed and recorded their every movement. A classic anecdote captures the essence of the workers' attitude:

"You're a very good worker," said the efficiency expert as he watched a carpenter plane a piece of wood. "Now if we can just stick a buffer on your elbow you could plane and buff the wood with the same motion."

"Yeah," the carpenter responded, "and if you'd stick a broom-stick up your ass you could take your notes and sweep the floor at the same time."[19]

Taylor published his *Principles of Scientific Management* in 1911, but its essential idea had been expressed much earlier. The "grand object of the modern manufacturer," Andrew Ure explained in 1835, is to reduce the human worker to "a component of a mechanical system."[20] This was a prescription for the dehumanization and robotization of labor. One perceptive critic, Harry Braverman, referred to the fragmentation of work without regard to human capabilities as "the subdivision of the individual" and called it "a crime against the person and against humanity."[21]

Although Scientific Management was a quintessentially capitalist movement, the economic benefits it promised proved irresistible even to sworn enemies of capitalism. In the first year of the Russian Revolution, Lenin declared that Taylorism,

> like all capitalist progress, is a combination of the refined brutality of bourgeois exploitation and a number of the greatest scientific achievements in the field of analyzing mechanical motions during work, the elimination of superfluous and awkward motions, the elaboration of correct methods of work, the introduction of the best system of accounting and control, etc. . . . We must organise in Russia the study and teaching of the Taylor system and systematically try it out and adapt it to our ends.[22]

STAKHANOVITES AND LYSENKOISTS

LENIN'S QUESTIONABLE ASSUMPTION that a government devoted to the workers' interests could practice a form of Taylorism without the "brutality of bourgeois exploitation" was not put to the test. The industrialization of the Soviet Union proceeded under Stalin's aegis, and only the most unreconstructed Stalinist could believe that his policies were designed with the well-being of individual workers in mind. The Stakhanovite movement of the 1930s and '40s was promoted as a sort of "Taylorism from below," with workers supposedly imposing "scientifically" determined work methods on themselves, but like all of Stalin's social initiatives, its claims were bogus, and it was implemented with brutality aplenty.

The most notorious example of Stalin's science policies was the support he gave to an agronomist, Trofim Lysenko, in a bitter conflict over genetics. Lysenkoism was a reversion to Lamarck's idea of the heritability of acquired characteristics, which had been definitively disproved long before Lysenko appeared on the scene. Stalin's ideologues used axioms of dialectical materialism to "prove" Lysenkoism's superiority to Mendelian genetics, an *a priori* method that Marx himself would certainly have vigorously condemned. The upshot was the destruction of a whole generation of Soviet geneticists and a devastating setback to the science of genetics in the Soviet Union.[23]

DEPRESSION, RADICALIZATION, AND PEOPLE'S SCIENCE

AT THE BEGINNING of the 1930s, the West was rapidly sliding into the Great Depression. The planned economy of the Soviet Union, by contrast, was robust and growing, and the social gains of the 1917 Russian Revolution appeared extremely attractive to a significant layer of Western intellectuals. In that rapidly radicalizing context, an international convocation was held in London in 1931 that was arguably the point of origin of the notion of people's history of science contained in this book.

It was at that conference—the second International Congress

of the History of Science and Technology—that Boris Hessen presented the seminal paper on Newton's *Principia* that was discussed in a previous chapter. Hessen was a member of a delegation of scientists and historians from the Soviet Union whose appearance at the conference represented a concerted effort on the part of the Soviet state to gain a hearing for Marxist ideas at the highest levels of international intellectual discourse. The congress's organizers were astonished by the size of the Soviet delegation that showed up in London:

> For months they had been assuming that only one Russian— a Professor Zavadovsky—would be participating. Now they were confronted with a small battalion of politicians, administrators, scientists, historians and philosophers, all of them armed with lengthy and detailed addresses that they wanted to deliver to the congress (and the world).[24]

The clash of ideologies at this conference had a profound impact on how the history of science would henceforth be studied and understood.

The Soviet delegation was headed by none other than the Bolshevik leader Nikolai Bukharin. In addition to his credentials as a revolutionary, Bukharin also happened to be among the Soviet Union's most able intellectuals:

> Bukharin, who was simultaneously the head of the Academy of Science's section on the history of science and the Director of Industrial Research for the Supreme Economic Council, was the obvious choice to select and lead the delegation. He picked, among others, the U.S.S.R.'s leading physicist (A. F. Joffe), its best-known biologist (N. I. Vavilov) and an obscure historian and physicist named Boris Hessen.[25]

Although no one anticipated that Hessen would be the star of the show, it was his paper on Newton's *Principia* that was to generate the most controversy and have the most enduring impact.

Stalin had begun his drive to power in the Soviet Union but had not yet succeeded in imposing total control. Although few outsiders were aware of it at the time, Bukharin and Stalin were factional rivals locked in mortal combat. Bukharin's aim at the London conference, then, was not only to demonstrate the value of Marxist theory to the external world but to convince his peers in the Soviet leadership that he was its most able defender.

The Western participants in the conference were, for the most part, "a collection of academics" who were "unlikely to view science as anything other than a body of ideas fathered by a succession of scientific geniuses."[26] A significant minority, however—including two of the congress's organizers, Joseph Needham and Lancelot Hogben—was favorably disposed toward the Marxist outlook propounded by Bukharin and his colleagues. The stage was thus set for a lively exchange of views.

Hessen's presentation, however, was so utterly incompatible with traditional history of science that there was scarcely any common ground that could serve as a basis of dialogue. Most of the Western participants in the congress were unable to perceive Hessen's paper as anything other than an exercise in Soviet propaganda or Marxist dogma. Hessen and his colleagues made their ideas easier to attack by presenting them somewhat stridently and in Marxist jargon, thus reinforcing the Westerners' prejudices.

As one British journalist observed, the Soviet delegation "would advance its cause more quickly and receive more sympathetic understanding if it would explain its ideas in the idiom of other languages, instead of merely translating expositions cast in the new Russian turn of thought."[27] But Bukharin and Hessen were bound by explicit instructions of the Soviet Politburo "to emphasize Marxism in their reports."[28] They were therefore no doubt more concerned with how their papers would be received at home than they were with impressing Western academics.

The congress resulted in an enduring polarization among professional historians of science. The traditional outlook remained dominant for several decades, but Hessen had planted the seed of an opposing viewpoint based on taking social context and factors external to science into account, and it began to germinate. The political and social radicalization of the 1960s gave fresh impetus to the "externalist" or "contextualist" challenge, which has increasingly guided the research agendas of historians of science ever since.

Ironically, in the immediate wake of the congress, Hessen's thesis linking science to the rise of capitalism fared no better in the Soviet Union than it did in the West. It was doomed by its factional associations; by the end of the decade, both Hessen and Bukharin had perished as victims of Stalin's purges. Before the reawakening of the 1960s, those most capable of appreciating Hessen's seminal insight were British scientists with Marxist inclinations such as Joseph Needham, J. D. Bernal, J. B. S. Haldane, Lancelot Hogben, and Hyman Levy.[29] Two important echoes of the Hessen thesis also occurred in the work of Edgar Zilsel (discussed in chapter 6) and in that of Robert Merton.

As an explicitly Marxist analysis, Zilsel's work received a cold shoulder in Cold War academic circles. By contrast, Robert Merton's *Science, Technology and Society in Seventeenth-Century England*, which "relied quite heavily on Hessen," was accorded a better reception. "Somewhat surprisingly for the man who was to become one principal representative of mainstream American sociology, both the overall organization of his pertinent account and a substantial number of factual data were taken over from Hessen, with due reference but rather uncritically."[30]

Merton sanitized the Hessen thesis for Western consumption by replacing the Marxist jargon with academic sociological jargon and by qualifying every statement almost to the point of meaninglessness. But while bending to the conservatizing pressures of Cold War American academia, Merton lent a measure of respectability to the Hessen thesis that allowed it a hearing it would not otherwise have received.[31]

STALIN'S "PROLETARIAN SCIENCE"

BECAUSE LYSENKOISM HAS become the paradigmatic case of the destructiveness of political interference with scientific practice, Stalin's claim that it was a proletarian science may seem to reflect badly on the very idea of people's science. The similarity of the two terms, however, is purely nominal. Lysenko's proletarian science *counterposed* itself to solidly established knowledge in the field of genetics. The people's science of this book, by contrast, refers to the participation of many kinds of people in *creating* the established knowledge in genetics and all other fields of science. The recognition that modern scientific knowledge has developed in specific social contexts does not negate its essentially universal nature.[32]

The Stalinist opposition of "proletarian science" to "bourgeois science" is akin to the various alternative sciences promoted by cultural nationalists. The Nazis notoriously counterposed an "Aryan science" to "Jewish science." Hindu traditionalists espouse a "Hindu science" and Islamic fundamentalists an "Islamic science" as weapons against "decadent Western science." But like Stalin and Lysenko, the various cultural movements that have sought to distance themselves from the universalism of modern science have always been more interested in exercising ideological authority than in gaining knowledge of nature.

Meera Nanda, an Indian scientist and social activist, has urged non-Western peoples to recognize the ethnocentric sciences as "untested and untestable cosmologies that are used to justify . . . the despotism of some of our own cultural traditions." She warned that "a cultural nationalism that turns against the internationalism of science is completely devoid of any progressive impulse, and for all its populist rhetoric, can only keep the people it claims to speak for in the bondage of age-old oppressions justified by ancient superstitions."[33] The despotism of which she writes has traditionally been most damaging to women.

Meera Nanda's point is well taken, but unfortunately, the benefits of universal science have not been universally shared.

In fact, the uses to which modern science has been put have more often than not proven oppressive to the people of the neo-colonial world. From the cultural nationalists' perspective, "Western" science has added insult to injury by first forging the chains of foreign domination and then flaunting itself as the symbol of cultural superiority. Furthermore, the claims that modern science could improve the conditions of life in the less affluent parts of the world have not been fulfilled. Its most celebrated promise was that it could solve the most fundamental of all social problems: world hunger.

THE GREEN REVOLUTION: SCIENCE FOR THE PEOPLE?

WIDESPREAD MALNUTRITION IN poor countries underlies diseases responsible for *tens of thousands of deaths every day*. The existence of malnutrition on so vast a scale means not enough food is being produced for everyone to have enough to eat, does it not? And if that is the problem, then the obvious solution is to increase food production, is it not? Based on that commonsense logic, certain Western institutions set out several decades ago to bring the resources of modern science to bear on the curse of world hunger by producing knowledge that could help poor farmers in Asia, Africa, and Latin America grow more food.

It began in 1944 with a Rockefeller Foundation program designed to increase agricultural productivity in Mexico. Its success in increasing crop yields was so dramatic that the desire to extend its benefits to the rest of the world was irresistible. That was the debut of the Green Revolution, modern science's answer to global hunger and poverty. It would, its proponents hoped, render the violent cataclysms of "red revolution" obsolete as the prerequisite to breaking the chains of poverty.

By the 1970s, new strains of wheat, rice, and corn developed by Rockefeller and Ford Foundation research institutes had been propagated throughout the world. The "miracle seeds" of the Green Revolution were accompanied by new farming practices that replaced the traditional methods of millions of poor farmers:

> By the 1990s, almost 75 percent of Asian rice areas were sown with these new varieties. The same was true for almost half of the wheat planted in Africa and more than half of that in Latin America and Asia, and about 70 percent of the world's corn as well. Overall, it was estimated that 40 percent of all farmers in the Third World were using Green Revolution seeds, with the greatest use found in Asia, followed by Latin America.[34]

On its own terms, the Green Revolution represented a brilliant accomplishment. "The production advances of the Green Revolution are no myth," even its severest critics acknowledge; "Thanks to the new seeds, tens of millions of extra tons of grain a year are being harvested." This was a good thing—but not good enough. The great, damning paradox of the Green Revolution is "that increased food production can—and often does—go hand in hand with greater hunger." A statistical analysis of the achievements of the Green Revolution from 1970 to 1990 revealed that in South America,

> while per capita food supplies rose almost 8 percent, the number of hungry people also went up, by 19 percent. In south Asia, there was 9 percent more food per person by 1990, but there were also 9 percent more hungry people. [It was not] increased population that made for more hungry people. The total food available per person actually increased.[35]

More food and yet more hunger? How can that be? "In a nutshell, if the poor don't have the money to buy food, increased production is not going to help them." The real problem is that "far too many people do not have access to the food that is already available because of deep and growing inequality." The Green Revolution, unfortunately, tended not to meliorate but to exacerbate social inequality. By introducing it into "a social system stacked in favor of the rich and against the poor, without addressing the social questions of access to the technology's benefits," it widened the gulf between larger and smaller producers.[36]

To work their miracle, the new varieties of wheat, rice, and corn required massive inputs of fertilizers and pesticides, which most peasant farmers could not afford. The big growers who could afford them prospered, whereas the poorer farmers had their livelihood destroyed. The big landowners then grew the most profitable cash crops for export rather than staple crops to feed the home population. And the large-scale "farmers" who benefited most of all from the new technology were not in Asia, Africa, or Latin America; they were agribusiness giants such as Del Monte, Anderson Clayton, and Standard Brands.

The naïve belief that increasing the food supply would resolve the hunger problem stemmed from a profound ignorance of agricultural economic history. Until little more than a century ago, food production was insufficient to meet the basic needs of the world's population. In that context Malthus's 1798 prediction that the global food supply would become ever more inadequate did not seem unreasonable. But how utterly wrong he was! By the last quarter of the nineteenth century, agricultural productivity had increased to the point that *too much* food was being produced—too much, that is, for the market to absorb. Crop prices plummeted, and growing numbers of farmers went bankrupt. A *permanent crisis of overproduction in agriculture* set in that has persisted to the present.

By the time of the Great Depression of the 1930s, the growing tide of "surplus" food required government intervention to avoid the total paralysis of agriculture, producing one of history's most glaring absurdities. In a hungry world, agricultural policy in the advanced, market-ruled economies has long been devoted to *reducing food production* in the name of "price support." Mountains of wheat and other grains have been burned or put into storage to keep them off the market, and farmers are paid to withhold tens of millions of acres of cropland from production. This state of affairs has been called "agricultural

Malthusianism" in ironic tribute to Malthus's wrong-way prognosis. Is it any wonder, then, that the Green Revolution's strategy of increasing food production has not succeeded in alleviating world hunger?

Ultimately, "hunger is not caused by a shortage of food, and cannot be eliminated by producing more."[37] Meanwhile, the increasing dependence of agriculture on costly petrochemicals strengthened international corporate control over the world's food supply. But as their chemical fertilizers and pesticides ran afoul of the law of diminishing returns, the same corporations began promoting genetic engineering as another scientific quick fix. "We must be skeptical," critics of the Green Revolution warn, "when Monsanto, DuPont, Novartis, and other chemical-cum-biotechnology companies tell us that genetic engineering will boost crop yields and feed the hungry."[38]

The dangers of relying on large-scale pesticide production to boost food production were most tragically demonstrated by the worst industrial accident in history: the 1984 leak of methyl isocyanate gas from a pesticide factory in Bhopal, India, that caused 20,000 deaths and left 100,000 people with chronic illnesses. Twenty years after the disaster, Amnesty International issued a scathing report condemning the Bhopal plant's owners for evading human rights responsibilities. The Union Carbide Corporation and its parent, Dow Chemical, have refused to accept legal responsibility for the accident and have denied meaningful compensation to its victims. Amnesty's report also charged Union Carbide with withholding information critical to the victims' treatment.[39]

The Green Revolution was advertised as the supreme example of "science for the people," but instead it illustrated how thoroughly science has been subordinated to the dictates of the dominant global economic system. Despite the brilliant advance of modern science and the technological wonders that have accompanied it, the world today is still a place where small pockets of glittering opulence mock the condition of billions of human beings mired in hunger, disease, oppression, and grinding poverty.

Small wonder that the late twentieth century witnessed the rise of widespread distrust of modern science. Scientists are now widely perceived to be fallible—and arrogant in their unwillingness to acknowledge their fallibility. They are also frequently thought of as paid apologists for the interests of giant corporations or governmental bureaucracies. Welcome to the era of "postmodern science."

This sea change in the popular attitude toward science was a reflection of the tragic course of twentieth-century history, most strikingly exemplified by the nuclear incineration of the populations of Hiroshima and Nagasaki. Prophets of doom had previously warned of the dark potential of modern science, but the fears they projected were for the most part ignored by international public opinion until the atomic bombs demonstrated just how dangerous knowledge of nature could become.

RACHEL CARSON
AND THE ENVIRONMENTAL MOVEMENT

Big Science's popularity had the advantage of inertia; it did not decline immediately. But cracks in the scientific monolith began to appear in the late 1950s when an organization founded by Barry Commoner and other politically conscious scientists, the St. Louis Committee for Nuclear Information, warned of the dangers of radioactive fallout from atom-bomb tests in Nevada. Commoner and his allies spearheaded a "science information movement" to promote the idea that a scientist's primary obligation is not to governmental policymakers but to the people. It is the moral duty of scientists, they maintained, to inform the general public fully and directly about the scientific aspects of social issues.[40]

But the most significant dissident voice by far was Rachel Carson's. The publication of her *Silent Spring* in 1962 gave rise to the environmentalist movement and sparked a debate that forced a thoroughgoing reconsideration of the place of science in human affairs.[41] She declared that the massive modern use of pesticides and other synthetic chemicals to produce food posed a serious threat to humankind and ultimately to the continued existence of all life on earth. By linking an impending environmental crisis to the blind drive for profits by agribusiness and the chemical industry, *Silent Spring* issued a fundamental challenge to the "union of capital and science."

Not unexpectedly, the industries she targeted, in collaboration with the U.S. Department of Agriculture, launched a ferocious assault against her and her book. Major chemical companies and their trade associations "spent a quarter of a million dollars to discredit her research and malign her character."[42] The *New York Times* reported, "Some agricultural chemicals concerns have set their scientists to analyzing Miss Carson's work, line by line. Other companies are preparing briefs defending the use of their products. Meetings have been held in Washington and New York. Statements are being drafted and counter-attacks plotted."[43]

Because "in postwar America, science was god, and science was male," it was inevitable that the author's gender would be a conspicuous element of the campaign against *Silent Spring*. The chemical industry's flacks portrayed Carson as

> a hysterical woman whose alarming view of the future could be ignored or, if necessary, suppressed. She was a "bird and bunny lover," a woman who kept cats and was therefore clearly suspect. She was a romantic "spinster" who was simply overwrought about genetics. In short, Carson was a woman out of control. She had overstepped the bounds of her gender and her science.[44]

Scientists on the payroll of the polluting corporations believed they could dismiss her arguments on the grounds that she was "an outsider who had never been part of the scientific establishment. . . . Her career path was nontraditional; she had no academic affiliation, no institutional voice." Most damning in their eyes was that "she deliberately wrote for the public rather than for a narrow scientific audience."[45] But in spite of the scientific elite's attempts to marginalize her, this "people's author" ignited a momentous social movement in defiance of Big Science. "We live in a scientific age," she declared, "yet we assume that knowledge of science is the prerogative of only a small number of human beings, isolated and priestlike in their laboratories. This is not true. The materials of science are the materials of life itself."[46]

Carson put forward "her own, alternative scientific method: people's observations and interpretations were as important as those of scientists, and community ethics served as the standard for making decisions about environmental risks."[47] As for her influence on the practice of science itself, by redirecting interest toward ecology and away from traditional mechanistic and reductionist approaches, *Silent Spring* had a major impact on the way biological knowledge would henceforth be pursued.

When the chemical industry and its kept scientists realized that public concern for the natural environment was a genie that could not be stuffed back in its bottle, a change in strategy became necessary. Rather than trying to combat the movement head-on, they co-opted it. The result was that "environmentalism became a matter of political consensus dominated by professional environmentalists." The scientific elite "embraced the mantra of ecology," which

> became one of the conceptual cornerstones of mainstream environmentalism. But it was not a subversive ecology that questioned fundamental values of economics, consumer habits, and techno-scientific control. It represented an engineering mentality in which problems of waste, pollution, population, biodiversity and the toxic environment could be solved scientifically.[48]

In addition to its mainstream, environmentalism includes more radical elements such as ecofeminism and the environmental justice movement. The latter represents, in the words of a spokesman, "people of color, working class people, and poor people," who are just as much concerned about wetlands, birds and wilderness areas, but we're also concerned with urban habitats, where people live in cities, about reservations, about things that are happening along the U.S.–Mexican border, about children that are being poisoned by lead in housing and kids playing outside in contaminated playgrounds.[49]

Despite the environmental justice movement, the ecofeminists, and other radicals, the corporate strategy of co-optation has for the most part been successful in resisting fundamental change. As a consequence, although some environmental problems have been addressed by governmental action and some limited reforms have been implemented, the overall picture remains bleak. "In spite of decades of environmental protest and awareness, and in spite of Rachel Carson's apocalyptic call alerting Americans to the problem of toxic chemicals," her biographer lamented, "reduction of the use of pesticides has been one of the major policy failures of the environmental era. Global contamination is a fact of modern life."[50]

Meanwhile, another genie had escaped that could not be rebottled. Science had long been touted by its proponents as a superior form of knowledge, with its truth value guaranteed by the impartiality and disinterestedness of its practitioners. But the attack on *Silent Spring* revealed—to all but the willfully blind—a highly partisan Big Science acting as the mouthpiece for corporate interests. The "union of capital and science" had long exerted a corrupting influence on science, but now the mechanisms of corruption had been exposed and were more obvious than ever. Despite their protestations of objectivity, scientists in corporate employ are increasingly perceived as having sold their souls to Mammon.

FEMINISM VERSUS MEDICAL SCIENCE

THE MOST THOROUGHGOING popular challenge to a branch of orthodox science occurred when the women's liberation and science information movements converged in the late 1960s to create a medical self-help movement. To escape the clutches of "condescending, paternalistic, judgmental and non-informative" physicians, large numbers of women began to educate themselves about female anatomy and physiology, to master techniques of self-examination, and to force the male-dominated medical profession to treat them with respect as knowledgeable healthcare consumers. So commonplace has much of the feminists' critique become that their immense contribution to the transformation of medical practice tends to be overlooked and underestimated.

A milestone in the campaign was the 1973 publication of *Our Bodies, Ourselves* by the Boston Women's Health Book Collective. A revised edition (1984) proclaimed the movement's goals:

> We want to reclaim the knowledge and skills which the medical establishment has inappropriately taken over. We also want preventive and nonmedical healing methods to be available to all who need them. We are committed to exposing how the medical establishment works to suppress these alternatives (home birth and nurse-midwifery, for example).[51]

Challenging the "myth" that "medicine is a science," the authors declared,

> Fact: Medicine has prestige largely because it is associated with science in the public mind, and claims objectivity and neutrality. Much of the theory on which modern medical-practice is based actually derives from the untested assumptions and prejudices of earlier generations of medical leaders. . . . Consider the ugly pronouncements medical authorities have made over the years, in the name of "science," about so-called "inferior" groups of people like women and blacks, and the way these pseudo-scientific statements have been used to deprive these groups of social control and political power.[52]

"Ugly pronouncements" with regard to gender have been the rule rather than the exception in the history of medical science. In the nineteenth century,

> medical "scientific" theory stated first that women were ill because they were women, and second that they became ill if they tried to do anything beyond their conventional female roles, a Catch-22 situation. The "illness" almost always related to the woman's reproductive system which, it was thought . . . was by its very nature pathological.[53]

The authors of *Our Bodies, Ourselves* asked and answered a crucial question: "Why doesn't the present U.S. medical system provide affordable services and emphasize prevention and primary care? The profit motive has to be the main reason."

> There has been a phenomenal rise in the "corporatization" of medical care; that is, profit-making medical care chains which operate laboratories, "emergency rooms," mobile CAT scanners, etc., in addition to hospitals and nursing homes—a proliferation of separate services purely for profit. Academic centers, medical schools and teaching hospitals have themselves created new and unprecedented arrangements amounting to millions of dollars with profit-making drug firms, all in the name of "scientific research."[54]

The feminist critique of the corporatization of medical science is applicable to modern science in general.

THE RISE OF THE SCIENTIFIC-INDUSTRIAL COMPLEX

IN 1961, PRESIDENT Eisenhower left office with an oft-quoted warning against "the potential for the disastrous rise of misplaced power" by "the military-industrial complex."[55] Science in the United States today is both the handmaiden of that complex and a subcomplex unto itself: an intricate entwinement of university, government, and big business. Federal research grants over the years have systematically favored a limited group of institutions, which "guaranteed that the rich got richer."[56]

"Two-thirds of research is now financed by companies. And much of this 'privatized' science is falling into the hands of ever fewer—and ever bigger—global corporations," the *New Scientist* reported in 2002.[57] The corruption of science is not confined to researchers who work directly for corporations. No less complicit are the university laboratories and government agencies whose ties to industry have bound them ever more tightly into a single entity. "Conflict of interest in science has become the norm of behavior, rather than the exception," social critic Sheldon Krimsky has observed; "The academic research milieu is generally acknowledged to be permanently and unabashedly linked to the private sector."[58]

The privatization of science was greatly facilitated by the passage of the Bayh-Dole Act in 1980, which allowed universities and small businesses to patent the findings produced by federally funded research. As surely as night follows day, major corporations were certain eventually to gain the same privileges, which they did in 1987. The boundary lines separating government, industrial, and academic research have increasingly blurred. The upshot is that public dollars pay universities to produce knowledge that becomes the private property of corporations.

"BIG PHARMA" AND THE MEDICAL SCIENCES

THE UNTRUSTWORTHINESS OF research findings generated by scientists on pharmaceutical industry payrolls is of crucial concern to all who ingest the medicines they produce. "About a quarter of scientists working in medical research have some sort of financial relationship with industry," one critic reports. "And, not surprisingly, there is a strong association between commercial sponsorship and the conclusions scientists draw from their findings."[59] A prestigious voice of the medical profession, the *Journal of the American Medical Association* (*JAMA*), has acknowledged the problem: "In collusion with money-seeking academic researchers, *JAMA* tells us, some pharmaceutical firms, large and small, have misrepresented and inflated scientific data to gain, first, regulatory approval, and then to influence physicians to prescribe their products to an unwitting public."[60]

JAMA, however, neglected to mention its own culpability. As editor of *The Lancet*, Richard Horton is competent to assess the current state of medical journalism. "The process of publication," he charged, "has been reduced to marketing dressed up as legitimate science." Medical journals have "devolved into information-laundering operations for the pharmaceutical industry," thus becoming an "obstacle to scientific truth-telling." Their posture as "neutral arbiters" is contradicted by their being "owned by publishers and scientific societies that derive and demand huge earnings from advertising by drug companies." Their pronouncements cannot be trusted. "Opinions are rented out to the highest bidder," Horton concluded. "Knowledge is just one more commodity to be traded."[61]

Just because a scientist's name is listed as an author of an article in a science journal, it cannot be assumed that he or she had anything to do with actually writing it, because there is now "a ghostwriting industry in science and medicine."[62] In May 2002, for example, the *New York Times* reported, in a story about Neurontin, a drug approved for treating epilepsy: "Warner-Lambert also hired two marketing firms to write articles about the unapproved uses of Neurontin and found doctors willing to sign their names to them as authors."[63]

Exerpta Medica is a medical publishing company in New Jersey that provides pharmaceutical companies "an invaluable tool: ready made scientific articles placed in leading medical journals, and carrying the imprimatur of influential academic leaders."[64]

> It works like this: Excerpta is contracted by a company to find a distinguished academic scholar to agree to have his or her name placed on a commentary, editorial, review, or research article, which has been written by someone either from the

company or someone selected by Excerpta. [In one particular case], the writer of the article was a freelancer who was paid $5,000 to research and write the article under company standards. The university scientist, whose name appeared as the author, was paid $1,500. Representatives of the pharmaceutical industry claim that it is a common practice to have articles that appear in journals ghostwritten by freelancers.[65]

In some cases, "the scientists named as authors will not have seen the raw data they are writing about—just tables compiled by company employees."[66]

Scientific information directed at laypeople is likewise often tainted by hidden and unacknowledged conflicts of interest. In 1966, a widely read book entitled *Feminine Forever* promoted estrogen as a miracle drug that would help women maintain their youth and beauty. Authored by an M.D., the book appeared medically authoritative, but it was later revealed to have been financed by Wyeth, a pharmaceutical company marketing hormone replacement therapy drugs.[67]

WON'T THE GOVERNMENT'S SCIENTISTS PROTECT US?

UNFORTUNATELY, GOVERNMENTAL OVERSIGHT of Big Pharma's science cannot be relied on. In the United States, Food and Drug Administration (FDA) scientists are charged with protecting the public from unsafe drugs. The FDA, however, has devolved into "an agency where conflicts of interest have become normalized in the process of drug evaluation."[68] In 1998, the FDA put its stamp of approval on a vaccine that soon thereafter had to be pulled from the market because of reports linking it to severe bowel obstruction in children. It was subsequently revealed that "the advisory committees of the FDA and the Centers for Disease Control were filled with members who had ties to the vaccine manufacturers."[69] In 2000, an investigation disclosed that "more than half of the experts hired to advise the government on the safety and effectiveness of medicine have financial relationships with the pharmaceutical companies that will be helped or hurt by their decisions."[70]

In November 2004, David Graham, an FDA safety officer for twenty years, testified before the U.S. Senate that his supervisors had tried to "silence him and pressure him to limit his criticism of the safety of some drugs." Although FDA officials denied Graham's allegations, the latter were corroborated one month later by a report of the Health and Human Services Department's inspector general. The report stated that "the work environment at the FDA's Center for Drug Evaluation and Research either allowed little dissent or stifled scientific dissent entirely." Of three hundred sixty FDA scientists surveyed, sixty-three reported having been "pressured to approve or recommend approval for a (new drug application) despite reservations about the safety, efficacy or quality of the drug." The federal agency did not intend to make the results of its investigation public; the report was forced into the open through the efforts of two science information groups: Public Employees for Environmental Responsibility and the Union of Concerned Scientists.[71]

Known cases of scientific conflict of interest and misconduct abound, although, as Krimsky pointed out, those that come to public notice are but "the tip of the proverbial iceberg." Remember the Alar scandal? In 1989, the American public was informed by the popular television program *60 Minutes* that this widely used pesticide was a potentially dangerous carcinogen.[72] In fact, four years earlier, a committee of Environmental Protection Agency (EPA) scientists had reached that conclusion, but the EPA was required to have its findings reviewed by an outside advisory panel of supposedly disinterested experts. That panel determined Alar (the trade name for the chemical daminozine) to be acceptably safe, thus allowing it to be put on the market. It was later discovered, however, that seven of the eight members of the advisory panel had served as paid consultants for Uniroyal, the sole manufacturer of daminozine.[73]

Toxicological studies that minimize the health hazards of cigarette smoking constitute perhaps the most transparent abuse of the authority of science. A World Health Organization report in the year 2000 revealed how "tobacco companies set up front scientific organizations and funded advocacy science to dispute responsible studies linking tobacco companies to cancer."[74] In one particularly blatant case, John Graham, director of Harvard University's Center for Risk Analysis, "solicited financial contributions from a cigarette company while also downplaying the risks of secondhand smoke."[75]

The Harvard policy center (and others like it) "serve unabashedly as legitimating voices for well-endowed corporations." Its financial support comes from "more than one hundred large corporations and trade associations, including Dow, Monsanto, DuPont, as well as major trade groups such as the Chlorine Chemistry Council and the Chemical Manufacturers Association." In addition to its services rendered to the tobacco industry, the center "downplayed the risks to children from exposures to pesticides and plasticizers (bisphenol-A and phthalates) without alerting the readers that the center is funded by the manufacturers of those chemicals."[76]

All too rarely do disclosures of scientific transgressions emerge from reports of governmental agencies or congressional hearings. More often they are brought to light by nongovernmental organizations such as the Union of Concerned Scientists, the Association for Science in the Public Interest, the Natural Resources Defense Council, and Public Citizen. But the most prolific source of such revelations is an occupational group that would not ordinarily be thought of as deserving mention in a history of science: investigative journalists. It is they who have most frequently and most successfully exposed the pervasive conflicts of interest that have undermined the integrity of modern science.

BIG SCIENCE AND BIG MONEY
IN THE UNITED STATES TODAY

ALTHOUGH BIG SCIENCE is international, any discussion of it must focus on its American component, because for the past half-century the United States has been the dominant player in the game. According to a 1998 report of the National Science Foundation, more money is spent on research in the United States than in Japan, Germany, England, France, Italy, and Canada *combined*. As for Russia and the other spin-offs of the former Soviet Union, science in those countries "barely survives, relying heavily on foreign charity, and no longer enters the list."[77] The money magnet has caused a "brain drain" that continuously enriches American science with talent from all over the world.

Throughout the nineteenth century and the early part of the twentieth, "science and government kept a mutually wary distance" in the United States. In those comparatively innocent days, basic research was "an elite activity, largely financed by private philanthropy and concentrated in a handful of universities and independent research institutes."[78] World War II changed all that, as the exigencies of modern military conflict compelled the American government to take command of the scientific enterprise. The centralized mobilization of research and the great increase in resources devoted to it produced a diversity of important results, from radar to antimalarial drugs, but its defining triumph was the nuclear fission weapon created by the Manhattan Project.

Competent scholars have convincingly argued that American policymakers ordered the atomic incineration of two Japanese cities not to defeat Japan but to hasten the war's end in order to exclude the Soviet Union from postwar spoils in that part of the world.[79] Be that as it may, the Cold War provided the rationale for the U.S. government's rapidly growing role in science after the war. Federal spending on academic research and development rose from less than $150 million in 1953 to almost $10 billion in 1990.[80]

But the governmental stewardship of science did not diminish with the end of the Cold War; the much-anticipated "peace dividend" never materialized.[81] The stated reason that federal research and development (R&D) budgets continued to increase year by year was the need to remain economically competitive with rival industrial nations, but the permanent war economy in the United States is difficult to disguise. The portion of the R&D budget allocated to what is euphemistically called "defense" amounted to almost exactly half of the total of $75.4 billion in fiscal 2000.[82] Another $8.4 billion went for space research officially categorized as civilian but ultimately motivated by considerations of its potential military applications. In the same spirit, "the Department of Energy, descended from the Manhattan Project, provides over $2 billion annually for physics, the nuclear sciences, and other disciplines in science and engineering."[83]

KEYNES AND THE PERMANENT CRISIS OF OVERPRODUCTION

BECAUSE THE FORM, content, and direction of science have been so strongly influenced by immense expenditures on war-related research, knowing why those expenditures are made is essential to understanding the place of science in contemporary America. It certainly has nothing to do with preparedness to combat a genuine military threat. When the bugaboo of "international communism" evaporated with the collapse of the Soviet Union, another specious justification for maintaining the several-hundred-billion-dollar war budget—"international terrorism"—was quickly conjured up. Random acts of terrorism pose a real (if statistically miniscule) threat to some urban populations, but to think that the architects of American imperialism really fear raggedy groups of Islamic radicals is equivalent to belief in the bogeyman.

An observation by Dr. Helen Caldicott renders the deceit apparent: the U.S. Department of Energy is currently engaged in "a massive scientific undertaking costing 5 to 6 billion dollars annually for the next ten to fifteen years, to design, test, and develop new nuclear weapons," but "the largest nuclear stockpile in the world can accomplish little in the face of terrorists armed with box cutters."[84]

Neither is the war spending primarily motivated by a desire

for weapons for offensive purposes. Most of all, it is necessary to keep the wheels of the American economy from rapidly grinding to a halt. The Great Depression revealed that the capitalist system, left to its own devices, has become so productive that it is no longer capable of generating enough purchasing power to absorb all the products with which it continuously floods the market. John Maynard Keynes explained to Franklin Roosevelt that to create enough "aggregate demand" to keep the economy from freezing up, governments would henceforth have to create new purchasing power (i.e., new jobs) by engaging in massive deficit spending.[85]

It would not suffice to merely "prime the pump" and then step back to allow the invisible hand of supply and demand to reestablish economic equilibrium. Government deficit spending was destined to become a *permanent* condition, with deficits *continuously increasing*. When questioned as to what would happen "in the long run" as governments continued endlessly piling up mountains of debt, Keynes's famous riposte was, "In the long run we are all dead."[86]

Not all deficit spending, it was discovered, is equally effective in preventing economic gridlock. Using government money to produce useful things such as schools or housing or highways does not help because it competes with private capital, which puts downward pressure on the number of jobs in the private sector and on the purchasing power they represent. The most effective of Roosevelt's public works programs were those that produced nothing, most notoriously exemplified by legions of workers with shovels digging holes and then filling them back in again.[87] As useless as such activity would seem to be, it gave workers paychecks that allowed them to buy some of the surplus production without having them create more surplus products. But the apparent wastefulness was an insult to reason, and it was impossible in the American political context to explain that the paradox was an inescapable feature of the capitalist economic system.

In any event, the deficit spending represented by Roosevelt's public works programs was far from adequate to lift the American economy out of the mire. What ended the Great Depression was the truly massive military expenditure in the run-up to World War II.

After the war, the rebuilding of Europe through the Marshall Plan eased the problem of insufficient aggregate demand, but that was a temporary fix. To prevent the world economy from once again lapsing into a terminal crisis of overproduction, governments would continuously have to spend enormous amounts on utterly useless production—industrial output that would not house, feed, clothe, or otherwise benefit anybody in any way. But how could that be justified? The answer was found in weapons systems deemed necessary (wink, wink) for national security. Thus was born the ever-increasing "defense" budget, which has been the primary source of science funding ever since. It is sad to have to conclude that the major portion of Big Science's attention has been and is still being directed toward a vast exercise in deliberate waste.

The most egregious example of Big Science's planned wastefulness is the Strategic Defense Initiative (SDI), popularly known as "Star Wars." The Reagan administration's 1983 announcement of the intention to create a "shield" in outer space that could protect the United States from incoming missiles raised the prospect of massive federal investment in scientific research. The immense contracts at stake were a powerful inducement to corporate and university laboratories, but the rise of significant opposition to the program on the part of scientists was an unexpected development.

The Union of Concerned Scientists produced a detailed report entitled *The Fallacy of Star Wars*. "Nationwide, some 2,300 university researchers were doing the unthinkable, pledging they would not apply for or accept the bountiful funds that the Strategic Defense Initiative Organization wished to infuse into academic research." But the negative findings of independent scientists were no match for the power of government money. SDI officials "reported over 3,000 applications for funds from university scientists willing to do business with the missile-defense program."[88]

And so SDI flourished, and a great deal of junk science was paid for and produced over many years in the effort to provide it with credibility. After Reagan and the first President Bush, the Clinton administration changed the program's name to the Ballistic Missile Defense Organization but continued to fund it. "From the birth of Star Wars, in 1984, to the end of the century, missile defense consumed over $60 billion. The enormous expenditures," however, "have produced negligible results."[89] Bush the Younger has continued down the path toward the militarization of space; his administration's National Missile Defense program has been aptly nicknamed "Son of Star Wars."

Unfortunately, the waste of resources represented by trillions of dollars in misguided research and unused weapons is not the worst part of the story. By the measure of human suffering, much more costly than bombs that have never been dropped are the ones that have: "Millions and millions of tons of bombs dropped; millions and millions of dead, mostly, of course, civilians."[90]

THE NUCLEAR THREAT AND
THE MOVEMENT OPPOSING IT

The undropped bombs, however, cannot be ignored. During the Cold War, they were used to maintain a "balance of terror" between the United States and the Soviet Union, a strategy appropriately named MAD, an acronym for "Mutually Assured Destruction." The end of the Cold War did not reduce the potential danger. This description of the dimensions of the still-existing international nuclear arsenal was published in 2004:

> The U.S. currently has 2,000 intercontinental land-based hydrogen bombs, 3,456 nuclear weapons on submarines roaming the seas 15 minutes from their targets, and 1,750 nuclear weapons on intercontinental planes ready for delivery. Of these 7,206 weapons, roughly 2,500 remain on hair-trigger alert, ready to be launched at the press of a button.

Russia has a similar number of strategic weapons, with approximately 2,000 on hair-trigger alert. In total there is now enough explosive power in the combined nuclear arsenals of the world to "overkill" every person on earth 32 times.[91]

Nuclear proliferation is the most perilous consequence of Big Science out of social control. And weapons are but the half of it. "Atoms for Peace" power plants that use nuclear reactors to generate energy also produce the greatest environmental hazard of all: radioactivity that finds its way into the earth's atmosphere and waters. Dr. Caldicott, a leading figure in the campaign to educate the public about the nuclear dangers, warned that "if present trends continue, the air we breathe, the food we eat, and the water we drink will soon be contaminated with enough radioactive pollutants to pose a potential health hazard far greater than any plague humanity has ever experienced."[92]

Added to the possibility of spectacular accidents that could release large quantities of radioactive materials is the accumulation day in and day out of nuclear waste, the by-product of uranium mining and processing and of the routine operation of power plants.[93] The radioactivity, which remains potent for hundreds of thousands of years, causes cancer and alters reproductive genes, "resulting in an increased incidence of congenitally deformed and diseased offspring, not just in the next generation, but for the rest of time."[94]

In response to these dangers, grassroots movements demanding nuclear disarmament and the dismantling of the power plants proliferated in the 1960s and '70s. A series of nuclear accidents—capped by the near-total meltdown of a reactor at Three Mile Island in Pennsylvania in March 1979—heightened the public's fear and distrust. The extent of popular opposition was most dramatically displayed in June 1982, when an estimated one million people participated in an antinuclear protest rally in New York City, the largest demonstration in American history. Antinuclear sentiment remained strong throughout the 1980s, especially after the April 1986 explosion of the Chernobyl nuclear plant, which scattered radioactive material over much of Europe.

In the United States, although a core of dedicated activists continued the antinuclear campaign, their efforts met with diminishing returns during the 1990s as the nuclear industry kept a low profile. Nonetheless, nuclear power production was quietly being stepped up; by the end of the century, 103 commercial power reactors accounted for about 20 percent of the nation's electricity. At the turn of the twenty-first century, the nuclear industry felt secure enough to return openly to the offensive by exploiting its close ties with a particularly accommodating government:

> Since taking office, [George W.] Bush and his administration have proposed expedited licensing for nuclear reactors employing new technology; favored expanding the generating capacity of existing nuclear plants; encouraged the relicensing of older plants scheduled to be shut down and mothballed; extended and expanded a government insurance program that provides liability coverage to nuclear plant operators for catastrophic accidents; even promoted nuclear power as an environmentally friendly power source.[95]

Antinuclear activists have been aided by independent scientists in making their case; the organization Physicians for Social Responsibility merits special commendation. But the nuclear industry and its governmental allies have managed to confuse the debate by hiring as their mouthpieces the finest scientists money can buy. Regardless of the quality of their individual or collective research, the pronouncements of paid apologists for commercial interests cannot be taken at face value and must be considered suspect.

It is profoundly ironic that physics is held up as the paragon of scientific objectivity while nuclear physicists have been among the worst offenders in allowing their science to be falsified and used by self-serving external interests. Their most prominent exemplar is the late Edward Teller—often misleadingly identified as "the father of the hydrogen bomb"—who served as an aggressive advocate for the military-industrial complex.[96]

Equally disturbing is the way major representatives of medical science have prostituted themselves to pronuclear interests. "Even the American Medical Association (AMA) is an apologist for nuclear power," Dr. Caldicott declared. A 1989 AMA position paper concluded that nuclear power generation in the United States is acceptably safe. "The Council of Scientific Affairs, which formulated the AMA document, was staffed by staunch advocates and employees of the nuclear industry."[97]

The apparently unchallengeable dominance of Big Science points toward the story of people's science ending like all biographies: with its subject dead. But wait! Cancel the funeral! The scientific creativity of ordinary human beings has not yet expired.

SCIENTISTS IN GARAGES

THE OBJECTS IN outer space known as black holes were once thought to be so massive that nothing—not a single subatomic particle or wavelet—could escape their gravitational fields. When subsequent observations tended to show otherwise, "wormholes" in space-time were postulated to explain how hapless little bits of matter or energy could worm their way out of the overpowering grasp of the gargantuan black holes. Analogously, it seems that the feeble but persistent spirit of people's science found a way, late in the twentieth century, to wiggle free of the clutches of Big Science's omnipotent force field. Just when it seemed that significant advances could never occur anywhere other than in elite professional settings, scientific innovations of the first magnitude began to emerge from the garage and attic workshops of college dropouts, initiating a tidal wave of scientific creativity outside the corporate-academic complex.

The first electronic digital computers were by-products of the Cold War; they were designed and utilized for military purposes. Gigantic in size and extremely expensive, the computers of the 1950s and '60s "were the very symbols of entrenched and centralized power—arrogant, haughty, impersonal, inefficient, and inaccessible."[98] In the 1970s, however, numerous self-taught amateur electronics hobbyists with meager resources at their disposal—many of them very young—democratized the new technology by bringing it within reach of millions of ordinary individuals.

In 1975, the birth of the "home computer" was announced when the mass-market magazine *Popular Electronics* featured a small programmable device—the Altair 8800—that could be bought in kit form for $395.[99] The creators of the Altair were three Air Force engineers, Edward Roberts, William Yates, and Jim Bybee. Their company, "a small model-rocket hobby shop in Albuquerque," operated out of the garage at Roberts's home. Improbable as it may seem, their microcomputer "managed to bring down the mighty houses of IBM, Wang, UNIVAC, Digital, and Control Data Corporation."[100]

Credit for creating the personal computer cannot be ascribed to Roberts, Yates, and Bybee alone, because their design depended on the prior work of many others—those who invented and developed the transistor, the integrated circuit, and the microprocessor, to mention a few. The Altair, however, was a breakthrough that attracted legions of young innovators to the field. Among them were Bill Gates, Paul Allen, and Monte Davidoff, who developed the coded programs that the Altair needed to function as a computer. Gates and Davidoff were students at Harvard; Allen had previously dropped out of Washington State University. Gates, nineteen years old, quit school and joined Allen in founding a company to market their software.

Earlier, in 1971, an auspicious collaboration had begun between another college dropout, Stephen Wozniak, and a high school student, Steven Jobs. The appearance of the Altair inspired the proliferation of computer clubs across the United States—a social movement of sorts. Wozniak became an ardent member of the Homebrew Computer Club, which originally met in an engineer's garage in Menlo Park, California. Drawing encouragement from that milieu, he produced an original design for a personal computer, and together with Jobs formed the Apple Computer Company to manufacture and market it.

To fund their venture, they raised a little over $6,000 by selling some of their personal belongings and borrowing money from friends. "Apple's humble distribution center, sales office, and world headquarters" was set up—where else?—in the Jobs family's garage.[101] The subsequent meteoric rise of Apple and the social impact of its Apple II and Macintosh in making computers "user-friendly" for ever-increasing numbers of people are familiar stories.[102]

As central as Wozniak and Jobs were to the creation of the Apple II and the Macintosh, the contributions to computer science made by their many collaborators also deserve recognition. A comprehensive history would describe the roles of Rod Holt, Andy Herzfeld, Bill Atkinson, Bud Tribble, Burrell Smith, Jerry Manock, and Randy Wigginton, to name but a few. Holt, one of the principal designers of the Apple II, complained that most histories of the subject begin with the Macintosh, but that, he says, "is like starting the history of the Russian Revolution with the ascendance of Stalin." First of all, he declares, "the Mac was later—much later. Apple dominated the world market long before Mac." And he makes a strong case for the centrality of the Apple II to the development and democratization of computer science:

> The Apple II was not a typewriter; it was an instrument for the intellect wrestling with beautiful algorithms. People did not buy it because they wanted to run programs. They bought it to write their own software. It is impossible to teach "how a computer works," or code-writing, with today's machines with their hundreds of megabytes of inscrutable code. But in 1976, kids—twelve-year-olds, fourteen-year-olds—could fix a machine that wasn't working.[103]

Among many important software innovations that the Apple II facilitated was VisiCalc, the first electronic spreadsheet.

VisiCalc was the creation of Daniel Bricklin and Robert Frankston, "a Harvard Business School student and his buddy, outsiders to computing."[104] Bricklin and Frankston operated out of Frankston's Arlington, Massachusetts, attic (because, one historian wryly observed, "the Boston area has fewer garages than in Silicon Valley").[105]

The institutions of Big Science at first ignored and even disparaged the innovations that were emerging from garages and attics across America, but eventually the upstarts got the attention of the establishment. And then it was not long before the upstarts *became* the establishment—or part of it, at least. Some were hired by corporate giants that entered the personal computer field, whereas others saw their garage-based enterprises grow to become major firms in their own right. The union of capital with this fledgling science occurred at warp speed. Within six years of its incorporation, Apple had some 4,700 employees and $983 million in sales.[106] The transformation of Gates's and Allen's tiny software company, meanwhile, into one of the world's mightiest corporations is legendary.

IBM had initially resisted the personal computer revolution. Recognizing that the movement could not be defeated, however, the company instead resolved to take it over. In 1981 the IBM PC entered the market and, backed by the financial power of one of the world's largest corporations, rapidly eclipsed its competitors. Although its PCs are used in homes, IBM's main concern—as its name suggests—was to transform the personal computer into a business machine. It succeeded in doing so, although eventually new competitors selling less expensive IBM "clones" would take a large part of the market away from IBM. The biggest winner in the computer profits sweepstakes was not a hardware producer but the software man, Bill Gates.

"ARTISANS OF THE INFORMATION AGE"

"PROGRAMMERS," ONE HISTORIAN of computing declared, "are the artisans, craftsmen, brick layers, and architects of the Information Age."[107] Because what they do looks more like brainwork than handwork, it may seem fallacious to identify programmers as artisans, yet they do indeed have much in common with the craftspeople who had been in the forefront of the Scientific Revolution. One of the creators of the FORTRAN programming language, John Backus, described his method as "innovation by iteration, a constant process of trial and error."[108] Insofar as that is the case, the creative core of computer science is empirical rather than theory-driven. Ken Thompson, a pioneer practitioner, explained the appeal of programming as "having all the craftsman's satisfactions of making things, without the cost and trouble of procuring all the materials."[109]

It was not apparent at first that programmer-artisans would constitute the vanguard of the new science. At the dawn of the computer era, "programming was an afterthought"; it "was considered more a technician's chore." The first of the enormous calculating machines, the ENIAC, "did not have software. Its handlers had to set up the machine by hand, plugging and unplugging a maze of wires and properly positioning row upon row of switches." Programming thus originated as a form of low-prestige *manual* work. To perform those tasks, "the government hired a handful of young women with math skills as trainees."[110] Those young women were the pioneers of programming, although the importance of their work was greatly underestimated by their contemporaries.[111]

Programming and software remained the province of scientific outsiders for several years:

> In the engineering culture of computing, programmers were long regarded askance by the hardware crowd; hardware was the real discipline, while programmers were the unruly bohemians of computing. The hardware people tended to come from the more established field of electrical engineering. There were EE departments in universities, and hardware behaved according to the no-nonsense rules of the "hard sciences" like physics and chemistry. Some mathematicians were fascinated by computers and programming, but their perspective was often from the high ground of theory, not

wrestling with code and debugging programs. It was not until the 1960s, with the formation of computer science departments, that programming began to be taken seriously in academia, and then only slowly.[112]

In the early days of computing, the programmers were scorned by the engineers, and the engineers were in turn looked down on by the "pure" mathematicians. Jean Sammet, a member of the team that created the COBOL language, had to overcome the intellectual snobbery imparted by her training in mathematics before she could embrace programming as a worthy pursuit. She and her fellow math students had shared a low opinion of computer work: "It's hard to describe how much contempt we had for the engineers in the computer center," she said.[113] Even after programming gained a measure of academic respectability, major innovations continued to come from outside the establishment. In the early years, writing computer code was a highly specialized craft. "Preparing an engineering or scientific problem so that it could be placed on a computer was an arduous and arcane task that could take weeks and required special skills. Only a small group of people had the mysterious knowledge of how to speak to the machine, as if high priests in a primitive society."[114] It was the advent of "higher-level" languages such as FORTRAN and COBOL that began to make computing accessible to growing numbers of nonspecialists.

FORTRAN was developed to allow scientists to bypass the programming priesthood and address their problems directly to the computers themselves. "The team that created FORTRAN" was "an eclectic bunch," all "outsiders to the industry establishment, which regarded their chances of success as slim to nil." They were "a young group, all still in their twenties and early thirties when FORTRAN was released."[115] Establishment foot-dragging notwithstanding, FORTRAN proved to be a brilliant success, and other higher-level languages followed in its wake. The one that was to have the most profound social impact was BASIC, which represented a major step toward computer science becoming a people's science. BASIC was created by two Dartmouth computer scientists, Thomas Kurtz and John Kemeny, but in spite of its university pedigree, most academics in the field looked down their noses at it. Many "disparaged BASIC as a toy language that fostered poor programming habits, and they refused to teach it."[116] BASIC's potential began to be realized when Gates, Allen, and Davidoff adapted it for use in the original microcomputers. But if Gates and his friends had been encumbered by formal education in computer science or mathematics, they might well have been rendered incapable of their singularly important achievement.

The variant of BASIC that Gates and Allen created for the Altair 8800 propelled them to the head of an entirely new industry. Their company, Microsoft, held the legal rights to the software that the Altair's successors also needed, famously turning Gates into "the richest man in the world."[117] Ironically, Microsoft's public image has become—not undeservedly—that of a villainous monopoly stifling the creativity of a new generation of people like Gates and Allen.

BUILDING THE INFORMATION SUPERHIGHWAY

DESPITE THE CORPORATE takeover of the personal computer industry and the crushing gravitational pull of Big Science, the democratization of computing continued to flourish into the 1990s. As more and more people became involved in computing, they discovered new ways to adapt their computers to perform new tasks. The most important social consequence was the computer's unanticipated transformation from a number-crunching machine into a major communications medium.

Although the Internet originated in the military-industrial complex, many of the crucial innovations that allowed it to grow into a global communications, information, commerce, and entertainment network were contributed by students and others outside the established institutions of computing. The World Wide Web, for example, "was invented at an unforeseen and totally unexpected place: the high-energy physics laboratory CERN, on the Swiss-French border. It did not come from the research laboratories of IBM, Xerox, or even Microsoft, nor did it come out of the famed Media Lab at MIT."[118]

Tim Berners-Lee, a physicist, happened to be working at CERN when he "articulated the vision" that gave rise to the World Wide Web. Although customarily credited with inventing the Web, Berners-Lee carefully noted that "many other people, most of them unknown, contributed essential ingredients."[119] CERN is certainly an august institution of Big Science, but aside from providing the birthplace, it seems otherwise to have played a minimal role—perhaps even a retarding role—in the birth of the Web. Berners-Lee, by his own account, pressed forward with his idea in spite of massive indifference on the part of the CERN bureaucracy. He "had to maintain a somewhat low profile," he recalled. "At any moment some higher-up could have questioned my time" and put an end to the informal project. He felt "more than a twinge of anxiety," because he feared "being bawled out for not sticking to CERN business."[120]

Without the World Wide Web, the Internet's social and cultural impact would have been far less profound. In general, "the carpenters and bricklayers of the Internet economy" were "the programmers building the transaction systems and industrial-strength Web sites for on-line commerce."[121] And without browsers and search engines to make the Web's massive volume of information accessible, its usefulness would have been minimal. Once again, students and garages came to the fore. The search engine Yahoo! was created by two Stanford students, Dave Filo and Jerry Yang, in 1994. That same year, two University of Illinois students, Marc Andreessen and Eric Bina, presented their browser, Mosaic, to the world. Two more Stanford students, Larry Page and Sergey Brin, produced a search engine they named Google. In 1998, Page and Brin formed a company in a Menlo Park, California, garage; their Google, Inc. rapidly became a multibillion-dollar enterprise.

"COMPUTER LIBERATION"

FOR EVERY BILL Gates and Paul Allen who saw computer science as a road to personal fortune, there was a Ted Nelson or Bob Albrecht who believed it was, or should be, a people's science. Nelson, originator of the concept of "hypertext," self-published a book that gained him recognition as "the iconoclastic evangelist for power-to-the-people 'computer liberation.'"[122]

Albrecht was a computer engineer who resigned from Control Data Corporation in the 1960s when he became "uneasy with the industry's emphasis on serving institutions and corporations instead of individuals." After moving to San Francisco, he "was soon at the center of the alternative computing culture of northern California. He started a tabloid, *People's Computer Company* (PCC), which spread the gospel of computing for the masses." Albrecht and his collaborators opened a storefront computer center, also named PCC, which became "the early home to a side of the computing culture that regarded machines and code as tools of liberation." PCC's members, generally speaking, were "advocates of free speech and were antiestablishment, anticorporate, and anti-Vietnam War."[123]

Albrecht and one of his key allies, Dennis Allison, produced a version of the BASIC language that greatly expanded the number of programmers doing innovative work with microcomputers. In contrast to Gates's and Allen's proprietary approach, Albrecht and Allison "published it for others to do with as they wanted. There was no thought of trying to make money from it; they believed in free speech and free software."[124]

Their vision was shared by Richard Stallman, founder of the Free Software Foundation. In the 1970s, Stallman had been a leading light of the Artificial Intelligence Lab at MIT:

> In January 1984, Stallman resigned from MIT to pursue his seemingly quixotic mission of freeing the software world. "The conclusion I reached was that proprietary software is wrong," he said. "It is based on dividing people and keeping

them helpless. I decided to fight it and try to bring about its downfall."

Stallman "waged his war against the proprietary regime with his only weapon—writing software and distributing it free."[125] His activities provided a major source of inspiration for the "open source software" movement that arose in the 1990s, a world-wide collaborative effort among programmers who freely share the source code they write.

In 1991, a student in Finland, Linus Torvalds, produced the framework of an operating system—a core program for control-ling a computer's basic operations—that became the open source movement's key project. Torvalds's creation was dis-tributed on the Internet and developed by legions of volunteer programmers into Linux, an operating system capable of chal-lenging Microsoft's virtual monopoly.

Although some proponents of the open source movement con-tinue to believe in its potential for radical social change, the fact that IBM has joined their crusade exposes those hopes as somewhat utopian. IBM's interest in promoting "shareware" stems from its competitive position vis-à-vis Microsoft and Sun Microsystems. The company's strategy was outlined in an internal report: "We recommend that IBM aggressively pursue a Linux-based applica-tion development platform and support Linux across all its prod-ucts. Doing so would disrupt the Sun-Microsoft stranglehold."[126]

MIDTERM REPORT CARD ON THE REVOLUTION(S)

THERE HAS BEEN no shortage of sweeping claims for the new technologies—see any issue of *Wired* magazine.[127] But how revolutionary—in the sense of decisively changing society for the better—have the computer revolution, the digital revolu-tion, the information revolution, or the Internet revolution real-ly proven to be? It is too early to write a history of the computer and its consequences; the story is still unfolding, but some pre-liminary conclusions are in order.

By the mid-1980s, computers in homes and workplaces numbered in the millions, and the cultural transformation they have wrought is beyond question. Computers have brought about drastic changes in the way we work and play and spend our money; they have altered our language, our understanding of life and mind and humanity, our ways of thinking about almost everything.

By decentralizing communications, the Internet and the World Wide Web have certainly had a democratizing effect. They allow ordinary people to transmit and receive information not under the control of governments, corporations, or other institutions. That the Internet has become the focus of a struggle for freedom of expression in dozens of countries throughout the world has been most dramatically illustrated by China's sojourn onto the information superhighway. The number of Internet users in China grew from zero in 1993 to a hundred thousand in 1997, to 17 million in 2000, and to 59 million in 2002, with no indications of the growth rate slowing down.[128]

The Chinese government, meanwhile, has been ambivalent about the new technology. It cannot afford to ignore the immense economic potential of Internet commerce, but dreads the political consequences of its citizens receiving uncensored news and communicating freely among themselves. In an attempt to maintain firm control over the flow of information, the authorities constructed a Great Firewall of China, channeling all Internet service through government servers that block access to Web sites of foreign news agencies, human rights groups, or Chinese dissidents. Many Chinese "netizens," however, found they could log onto anonymous proxy servers outside the Great Firewall that would then connect them to the forbidden sites. Government Internet monitors in turn identified new proxy servers as they appeared and added them to the list of banned sites.

This cat-and-mouse game has escalated to ever higher levels of technological sophistication, but the Chinese "Internet police," with much greater resources at their disposal, have generally managed to keep pace with the "cyberdissidents." That is not to say that the authorities have succeeded in maintaining their monopoly of information and communications.

The proliferation of Chinese chat rooms, bulletin boards, and unofficial Web sites has overwhelmed the censors, and the ability of tens of millions of Internet users to exchange personal e-mails has been augmented by text messaging between hundreds of millions of cell phone users.[129] Nevertheless, the risk of incurring Draconian penalties encourages Chinese netizens to practice a great deal of self-censorship. The bottom line is that although "China with the Internet is certainly a freer place than China without the Internet,"[130] the repressive government still retains the upper hand.

Not only in China but throughout the world, the effect of the spread of personal computers has turned out to be "less revolutionary than its proponents imagined." Computers "did not put ordinary individuals on an equal footing with those in positions of power." Their primary impact has been "in the office and not the home, as a tool that assisted the functions of the corporate workplace." Computers, Paul Ceruzzi, concluded, "came 'to the people,' but for a price: corporate control."[131]

The union of capital and computer science was consummated most visibly in the rapid commercialization of the World Wide Web, culminating in the meteoric rise and fall of the "dot-coms" early in the twenty-first century. It was a virtual parody of the boom-and-bust cycle of classic capitalism. Dreams of computer liberation gave way to a frenzied stampede of small technology companies—Microsoft wannabes—onto the information superhighway. All but a very few overextended themselves and went bankrupt. When the dust cleared, the dominance of the new medium by large corporate entities was more pronounced than ever. By 2004, the twenty leading Web sites were owned by AOL Time Warner, Disney, Viacom, Fox Broadcasting, and other media giants; "just 14 companies attract 60% of all the time Americans spend online."[132]

To expect that the proliferation of computers would somehow lead to "liberation" was asking too much of a technology. Science and technology are tools and weapons. The results they produce depend on how they are used, and that in turn depends on whose hands control them. In the current world system, computers and computer science for the most part serve major financial and corporate interests. If liberation means creating a global economy governed by principles of social justice rather than by blind market forces that empower wealthy individuals and nations, the computer revolution makes liberation all the more difficult.

That is not to say that the knowledge embodied in computer science—or indeed any branch of modern science—cannot be put to use by proponents of social change, but if liberation is to be achieved, it will be the work of political forces to which science and technology are totally subordinate.

IN CONCLUSION:
SCIENCE, THE PEOPLE, AND THE FUTURE

FROM THE HUNTER-GATHERERS' accumulation of natural knowledge to the Manhattan Project and beyond, science has always been a social activity requiring the combined contributions of large numbers of individuals. What has the human race to show for their collective efforts?

Modern science has succeeded in vastly increasing our knowledge of nature, from the microworld of subatomic particles to the vastness of intergalactic space. The proof is in the technological pudding. At the same time, it is contaminated by its untrustworthiness with regard to the issues of most importance to human beings—that is, anything to do with social, economic, or political matters. In spite of the lengthening of life expectancy (at least in affluent areas) and the massive proliferation of entertaining and "laborsaving" gadgets, a powerful argument has often been made that modern science has not improved the quality of most people's lives.

The efforts of the science information movement and the resistance offered by the environmental, antinuclear, feminist, and other popular movements have not proved capable of forcing Big Science onto the path of social responsibility. The root of the problem apparently lies in the subordination of knowledge production to the profit motive, and the union of capital and science shows no signs of weakening.

What are the prospects that the domination of science by capital will diminish? And if it does, would it matter? Attempting to answer these questions would be a speculative exercise in futurology that exceeds the responsibility of the historian. For what it is worth, however, I do have an opinion on the latter issue, and in closing—in the interest of full disclosure—will state it briefly.

I do not subscribe to the "Bernal thesis," which maintains that eliminating private property in the means of production and replacing the market system with planned economy would inevitably lead to the flowering of a science that would serve human needs. The experience of the Soviet Union under Stalin and of China under Mao Tse-Tung demonstrated that although those may be *necessary* steps to take, they are certainly not *sufficient*.

On the other hand, I also reject the notions of the "deep ecology" school of thought, which holds that Big Science, Big Technology, and Big Industry are *inherently* antisocial enterprises.[133] The more extreme theorists of deep ecology, in their admirable desire to protect the natural environment, have put forward an abstract notion of nature in which all species are assigned equal value, with the interests of human beings counting for no more than those of any other. If anyone really wants to champion that value system, no rational arguments will dissuade them, but as a member of *Homo sapiens*—the only species capable of making rational choices—I choose to defend a humanist ideology.

"Putting the Earth first" is an attractive abstraction, but insofar as it requires radical reductions in industrial production, the real consequence would be the extermination of billions of people. In spite of its proponents' gentle intentions, from the standpoint of human well-being, theirs is a far crueler ideology than social Darwinism. If I were to summarize my outlook in a slogan, it would not be "Earth first!" but "People first!"

Unfortunately, I must with all humility admit that my answer to how science can be transformed to serve the interests of the world's six or seven billion people is hardly less abstract, because it depends on a heretofore untested proposition. Modern science will continue to be blindly destructive as long as its operations are determined by the anarchism of market economic forces. The problem to be solved is whether science, technology, and industry can be brought under genuinely democratic control in the context of a global planned economy, so that all of us can collectively put our hard-won scientific knowledge to mutually beneficial use. I am confident this *can* be accomplished, but *will* it? If so, there is reason for optimism. If not . . . well, to paraphrase Keynes, "in the not-so-long run we're all dead."

NOTES

1. Muller left the USSR in 1937 in disillusionment over the Lysenko affair. (See "Stakhanovites and Lysenkoists" in this chapter.)
2. Quoted in Roy Porter, *The Greatest Benefit to Mankind*, p. 424.
3. Garland E. Allen, "Is a New Eugenics Afoot?" pp. 59–61.
4. Ibid.
5. Two excellent historical treatments of eugenics are Daniel J. Kevles, *In the Name of Eugenics*, and Garland E. Allen, "The Eugenics Record Office at Cold Spring Harbor, 1910–1940."
6. Porter, *Greatest Benefit to Mankind*, pp. 648–649.
7. Ibid., pp. 649–650.
8. Ibid., p. 650.
9. Ibid., p. 650.
10. Robert Jay Lifton, "Doctors and Torture."
11. Quoted in Neil A. Lewis, "Red Cross Finds Detainee Abuse in Guantánamo."
12. Edward O. Wilson has tried to extrapolate from what he has learned about ant behavior to explain human social organization. See E. O. Wilson, *Sociobiology*. For a detailed critique of biological determinism in general and sociobiology in particular, see Richard Lewontin, Stephen Rose, and Leon Kamin, eds., *Not in Our Genes*.

13. Steven Rose, *The Times* (London), November 9, 1976, p. 17.

14. See esp. Leon Kamin, *The Science and Politics of IQ;* and Leslie Hearnshaw, *Cyril Burt*. For a brief synopsis, see Lewontin, Rose, and Kamin, eds., *Not in Our Genes*, pp. 101–106. It should not be surprising, given the ideological stakes involved, that Burt still has his defenders; see, e.g., Ronald Fletcher, *Science, Ideology, and the Media*.

15. Richard J. Herrnstein and Charles Murray, *The Bell Curve*. For the antidote to this poisonous brew, see Russell Jacoby and Naomi Glauberman, eds., *The Bell Curve Debate*.

16. Adrian Desmond, *The Politics of Evolution*, p. 378.

17. Allen, "Is a New Eugenics Afoot?"

18. Frederick W. Taylor, *Scientific Management*, pp 133–134.

19. Mitchel Cohen, *Big Science, the Fragmenting of Work, and the Left's Curious Notion of Progress*.

20. See the epigraph at the head of this chapter.

21. Harry Braverman, *Labor and Monopoly Capital*, p. 73.

22. V. I. Lenin, "The Immediate Tasks of the Soviet Government," p. 664.

23. The thumbnail description of Lysenkoism to which I must limit myself here does not do justice to the complexity of the phenomenon. For an appreciation of that complexity, see "The Problem of Lysenkoism," in Richard Levins and Richard Lewontin, *The Dialectical Biologist*.

24. Gary Werskey, *The Visible College*, p. 139. The papers from the congress are collected in Anonymous, *Science at the Cross Roads*.

25. Werskey, *Visible College*, p. 139.

26. Ibid., p. 142.

27. J. G. Crowther, *Manchester Guardian*, July 7, 1931; quoted in Werskey, *Visible College*, p. 145.

28. Loren Graham, "The Socio-Political Roots of Boris Hessen," p. 713.

29. Werskey's *Visible College* is a collective biography of these five dissident scientists.

30. H. Floris Cohen, *The Scientific Revolution*, p. 334.

31. Robert K. Merton, *Science, Technology and Society in Seventeenth-Century England*, is an eclectic blend of ideas that originated with Hessen and Max Weber. Western scholars tended to focus on the part based on Weber and to ignore the part based on Hessen.

32. Some postmodernist academics claim that modern science has been "socially constructed" and deny that it has anything to do with gaining objective knowledge of the world of material reality. (The foundation text of this school of thought is Jean-François Lyotard, *The Postmodern Condition: A Report on Knowledge*.) Although I have stressed social factors involved in the development of science and have criticized narrow conceptions of objectivity, I contend that modern science has produced valid knowledge

of nature that is universal, which is to say that it does not vary from culture to culture. In brief, I prefer to describe modern science as having been "socially mediated" rather than "socially constructed."

33. Meera Nanda, "Against Social De(con)struction of Science," pp. 1, 4. See also Meera Nanda, *Prophets Facing Backward.*

34. Peter Rosset, Joseph Collins, and Frances Moore Lappé, "Lessons from the Green Revolution."

35. Ibid.

36. Ibid.

37. Ibid.

38. Ibid.

39. Amnesty International, "Clouds of Injustice," November 2004. See Saritha Rai, "Bhopal Victims Not Fully Paid, Rights Group Says," *New York Times,* November 30, 2004. For a concise analysis of Union Carbide's culpability, see Timothy H. Holtz, "Tragedy Without End."

40. Maril Hazlett and Michael Egan, "Technological and Ecological Turns."

41. *Silent Spring* was partially serialized in the *New Yorker* magazine beginning in June 1962 and published by Houghton Mifflin in September of that year.

42. Linda Lear, "Introduction" to the anniversary edition of *Silent Spring,* p. xvii

43. John M. Lee, "Silent Spring Is Now Noisy Summer."

44. Lear, "Introduction," pp. xi, xvii.

45. Ibid., p. xi.

46. Rachel Carson, *Lost Woods,* p. 91.

47. Hazlett and Egan, "Technological and Ecological Turns."

48. Gary Kroll, "Rachel Carson's *Silent Spring.*"

49. Robert Bullard, "Environmental Justice."

50. Lear, "Introduction," p. xviii.

51. Boston Women's Health Book Collective, *The New Our Bodies, Ourselves,* pp. xvii, 557. On this subject, see also Sandra Morgen, *Into Our Own Hands;* and Ellen Frankfort, *Vaginal Politics.* For a feminist critique of the psychological sciences, see Phyllis Chesler, *Women and Madness.*

52. Boston Women's Health Book Collective, *New Our Bodies,* pp. 558, 561.

53. Ann Dally, "The Development of Western Medical Science," p. 59.

54. Boston Women's Health Book Collective, *New Our Bodies,* pp. 563–564.

55. Dwight D. Eisenhower, "Farewell Address to the Nation," January 17, 1961.

56. Daniel S. Greenberg, *Science, Money, and Politics,* p. 100.

57. David Concor, "Corporate Science versus the Right to Know," *New Scientist,* March 16, 2002; quoted in Sheldon Krimsky, *Science in the Private Interest,* p. 80.

58. Krimsky, *Science in the Private Interest,* pp. 6, 51. See also Derek Bok, *Universities in the Marketplace,* chap. 4, "Scientific Research."

59. Richard Horton, "The Dawn of McScience," p. 9.
60. Greenberg, *Science, Money, and Politics*, pp. 349–350. Greenberg cited Drummond Rennie, "Fair Conduct and Fair Reporting of Clinical Trials," *Journal of the American Medical Association*, November 10, 1999.
61. Horton, "Dawn of McScience," pp. 7, 9.
62. Krimsky, *Science in the Private Interest*, p. 115.
63. Melody Peterson, "Suit Says Company Promoted Drug in Exam Rooms," *New York Times*, May 15, 2002; quoted in Krimsky, *Science in the Private Interest*, pp. 116–117.
64. Mathew Kaufman and Andrew Julian, "Scientists Helped Industry to Push Diet Drug," *Hartford Courant*, April 10, 2000; quoted in Krimsky, *Science in the Private Interest*, p. 115.
65. Krimsky, *Science in the Private Interest*, p. 116.
66. Sarah Bosely, "Scandal of Scientists Who Take Money for Papers Ghostwritten by Drug Companies," *Guardian Weekly* (UK), February 7, 2002; quoted in Krimsky, *Science in the Private Interest*, p. 116.
67. Krimsky, *Science in the Private Interest*, p. 173.
68. Ibid., p. 99.
69. Ibid., pp. 9, 23–24.
70. Dennis Cauchon, "Number of Experts Available Is Limited," *USA Today*, September 25, 2000; quoted in Krimsky, *Science in the Private Interest*, p. 96.
71. Marc Kaufman, "Many Workers Call FDA Inadequate at Monitoring Drugs."
72. The *60 Minutes* segment was based on a report released by a leading environmental group, the Natural Resources Defense Council, entitled "Intolerable Risk: Pesticides in our Children's Food," February 27, 1989. Cited in Krimsky, *Science in the Private Interest*, p. 102.
73. Krimsky, *Science in the Private Interest*, pp. 101–102.
74. Ibid., p. 51, citing World Health Organization, *Tobacco Company Strategies to Undermine Tobacco Control Activities at the World Health Organization*, July 2000.
75. Krimsky, *Science in the Private Interest*, p. 39, citing Public Citizen, *Safeguards at Risk: John Graham and Corporate America's Back Door to the Bush White House*, March 2001.
76. Ibid., p. 39.
77. Greenberg, *Science, Money, and Politics*, p. 74. His source is National Science Foundation, *Science & Engineering Indicators, 1998*.
78. Greenberg, *Science, Money, and Politics*, p. 43.
79. See esp. Gar Alperovitz, *Atomic Diplomacy;* and Alperovitz, *The Decision to Use the Atomic Bomb and the Architecture of an American Myth*.
80. National Science Foundation, *National Patterns of R&D Resources*, reproduced in the Appendix to Greenberg, *Science, Money, and Politics*, Table 1.

81. Between 1990 and 2000 the total R&D budget rose steadily from $63.8 billion to $75.4 billion. See Table 5 in the Appendix to Greenberg, *Science, Money, and Politics*.

82. Since World War II, the American military machine has been deployed throughout the world not to defend against attacks on United States territory but to protect and extend the economic dominance of American corporations. The proper word to describe it is not "defense" but "imperialism."

83. Greenberg, *Science, Money, and Politics*, p. 51.

84. Helen Caldicott, *The New Nuclear Danger*, pp. 4–5.

85. For an early example of Keynes's advice to Roosevelt, see his "Open Letter to President Roosevelt" (1933).

86. Keynes's first use of this phrase actually preceded the Great Depression, but he utilized it as an all-purpose response to deflect any and all questions concerning the "long run." John Maynard Keynes, *A Tract on Monetary Reform* (1923).

87. Keynes, in his most important work, alluded to the benefits of this kind of apparently absurd economic activity: "If the Treasury were to fill old bottles with banknotes, bury them at suitable depths in disused coalmines which are then filled up to the surface with town rubbish, and leave it to private enterprise . . . to dig the notes up again . . . the real income of the community, and its capital wealth also, would probably become a good deal larger than it is." Keynes, *The General Theory of Employment, Interest and Money*, chap. 10, sect. 6.

88. Greenberg, *Science, Money, and Politics*, p. 285.

89. Ibid., p. 332, citing *Physics Today*, July 1999.

90. Tom Engelhardt, "Icarus (Armed with Vipers) over Iraq."

91. Caldicott, *New Nuclear Danger*, p. 3.

92. Helen Caldicott, *Nuclear Madness*, pp. 21–22.

93. See Donald L. Barlett and James B. Steele, *Forevermore*.

94. Caldicott, *Nuclear Madness*, p. 24.

95. Kevin Bogardus, "The Politics of Energy."

96. Teller was a member of a team of physicists who were collectively responsible for the scientific knowledge that made thermonuclear weapons possible; he does not deserve full credit. His reputation as "father of the hydrogen bomb" arose from his tireless lobbying and propagandizing for its development. The subtitle of a recent biography is an accurate characterization: Peter Goodchild, *Edward Teller: The Real Dr. Strangelove*.

97. Caldicott, *Nuclear Madness*, pp. 146–147. The AMA paper, "Medical Perspective on Nuclear Power," was published in the *Journal of the American Medical Association*, November 17, 1989.

98. Stan Augarten, *Bit by Bit*, pp. 195, 253.

99. H. Edward Roberts and William Yates, "Exclusive! Altair 8800."

100. Paul E. Ceruzzi, *A History of Modern Computing*, pp. 226, 304.

101. Augarten, *Bit by Bit*, pp. 276–280.

102. Among the many books that tell the story of Apple and the Macintosh are Michael Moritz, *The Little Kingdom;* Owen Linzmayer, *Apple Confidential 2.0;* and Steven Levy, *Insanely Great.*

103. Personal communication from Rod Holt to the author, November 26, 2004.

104. Steve Lohr, *Go To*, p. 171. Although Bricklin and Frankston "had deep and varied experience in computing," they were nonetheless outsiders to institutionalized computer science.

105. Ceruzzi, *History of Modern Computing*, p. 267.

106. Augarten, *Bit by Bit*, p. 280.

107. Lohr, *Go To*, p. 7.

108. Ibid., p. 29.

109. Ibid., p. 68. Thompson was one of the creators of the Unix operating system.

110. Ibid., pp. 3, 7.

111. The word "computer" in the pre-electronic age referred not to machines but to low-paid, low-status employees—most often women—who performed tedious calculations for scientists by hand. A recent biography tells the story of one such human computer, Henrietta Swan Leavitt, who was a member of a team of women hired by the Harvard College Observatory in the early years of the twentieth century to do numerical calculations of the brightness of stars in astronomical photographs. Leavitt's familiarity with that data led her, in 1912, to a scientific breakthrough of supreme importance to astronomy and cosmology: the discovery that measurements of the cyclically varying brightness of certain stars (Cepheid variables) provide a way to measure the vast distances of intergalactic space. Her accomplishment deserves far more than a footnote; fortunately, it has been admirably described by George Johnson, *Miss Leavitt's Stars.*

112. Lohr, *Go To*, p. 6.

113. Quoted in Lohr, *Go To*, pp. 47–48. Sammet was a leading member of the team that created the COBOL language.

114. Lohr, *Go To*, p. 13.

115. Ibid., pp. 13–14.

116. Ceruzzi, *History of Modern Computing*, p. 232.

117. *Fortune* magazine's 2004 list of wealthiest Americans had Gates at the top for the eleventh year in a row; his wealth was estimated at $48 billion.

118. Ceruzzi, *History of Modern Computing*, p. 301.

119. Tim Berners-Lee, *Weaving the Web*, p. 2.

120. Ibid., pp. 31–32, 42–43, 55.

121. Lohr, *Go To*, p. 201.

122. Ibid., p. 177. See Ted Nelson, *Computer Lib.*

123. Lohr, *Go To*, p. 88.

124. Ibid., p. 89.

125. Ibid., p. 212.

126. The report, written by Nick Bowen and dated December 20, 1999, was endorsed by Samuel Palmisano, an IBM senior vice president and later IBM president, in a January 7, 2000, e-mail to IBM senior management. Quoted in Lohr, *Go To*, p. 218.

127. "*Wired* magazine was launched in January 1993 to cover the Digital Revolution, a term popularized by the Company to describe the profound changes caused by the convergence of the computer, media, and communications industries. . . . *Wired* magazine is not a computer magazine; it is about people, companies, and ideas of the Digital Revolution." From a document that Wired Ventures, Inc., filed with the Securities and Exchange Commission in 1996.

128. Internet use figures from the China Internet Network Information Center (CNNIC) are cited in A. Lin Neumann, "The Great Firewall"; and the Chinese news service Xinhua (December 25, 2002). It should be kept in mind that 60 million Internet users represent only about 5 percent of the population of China.

129. As of April 2005, there were an estimated 350 million cellphone users in China. "A Hundred Cellphones Bloom, and Chinese Take To the Streets," *New York Times*, April 25, 2005.

130. Neumann, "The Great Firewall."

131. Ceruzzi, *History of Modern Computing*, pp. 280, 349.

132. John Pilger, "Australia's Samizdat."

133. See, e.g., David Watson, *Beyond Bookchin*.

BIBLIOGRAPHY

Ackerknecht, Erwin H. *A Short History of Medicine* (Baltimore, MD: Johns Hopkins University Press, 1982).

Aczel, Amir D. *The Riddle of the Compass* (New York: Harcourt, 2001).

Agricola, Georgius. *De re metallica*, trans. Herbert Clark Hoover and Lou Henry Hoover (New York: Dover, 1950).

Allen, Garland E. "The Eugenics Record Office at Cold Spring Harbor, 1910–1940." *Osiris* 2 (1986).

———. "Is a New Eugenics Afoot?" *Science* 294, 5540 (October 5, 2001).

Alperovitz, Gar. *Atomic Diplomacy: Hiroshima and Potsdam* (New York: Penguin, 1985).

———. *The Decision to Use the Atomic Bomb and the Architecture of an American Myth* (New York: Knopf, 1995).

Anonymous. *Calcoen: A Dutch Narrative of the Second Voyage of Vasco da Gama to Calicut, Printed at Antwerp circa 1504*, trans. J. Ph. Berjeau (London: Pickering, 1874).

———. *An Essay on the Usefulness of Mathematical Learning. In a Letter from a Gentleman in the City, to his Friend at Oxford*, 3rd ed. (London, 1745).

———. *Report of Dr. Benjamin Franklin and Other Commissioners Charged by the King of France, with the Examination of the animal Magnetism as Now Practised at Paris* (London, 1785).

———. *Science at the Cross Roads. Papers presented to the International Congress of the History of Science and Technology (London, 1931) by the delegates of the USSR* (London: Frank Cass, 1971; originally published in 1931).

Appel, Toby A. *The Cuvier-Geoffroy Debate: French Biology in the Decades Before Darwin* (Oxford: Oxford University Press, 1987).

Aristotle. *Metaphysics*, W. D. Ross, trans. (Chicago, IL: Great Books of the Western World, 1952).

_____. *Meteorology*, E. W. Webster, trans. (Chicago, IL: Great Books of the Western World, 1952).

_____. *On the Heavens*, J. L. Stocks, trans. (Chicago, IL: Great Books of the Western World, 1952).

_____. *Politics*, Benjamin Jowett, trans. (Chicago, IL: Great Books of the Western World, 1952).

Aronson, J. K. *An Account of the Foxglove and Its Medical Uses, 1785–1985* (London: Oxford University Press, 1985).

Ashton, T. S. *The Industrial Revolution, 1760–1830* (London: Oxford University Press, 1964).

Aubrey, John. *Aubrey's Brief Lives*, ed. Oliver Lawson Dick (Boston, MA: David R. Godine, 1999).

Augarten, Stan. *Bit by Bit: An Illustrated History of Computers* (New York: Ticknor & Fields, 1984).

Aveni, Anthony F. *Ancient Astronomers* (Washington, DC: Smithsonian Books, 1993).

Bachelard, Gaston. *The Psychoanalysis of Fire* (Boston, MA: Beacon Press, 1968).

Bacon, Francis. *The Essays, or Counsels Civil and Moral* (Oxford, UK: Oxford University Press, 1999).

_____. "The Great Instauration," in *The New Organon*.

_____. *The New Organon*, ed. Fulton H. Anderson (New York: Macmillan, 1960).

_____. "Of Seditions and Troubles," in *The Essays*.

_____. *The Works of Francis Bacon*, ed. James Spedding, Robert L. Ellis, and Douglas D. Heath (London: Longman, 1857–1874).

Bailey, Geoff, ed. *Hunter-Gatherer Economy in Prehistory: A European Perspective* (Cambridge, UK: Cambridge University Press, 1983).

Baker, Keith. *Inventing the French Revolution: Essays on French Political Culture in the Eighteenth Century* (Cambridge, UK: Cambridge University Press, 1990).

Balick, Michael J., and Paul Alan Cox, *Plants, People and Culture: The Science of Ethnobotany* (New York: Scientific American Library, 1996).

Barine, Arvède. *Bernardin de Saint-Pierre* (Chicago, IL: A. C. McClurg, 1893).

Barlett, Donald L., and James B. Steele. *Forevermore: Nuclear Waste in America* (New York: W. W. Norton, 1985).

Barzun, Jacques. *From Dawn to Decadence* (New York: HarperCollins, 2000).

Bazin, Hervé. *The Eradication of Smallpox* (San Diego, CA: Academic Press, 2000).

Beadle, George W. "The Ancestry of Corn," *Scientific American*, 242 (1980).

Beaglehole, J. C., ed. *The Endeavour Journal of Joseph Banks 1768–1771* (Sydney, Australia: Angus and Robertson, 1962).

Beall, Otho T., and Richard Shryock. *Cotton Mather: First Significant Figure in American Medicine* (Baltimore, MD: Johns Hopkins University Press, 1954).

Bedini, Silvio A. *Thinkers and Tinkers: Early American Men of Science* (New York: Scribner's, 1975).

————. *Patrons, Artisans and Instruments of Science, 1600–1750* (Brookfield, VT: Ashgate/Variorum, 1999).

Beeching, Jack. "Introduction" to Richard Hakluyt, *Voyages and Discoveries*.

Bennett, J. A. "The Challenge of Practical Mathematics," in Pumphrey, Rossi, and Slawinski, eds., *Science, Culture and Popular Belief in Renaissance Europe*.

————. "The Mechanics' Philosophy and the Mechanical Philosophy." *History of Science* 24 (1986).

Bergasse, Nicolas. *Considérations sur le Magnétisme animal, ou Sur la théorie du monde et des êtres organisés, d'après les principes de M. Mesmer* (La Haye, 1784).

Bergasse du Petit-Thouars, Aristide G. H. N. *Nicolas Bergasse: Un Défenseur des principes traditionnels sous la Révolution* (Paris: Librairie Académique, 1910).

Bergreen, Laurence. *Over the Edge of the World* (New York: William Morrow, 2003).

Bernal, J. D. *Science in History* (Cambridge, MA: MIT Press, 1971).

Bernal, Martin. "Animadversions on the Origins of Western Science," in Shank, ed., *The Scientific Enterprise in Antiquity and the Middle Ages*.

————. *Black Athena* (New Brunswick, NJ: Rutgers University Press, 1987).

————. "Response to Professor Snowden." *Arethusa* 22 (1989).

Bernardin de Saint-Pierre, Jacques Henri. *Etudes de la nature*, new ed. (Basel, Switzerland: Chez Tourneizen, 1797; originally published in 1784).

————. *Paul and Virginia* (London: Penguin, 1982).

Berners-Lee, Tim. *Weaving the Web: The Original Design and Ultimate Destiny of the World Wide Web* (New York: HarperBusiness, 2000).

Biagioli, Mario, and Peter Galison, eds. *Scientific Authorship: Credit and Intellectual Property in Science* (New York: Routledge, 2003).

Biot, Jean Baptiste. *Essai sur l'histoire générale des sciences pendant la Révolution française* (Paris, 1803).

Bivins, Roberta, "The Body in Balance," in Porter, ed., *Medicine: A History of Healing*.

Blake, John B. "The Inoculation Controversy in Boston: 1721–1722." *New England Quarterly* 25 (1952).

Blumenbach, Johann Friedrich. *On the Natural Varieties of Mankind*, trans. Thomas Bendyshe (New York: Bergman, 1969).

Blurton-Jones, Nicolas, and Melvin J. Konner. "!Kung Knowledge of Animal Behavior (or: The Proper Study of Mankind is Animals)," in Lee and De Vore, eds., *Kalahari Hunter-Gatherers*.

Bogardus, Kevin. "The Politics of Energy: Nuclear Power." December 11, 2003. http://www.icij.org/report.aspx?aid=122&sid=200

Bok, Derek. *Universities in the Marketplace: The Commercialization of Higher Education* (Princeton, NJ: Princeton University Press, 2003).

Boston Women's Health Book Collective. *The New Our Bodies, Ourselves* (New York: Simon & Schuster, 1984).

_____. *Our Bodies, Ourselves* (New York: Simon & Schuster, 1973).

Bougainville, Louis-Antoine de. *A Voyage Round the World, Performed by Order of His Most Christian Majesty, in the Years 1766, 1767, 1768, and 1769*, trans. John Reinhold Forster (London, 1772).

Boulding, Kenneth. "The Great Laws of Change," in Tang, Westfield, and Worley, eds., *Evolution, Welfare, and Time in Economics.*

Bourne, Edward Gaylord. "Prince Henry the Navigator," in *Essays in Historical Criticism* (New York: Scribner's, 1901).

Bowen, Catherine Drinker. *Francis Bacon: The Temper of a Man* (Boston, MA: Little, Brown, 1963).

Boyle, Robert. *An Account of Philaretus, during His Minority*, in *The Works of the Honourable Robert Boyle*, vol. 1.

_____. *A Continuation of New Experiments Physico-Mechanical, Touching the Spring and Weight of the Air, Second Part*, in *The Works of the Honourable Robert Boyle*, vol. 4.

_____. *The Excellency of Theology Compared with Natural Philosophy*, in *The Works of the Honourable Robert Boyle*, vol. 4.

_____. *New Experiments Physico-Mechanical, Touching the Spring of the Air*, in *The Works of the Honourable Robert Boyle*, vol. 1.

_____. *Some Considerations Touching the Usefulness of Experimental Natural Philosophy*, in *The Works of the Honourable Robert Boyle*, vol. 2.

_____. *That the Goods of Mankind May Be Much Increased by the Naturalist's Insight into Trades*, in *The Works of the Honourable Robert Boyle*, vol. 3.

_____. *The Works of the Honourable Robert Boyle* (London, 1772).

Braverman, Harry. *Labor and Monopoly Capital: The Degradation of Work in the Twentieth Century* (New York: Monthly Review Press, 1974).

Breasted, James H. *The Conquest of Civilization* (New York: Harper, 1926).

_____. *The Edwin Smith Surgical Papyrus Published in Facsimile and Hieroglyphic Transliteration with Translation and Commentary in Two Volumes* (Chicago, IL: University of Chicago Press, 1930).

Bronowski, Jacob. *The Ascent of Man* (London: British Broadcasting Corporation, 1973).

Brookes, Martin. *Extreme Measures: The Dark Visions and Bright Ideas of Francis Galton* (New York: Bloomsbury, 2004).

Brown, Lloyd A. *The Story of Maps* (New York: Dover, 1979).

Bullard, Robert. "Environmental Justice: An Interview with Robert Bullard." *Earth First! Journal* July 1999. http://www.ejnet.org/ej/bullard.html

Bullough, Vern L. *The Development of Medicine as a Profession* (New York: Hafner, 1966).

Burke, John G., ed. *The Uses of Science in the Age of Newton* (Berkeley, CA: University of California Press, 1983).

Burkert, Walter. *Lore and Science in Ancient Pythagoreanism* (Cambridge, MA: Harvard University Press, 1972).

Bush, Vannevar. "As We May Think," *Atlantic Monthly*, July 1945.

Caldicott, Helen. *The New Nuclear Danger: George W. Bush's Military-Industrial Complex* (New York: New Press, 2004).

_____. *Nuclear Madness: What You Can Do* (New York: W. W. Norton, 1994).

Camac, C. N. B., ed. *Classics of Medicine and Surgery* (New York: Dover, 1959).

Campbell, Tony. "Portolan Charts from the Late Thirteenth Century to 1500," in Harley and Woodward, eds., *The History of Cartography*, vol. 1.

Canny, Nicholas. *The Upstart Earl: A Study of the Social and Mental World of Richard Boyle, First Earl of Cork, 1566–1643* (Cambridge, UK: Cambridge University Press, 1982).

Carney, Judith Ann. *Black Rice: The African Origins of Rice Cultivation in the Americas* (Cambridge, MA: Harvard University Press, 2001).

Carson, Rachel. *Lost Woods: The Discovered Writing of Rachel Carson*, ed. Linda Lear (Boston, MA: Beacon Press, 1998).

_____. *Silent Spring* (Boston, MA: Houghton Mifflin, 1962).

_____. *Silent Spring*, anniversary ed. (Boston, MA: Mariner, 2002).

Cartier, Jacques. *The Voyages of Jacques Cartier*, ed. H. P. Biggar (Toronto: University of Toronto Press, 1993).

Casson, Lionel. *The Ancient Mariners: Seafarers and Sea Fighters of the Mediterranean in Ancient Times* (New York: Macmillan, 1959).

_____. *Libraries in the Ancient World* (New Haven, CT: Yale University Press, 2001).

_____. *Ships and Seamanship in the Ancient World* (Princeton, NJ: Princeton University Press, 1971).

_____, ed. and trans. *The Periplus Maris Erythraei* (Princeton, NJ: Princeton University Press, 1989).

Ceruzzi, Paul E. *A History of Modern Computing* (Cambridge, MA: MIT Press, 2003).

Chaplin, Joyce. *An Anxious Pursuit: Agricultural Innovation and Modernity in the Lower South, 1730–1815* (Chapel Hill, NC: University of North Carolina Press, 1993).

Charbonnier, Georges. *Conversations with Claude Lévi-Strauss* (London: Jonathan Cape, 1969).

Chesler, Phyllis. *Women and Madness* (Garden City, NY: Doubleday, 1972).

Chiera, Edward. *They Wrote on Clay* (Chicago, IL: University of Chicago Press, 1966).

Childe, V. Gordon. *Man Makes Himself* (New York: New American Library, 1951).

_____. *What Happened in History* (New York: Penguin, 1954).

Christianson, John Robert. *On Tycho's Island: Tycho Brahe and His Assistants, 1570–1601* (Cambridge, UK: Cambridge University Press, 2003).

Clagett, Marshall. *The Science of Mechanics in the Middle Ages* (Madison, WI: University of Wisconsin Press, 1959).

_____, ed. *Critical Problems in the History of Science* (Madison, WI: University of Wisconsin Press, 1959).

Cobban, Alfred. *The Social Interpretation of the French Revolution* (Cambridge, UK: Cambridge University Press, 1964).

Cohen, H. Floris. *The Scientific Revolution: A Historiographical Inquiry* (Chicago, IL: University of Chicago Press, 1994).

Cohen, Mitchel. *Big Science, the Fragmenting of Work, and the Left's Curious Notion of Progress* (Brooklyn, NY: Red Balloon Collective, 2004).

Cohen, Morris R., and I. E. Drabkin, *A Source Book in Greek Science* (New York: McGraw-Hill, 1948).

Collins, K. St. B. "Introduction" to Taylor, *The Haven-Finding Art*.

Conner, Clifford D. "Jean Paul Marat and the Scientific Underground of the Old Regime." Unpublished dissertation, City University of New York, 1993.

Cooper, Lane. *Aristotle, Galileo and the Tower of Pisa* (Ithaca, NY: Cornell University Press, 1935).

Corney, Bolton Glanvill, ed. *The Quest and Occupation of Tahiti by Emissaries of Spain during the Years 1772–1776* (London: Hakluyt Society, 1915).

Cortesão, Armando. *The Mystery of Vasco da Gama* (Lisbon, 1973).

Crabtree, Adam. *Animal Magnetism, Early Hypnotism, and Psychical Research 1766–1925* (White Plains, NY: Kraus, 1988).

Creath, Richard. "The Unity of Science: Carnap, Neurath, and Beyond," in Galison and Stump, eds., *The Disunity of Science* (Stanford, CA: Stanford University Press, 1996).

Crellin, John. "Herbalism," in Porter, ed., *Medicine: A History of Healing*.

Crombie, A. C. *Augustine to Galileo* (Cambridge, MA: Harvard University Press, 1961).

_____. *Robert Grosseteste and the Origins of Experimental Science, 1100–1700* (Oxford: Clarendon Press, 1953).

_____. "Commentary on the Papers of Rupert Hall and Giorgio de Santillana," in Clagett, ed., *Critical Problems in the History of Science*.

Crosby, Alfred W. *Ecological Imperialism: The Biological Expansion of Europe, 900–1900* (Cambridge, UK: Cambridge University Press, 1986).

Cunliffe, Barry. *The Extraordinary Voyage of Pytheas the Greek* (New York: Walker, 2002).

Cunningham, Andrew, and Nicholas Jardine, eds. *Romanticism and the Sciences* (Cambridge, UK: Cambridge University Press, 1990).

Cuvier, Georges. *Rapport historique sur les progrès des sciences naturelles depuis 1789, et sur leur état actuel* (Paris: De l'Imprimerie impériale, 1810).

Dahlberg, Frances, ed. *Woman the Gatherer* (New Haven, CT: Yale University Press, 1981).

d'Alembert, Jean Le Rond. *Discours préliminaire de l'Encyclopédie* (1751), trans. Richard N. Schwab as *Preliminary Discourse to the Encyclopedia of Diderot* (Chicago, IL: University of Chicago Press, 1995).

Dales, Richard C. *The Scientific Achievement of the Middle Ages* (Philadelphia: University of Pennsylvania Press, 1973).

Dally, Ann. "The Development of Western Medical Science," in Porter, ed., *Medicine: A History of Healing*.

Dantzig, Tobias. *Number: The Language of Science* (New York: Macmillan, 1954).

Darnton, Robert. *The Literary Underground of the Old Regime* (Cambridge, MA: Harvard University Press, 1982).

_____. *Mesmerism and the End of the Enlightenment in France* (Cambridge, MA: Harvard University Press, 1968).

Darwin, Charles. *The Autobiography of Charles Darwin* (New York: W. W. Norton, 1993; originally published in 1876).

_____. *The Descent of Man* (London: John Murray, 1971).

_____. *The Life and Letters of Charles Darwin*, ed. F. Darwin (New York: Basic Books, 1959).

_____. *Origin of Species* (Oxford, UK: Oxford University Press, 1996; originally published in 1859).

da Vinci, Leonardo. *The Notebooks of Leonardo da Vinci*, trans. and ed. Edward MacCurdy (New York: George Braziller, 1955).

Davis, John. *The Seaman's Secrets* (London, 1607).

Davy, Humphry. "Discourse Introductory to a Course of Lectures on Chemistry," in John Davy, ed., *The Collected Works of Humphry Davy* (London: Smith, Elder, 1839–1840).

de Castro e Almeida, Virginia, ed., and Bernard Miall, trans. *Conquests and Discoveries of Henry the Navigator; Being the Chronicles of Azurara* (London: Allen & Unwin, 1936).

Debus, Allen G. *The Chemical Philosophy: Paracelsian Science and Medicine in the Sixteenth and Seventeenth Centuries* (New York: Science History Publications, 1976).

_____. *The English Paracelsians* (New York: Franklin Watts, 1966).

Denham, T. P., S. G. Haberle, C. Lentfer, R. Fullagar, J. Field, M. Therin, N. Porch, and B. Winsborough. "Origins of Agriculture at Kuk Swamp in the Highlands of New Guinea." *Science* July 11, 2003.

Descartes, René. *Oeuvres de Descartes*, Charles Adams and Paul Tannery, eds. (Paris: J. Vrin, 1996).

_____. *Rules for the Direction of the Mind*, Elizabeth S. Haldane and G. R. T. Ross, trans. (Chicago, IL: Great Books of the Western World, 1952).

Desmond, Adrian. *Archetypes and Ancestors: Palaeontology in Victorian London 1850–1875* (Chicago, IL: University of Chicago Press, 1982).

_____. *The Politics of Evolution: Morphology, Medicine, and Reform in Radical London* (Chicago, IL: University of Chicago Press, 1989).

Desmond, Adrian, and James Moore. *Darwin* (London: Michael Joseph, 1991).

De Vorsey, Louis. "Amerindian Contributions to the Mapping of North America: A Preliminary View," in Storey, William K., ed., *Scientific Aspects of European Expansion* (Brookfield, VT: Variorum, 1996).

Dhombres, Nicole, and Jean Dhombres. *Naissance d'un nouveau pouvoir: Sciences et savants en France, 1793–1824* (Paris: Editions Payot, 1989).

Diamond, Jared. *Guns, Germs, and Steel* (New York: W. W. Norton, 1998).

Dilke, O. A. W. "Cartography in the Ancient World: An Introduction," in Harley and Woodward, eds., *The History of Cartography*, vol. 1.

_____. "The Culmination of Greek Cartography in Ptolemy," in Harley and Woodward, eds., *The History of Cartography*, vol. 1.

_____. "Roman Large-Scale Mapping in the Early Empire," in Harley and Woodward, eds., *The History of Cartography*, vol. 1.

Dobbs, B. J. T. "Newton as Final Cause and First Mover," in Osler, ed., *Rethinking the Scientific Revolution*.

Dobell, Clifford, ed. *Antony van Leeuwenhoek and His "Little Animals": A Collection of Writings by the Father of Protozoology and Bacteriology* (New York: Dover, 1960).

Dorn, Harold. *The Geography of Science* (Baltimore, MD: Johns Hopkins University Press, 1991).

Dor-Net, Zvi. *Columbus and the Age of Discovery* (New York: William Morrow, 1991).

Drake, Stillman. *Cause, Experiment and Science* (Chicago, IL: University of Chicago Press, 1981).

_____. *Galileo at Work: His Scientific Biography* (Chicago, IL: University of Chicago Press, 1978).

Drury, Shadia. *Leo Strauss and the American Right* (New York: Palgrave Macmillan, 1999).

_____. *The Political Ideas of Leo Strauss* (London: Macmillan, 1988).

_____. "Noble Lies and Perpetual War: Leo Strauss, the Neo-Cons, and Iraq." An interview with Shadia Drury by Danny Postel, October 16, 2003. http://www.informationclearinghouse.info/article5010.htm

Duveen, Denis I., and Herbert S. Klickstein, *Bibliography of the Works of Antoine Laurent Lavoisier, 1743–1794* (London: Wm. Dawson & Sons, 1954).

Eamon, William. *Science and the Secrets of Nature: Books of Secrets in Medieval and Early Modern Culture* (Princeton, NJ: Princeton University Press, 1996).

Easlea, Brian. *Witch Hunting, Magic and the New Philosophy* (Atlantic Highlands, NJ: Humanities Press, 1980).

Eisenhower, Dwight D. "Farewell Address to the Nation," January 17, 1961. http://www.eisenhower.archives.gov/farewell.htm

Eisenstein, Elizabeth L. *The Printing Press as an Agent of Change: Communications and Cultural Transformations in Early Modern Europe* (Cambridge, UK: Cambridge University Press, 1979).

Eisley, Loren. *The Man Who Saw Through Time* (New York: Scribner's, 1973).

Engelhardt, Tom. "Icarus (Armed with Vipers) over Iraq." December 6, 2004. Tom Dispatch, http://www.tomdispatch.com/index.mhtml?pid=2047

Engels, Frederick. *The Part Played by Labor in the Transistion from Ape to Man* (Moscow: Progress Publishers, 1934; first published in 1876).

Farrington, Benjamin. *Francis Bacon: Philosopher of Industrial Science* (New York: Farrar, Straus and Giroux, 1979).

_____. *Greek Science* (Harmondsworth, UK: Penguin, 1969).

_____. *Science and Politics in the Ancient World* (London: Allen & Unwin, 1939).

_____. *Science in Antiquity* (Oxford, UK: Oxford University Press, 1969).

Feinberg, Richard. *Polynesian Seafaring and Navigation* (Kent, OH: Kent State University Press, 1988).

Fenn, Elizabeth Anne. *Pox Americana: The Great Smallpox Epidemic of 1775–82* (New York: Hill & Wang, 2001).

Field, J. V. "Mathematics and the Craft of Painting: Piero della Francesca and Perspective," in Field and James, eds., *Renaissance and Revolution: Humanists, Scholars, Craftsmen and Natural Philosophers in Early Modern Europe*.

_____ and Frank A. J. L. James, eds. *Renaissance and Revolution: Humanists, Scholars, Craftsmen and Natural Philosophers in Early Modern Europe.* (Cambridge, UK: Cambridge University Press, 1993).

Finley, M. I. *Ancient Slavery and Modern Ideology* (Harmondsworth, UK: Penguin, 1980).

_____. *The Ancient Greeks* (Harmondsworth, UK: Penguin, 1979).

Fletcher, Ronald. *Science, Ideology, and the Media: The Cyril Burt Scandal* (London: Transaction Publishers, 1991).

Frankfort, Ellen. *Vaginal Politics* (New York: Quadrangle Books, 1972).

Franklin, Benjamin. "Chart of the Gulf Stream," in American Philosophical Society, *Transactions* 28 (Philadelphia, 1786), opposite p. 315.

_____. *The Ingenious Dr. Franklin: Selected Scientific Letters of Benjamin Franklin*, ed. Nathan G. Goodman (Philadelphia: University of Pennsylvania Press, 1931).

Galilei, Galileo. *Dialogue Concerning the Two Chief World Systems*, trans. Stillman Drake (Berkeley: University of California Press, 1967).

_____. *Dialogues Concerning Two New Sciences*, trans. Henry Crew and Alfonso de Salvio (New York: Dover, 1954).

_____. *Il Saggiatore* [The Assayer], trans. Stillman Drake, in *The Controversy on the Comets of 1618* (Philadelphia: University of Pennsylvania Press, 1960).

_____. *Istoria e dimostrationi intorno alle macchie solari e loro accidenti* (Bologna, 1655; originally published in 1613).

_____. *Sidereus nuncius* [The Starry Messenger] (1610), trans. Albert Van Helden (Chicago, IL: University of Chicago Press, 1989).

Galison, Peter, and Bruce Hevly, eds. *Big Science: The Growth of Large-Scale Research* (Stanford, CA: Stanford University Press, 1992).

Galison, Peter, and David J. Stump, eds. *The Disunity of Science* (Stanford, CA: Stanford University Press, 1996).

Galton, Francis. *English Men of Science: Their Nature and Nurture* (London: Macmillan, 1874).

_____. *Hereditary Genius: An Inquiry into Its Laws and Consequences* (London: Macmillan, 1869).

_____. "Hereditary Improvement." *Fraser's Magazine* 7 (1865).

_____. "Hereditary Talent and Character," *Macmillan's Magazine* 12 (1865).

Gerard, John. *The Herball, or General Historie of Plantes* (London, 1597).

Gilbert, William, *De magnete magneticisque corporibus et de magno magneto tellure physiologia nova* (1600), trans. P. Fleury Mottelay as *On the Loadstone and Magnetic Bodies and on the Great Magnet the Earth* (Chicago, IL: Great Books of the Western World, 1952).

Gillispie, C. C. *The Edge of Objectivity* (Princeton, NJ: Princeton University Press, 1960).

_____. "The *Encyclopédie* and the Jacobin Philosophy of Science," in Clagett, ed., *Critical Problems in the History of Science*.

Ginzburg, Carlo. "Clues: Roots of an Evidential Paradigm," in *Myths, Emblems, Clues*.

_____. *Myths, Emblems, Clues* (London: Radius, 1990).

Glacken, Clarence J. *Traces on the Rhodian Shore* (Berkeley: University of California Press, 1967).

Gladwin, Thomas. *East Is a Big Bird: Navigation and Logic on Puluwat Atoll* (Cambridge, MA: Harvard University Press, 1970).

Glanvill, Joseph. *Plus Ultra, or, The Progress and Advancement of Knowledge since the Days of Aristotle* (Gainesville, FL: Scholars' Facsimiles & Reprints, 1958; originally published in 1668).

Goethe, Johann Wolfgang von. *Faust*, trans. Philip Wayne (Harmondsworth, UK: Penguin, 1962).

Golino, Carlo L., ed. *Galileo Reappraised* (Berkeley: University of California Press, 1966).

Gomperz, T. *Greek Thinkers* (London: J. Murray, 1901–1912).

Goodchild, Peter. *Edward Teller: The Real Dr. Strangelove* (Cambridge, MA: Harvard University Press, 2004).

Goodenough, Ward H. *Native Astronomy in the Central Carolines* (Philadelphia: University Museum, 1953).

Gooding, David, Trevor Pinch, and Simon Schaffer, eds. *The Uses of Experiment: Studies in the Natural Sciences* (Cambridge, UK: Cambridge University Press, 1989).

Gould, Stephen Jay. *Ever Since Darwin* (New York: W. W. Norton, 1977).

_____. *The Mismeasure of Man* (New York: W. W. Norton, 1981).

_____. "Posture Maketh the Man," in Gould, *Ever Since Darwin*.

Graham, Loren. "The Socio-Political Roots of Boris Hessen: Soviet Marxism and the History of Science," *Social Studies of Science* 15 (1985).

Greenberg, Daniel S. *Science, Money, and Politics: Political Triumph and Ethical Erosion* (Chicago, IL: University of Chicago Press, 2001).

Greene, Mott T. *Geology in the Nineteenth Century* (Ithaca, NY: Cornell University Press, 1982).

————. "History of Geology," *Osiris* 1 (1985).

Grimé, William. *Botany of the Black Americans* (St. Clair Shores, MI: Scholarly Press, 1976).

Guedj, Denis. *La Révolution des Savants* (Paris: Gallimard, 1988).

Guerlac, Henry. *Newton on the Continent* (Ithaca, NY: Cornell University Press, 1981).

Gunther, R. T. "The Great Astrolabe and Other Scientific Instruments of Humphrey Cole," *Archaeologia* 76 (1926–1927).

Hahn, Roger. *The Anatomy of a Scientific Institution: The Paris Academy of Sciences, 1666–1803* (Berkeley: University of California Press, 1971).

————. *Laplace as a Newtonian Scientist* (Los Angeles: University of California Press, 1967).

Hakluyt, Richard. *Voyages and Discoveries* (London: Penguin, 1985).

Hale, J. R. *Renaissance Exploration: An Authoritative Survey of the Great Age of European Discovery* (New York: W. W. Norton, 1968).

Hall, A. Rupert. *From Galileo to Newton* (New York: Harper & Row, 1963).

————. "Gunnery, Science, and the Royal Society," in Burke, *The Uses of Science in the Age of Newton*.

————. "The Scholar and the Craftsman in the Scientific Revolution," in Clagett, *Critical Problems in the History of Science*.

Hall, Marie Boas. *Robert Boyle and Seventeenth-Century Chemistry* (Cambridge, UK: Cambridge University Press, 1958).

Hamm, E. P. "Knowledge from Underground: Leibniz Mines the Enlightenment," *Earth Sciences History* 16 (1997).

Harding, Sandra. *The Science Question in Feminism* (Ithaca, NY: Cornell University Press, 1986).

Harley, J. B. "New England Cartography and the Native Americans," in J. B. Harley, *The New Nature of Maps: Essays in the History of Cartography* (Baltimore, MD: Johns Hopkins University Press, 2001).

Harley, J. B., and David Woodward, "The Growth of an Empirical Cartography in Hellenistic Greece," in Harley and Woodward, eds., *The History of Cartography*, vol. 1.

————, eds. *The History of Cartography*, vol. 1: *Cartography in Prehistoric, Ancient, and Medieval Europe and the Mediterranean* (Chicago, IL: University of Chicago Press, 1987).

Harris, L. E. *The Two Netherlanders: Humphrey Bradley and Cornelius Drebbel* (Leiden, Netherlands: Brill, 1961).

Harris, Seale. *Banting's Miracle: The Story of the Discoverer of Insulin* (Philadelphia, PA: Lippincott, 1946).

Hawke, David Freeman. *Nuts and Bolts of the Past: A History of American Technology, 1776–1860* (New York: Harper & Row, 1988).

Hawkins, Gerald. "Stonehenge: A Neolithic Computer," *Nature*, January 27, 1964.

_____. *Stonehenge Decoded* (New York: Dell, 1965).

Hazlett, Maril, and Michael Egan. "Technological and Ecological Turns: Science and American Environmentalism." Paper presented at the History of Science Society Annual Conference, November 2003.

Hearnshaw, Leslie. *Cyril Burt: Psychologist* (Ithaca, NY: Cornell University Press, 1979).

Heilbron, J. L. *Elements of Early Modern Physics* (Berkeley: University of California Press, 1982).

Henry, John. "Doctors and Healers: Popular Culture and the Medical Profession," in Pumphrey, Rossi, and Slawinski, eds., *Science, Culture and Popular Belief in Renaissance Europe.*

Herodotus. *The History*, trans. George Rawlinson (Chicago, IL: Great Books of the Western World, 1952).

Herrnstein, Richard J., and Charles Murray, *The Bell Curve: Intelligence and Class Structure in American Life* (New York: Free Press, 1994).

Hessen, Boris. *The Social and Economic Roots of Newton's "Principia"* (New York: Howard Fertig, 1971).

Hill, Christopher. "Newton and His Society," in Robert Palter, ed., *The Annus Mirabilis of Sir Isaac Newton, 1666–1966* (Cambridge, MA: MIT Press, 1970).

_____. "Science and Magic," in *The Collected Essays of Christopher Hill*, vol. 3: *People and Ideas in Seventeenth-Century England* (Amherst: University of Massachusetts Press, 1986).

_____. *The World Turned Upside Down: Radical Ideas during the English Revolution* (New York: Penguin, 1975).

Hippocrates. *Hippocratic Writings* (Harmondsworth, UK: Penguin, 1986).

Hodges, Henry. *Technology in the Ancient World* (New York: Alfred A. Knopf, 1970).

Hogben, Lancelot. *Astronomer Priest and Ancient Mariner* (New York: St. Martin's Press, 1974).

_____. *Mathematics in the Making* (London: Macdonald, 1960).

Holtz, Timothy H. "Tragedy Without End: The 1984 Bhopal Gas Disaster," in *Dying for Growth: Global Inequality and the Health of the Poor*, Jim Yong Kim *et al*, eds.

Homer. *The Odyssey*, trans. Samuel Butler (Chicago, IL: Great Books of the Western World, 1952).

Hooke, Robert. *General Scheme or Idea of the Present State of Natural Philosophy* (1705), in *The Posthumous Works of Robert Hooke* (New York: Johnson Reprint Corp., 1969).

Hoover, Herbert Clark, and Lou Henry Hoover. "Introduction" to Agricola, *De re metallica.*

Hooykaas, Reijer. "The Portuguese Discoveries and the Rise of Modern Science," in Hooykaas, *Selected Studies in History of Science.*

_____. "The Rise of Modern Science: When and Why?" *British Journal for History of Science* 20, 4 (1987).

_____. "Science and Reformation," in Clagett, ed., *Critical Problems in the History of Science.*

_____. *Selected Studies in History of Science* (Coimbra: Por ordem da Universidade, 1983).

Horton, Richard. "The Dawn of McScience," *New York Review of Books*, March 11, 2004.

_____. "Myths in Medicine: Jenner Did Not Discover Vaccination." *British Medical Journal*, 310 (1995).

Hudson, Travis, and Ernest Underhay, *Crystals in the Sky: An Intellectual Odyssey Involving Chumash Astronomy, Cosmology, and Rock Art* (Socorro, NM: Ballena Press, 1978).

Hughes, M. J. *Women Healers in Medieval Life and Literature* (New York: King's Crown, 1943).

Huygens, Christiaan. *Horologium oscillatorium* (1673), trans. Richard J. Blackwell as *The Pendulum Clock* (Ames, IA: Iowa State University Press, 1986).

Ifrah, Georges. *The Universal History of Numbers* (New York: Wiley, 2000).

Isocrates. *Busiris*, in *Isocrates*, trans. George Norlin. Loeb Classical Library (New York: G. P. Putnam's Sons, 1928-1945).

Jackson, Myles W. "Can Artisans Be Scientific Authors?" in Biagioli and Galison, eds., *Scientific Authorship: Credit and Intellectual Property in Science.*

Jacob, James R. "'By an Orphean Charm': Science and the Two Cultures in Seventeenth-Century England," in P. Mack and M. C. Jacob, eds., *Politics and Culture in Early Modern Europe* (Cambridge, UK: Cambridge University Press, 1987).

_____. *Robert Boyle and the English Revolution: A Study in Social and Intellectual Change* (New York: B. Franklin, 1977).

_____. *The Scientific Revolution: Aspirations and Achievements, 1500–1700* (Atlantic Highlands, NJ: Humanities Press, 1998).

Jacob, James R., and Margaret C. Jacob, "Anglican Origins of Modern Science," *Isis* 71 (1980).

Jacob, Margaret C. *The Newtonians and the English Revolution, 1689–1720* (Ithaca, NY: Cornell University Press, 1976).

_____. *The Cultural Meaning of the Scientific Revolution* (New York: Alfred A. Knopf, 1988).

_____, ed. *The Politics of Western Science: 1640–1990* (Atlantic Highlands, NJ: Humanities Press, 1992).

Jacoby, Russell, and Naomi Glauberman, eds. *The Bell Curve Debate: History, Documents, Opinions* (New York: Times Books, 1995).

Jardine, Lisa. *The Curious Life of Robert Hooke* (New York: HarperCollins, 2004).

_____. *Ingenious Pursuits: Building the Scientific Revolution* (New York: Anchor Books, 1999).

Jardine, Lisa, and Alan Stewart. *Hostage to Fortune: The Troubled Life of Francis Bacon* (New York: Hill & Wang, 1999).

Johnson, Francis R. "Commentary on the Paper of Rupert Hall," in Clagett, ed., *Critical Problems in the History of Science.*

Johnson, George. *Miss Leavitt's Stars: The Untold Story of the Woman Who Discovered How To Measure the Universe* (New York: W. W. Norton, 2005).

Kamin, Leon. *The Science and Politics of IQ* (Potomac, MD: Lawrence Erlbaum Associates, 1974).

Kaufman, Marc. "Many Workers Call FDA Inadequate at Monitoring Drugs," *Washington Post,* December 17, 2004.

Kautsky, Karl. *Foundations of Christianity* (New York: Monthly Review Press, 1972; originally published in 1908).

Kearny, Hugh. *Science and Change: 1500–1700* (New York: McGraw-Hill, 1971).

Kelly, Robert L. *The Foraging Spectrum: Diversity in Hunter-Gatherer Lifeways* (Washington, DC, and London: Smithsonian Institution Press, 1995).

Kelly, Thomas. *George Birbeck: Pioneer of Adult Education* (Liverpool, UK: Liverpool University Press, 1957).

Kevles, Daniel J. *In the Name of Eugenics* (Berkeley: University of California, 1985).

Keynes, John Maynard. *The General Theory of Employment, Interest and Money* (London: Macmillan, 1936).

_____. "An Open Letter to President Roosevelt," *New York Times,* December 31, 1933. http://newdeal.feri.org/misc/keynes2.htm

_____. *A Tract on Monetary Reform* (London: Macmillan, 1923).

Kim, Jim Yong, Joyce V. Millen, Alec Irwin, and John Gershman, eds. *Dying for Growth: Global Inequality and the Health of the Poor* (Monroe, ME: Common Courage Press, 2000).

King, Henry C. *The History of the Telescope* (London: Griffin, 1955).

King, Ross. *Brunelleschi's Dome: How a Renaissance Genius Reinvented Architecture* (New York: Penguin, 2000).

Kirk, G. S., J. E. Raven, and M. Schofield, eds. *The Presocratic Philosophers* (Cambridge, UK: Cambridge University Press, 1983).

Kittredge, George Lyman, ed. "Lost Works of Cotton Mather," *Proceedings of the Massachusetts Historical Society* 45 (1912).

Kotzebue, Otto von. *A Voyage of Discovery, into the South Sea and Beering's Straits, for the Purpose of Exploring a North-east Passage, Undertaken in the Years 1815–1818* (London: Longman, Hurst, Rees, Orme, and Brown, 1821).

Koyré, Alexandre. *Metaphysics and Measurement* (Cambridge, MA: Harvard University Press, 1968).

Kraus, Paul. *Jabir ibn Hayyan: Contribution à l'histoire des idées scientifiques dans l'Islam,* in *Mémoires présentés à l'Institut d'Egypte,* tomes 44-45 (Cairo: Imprimerie de l'Institut Française d'Archéologie Orientale, 1942-1943).

Krimsky, Sheldon. *Science in the Private Interest* (Lanham, MD: Rowman & Littlefield, 2003).

Kroll, Gary. "Rachel Carson's *Silent Spring:* A Brief History of Ecology as a Subversive Subject." Online Ethics Center for Engineering and Science at Case Western Reserve University, http://onlineethics.org/moral/carson/kroll.html

Krupp, E. C. *Echoes of the Ancient Skies* (New York: Harper & Row, 1983).

————. *Skywatchers, Shamans & Kings* (New York: Wiley, 1997).

————, ed. *Archaeoastronomy and the Roots of Science* (Boulder, CO: Westview Press, 1984).

Kuhn, Thomas. "Alexandre Koyré and the History of Science," *Encounter* 34 (1970).

Landels, J. G. *Engineering in the Ancient World* (Berkeley: University of California Press, 1978).

Landes, David S. *Revolution in Time: Clocks and the Making of the Modern World* (Cambridge, MA: Harvard University Press, 1983).

————. *The Unbound Prometheus: Technological Change and Industrial Development in Western Europe from 1750 to the Present* (Cambridge, UK: Cambridge University Press, 1969).

Latour, Bruno. *The Pasteurization of France* (Cambridge, MA: Harvard University Press, 1988).

Lear, Linda. "Introduction" to Carson, *Silent Spring* (2002 edition).

Lee, John M. "Silent Spring Is Now Noisy Summer: Pesticides Industry up in Arms," *New York Times,* July 22, 1962.

Lee, R. B., and I. DeVore, *Man the Hunter* (Chicago, IL: Aldine, 1968).

————, eds. *Kalahari Hunter-Gatherers: Studies of the !Kung San and Their Neighbors* (Cambridge, MA: Harvard University Press, 1976).

Lefkowitz, Mary R. "Ancient History, Modern Myths," in Lefkowitz and Rogers, eds., *Black Athena Revisited.*

————. *Not Out of Africa* (New York: Basic Books, 1996).

Lefkowitz, Mary R., and Guy MacLean Rogers, eds. *Black Athena Revisited* (Chapel Hill: University of North Carolina Press, 1996).

Leicester, Henry M. *The Historical Background of Chemistry* (New York: Dover, 1971).

Lenin, V. I. "The Immediate Tasks of the Soviet Government" (April 1918), in *Selected Works*, vol. 2 (New York: International Publishers, 1967).

Levathes, Louise. *When China Ruled the Seas: The Treasure Fleet of the Dragon Throne, 1405–1433* (Oxford, UK: Oxford University Press, 1996).

Levins, Richard, and Richard Lewontin, *The Dialectical Biologist* (Cambridge, MA: Harvard University Press, 1985).

Lévi-Strauss, Claude. *The Savage Mind* (London: Weidenfeld and Nicolson, 1966).

Levy, Steven. *Insanely Great: The Life and Times of Macintosh, the Computer That Changed Everything* (New York: Penguin, 1995).

Lewis, David. *We the Navigators: The Ancient Art of Landfinding in the Pacific* (Honolulu: University of Hawaii Press, 1994).

Lewis, Neil A. "Red Cross Finds Detainee Abuse in Guantánamo," *New York Times,* November 30, 2004.

Lewontin, R. C., Steven Rose, and Leon J. Kamin, eds. *Not in Our Genes: Biology, Ideology and Human Nature* (New York: Pantheon, 1984).

Liebenberg, L. W. *The Art of Tracking: The Origin of Science* (Cape Town, South Africa: David Philip, 1990).

Liebermann, S. J. "Of Clay Pebbles, Hollow Clay Balls, and Writing," *American Journal of Archaeology* 84 (1980).

Lifton, Robert Jay. "Doctors and Torture," *New England Journal of Medicine* 351, 5 (July 29, 2004).

Lind, James. *A Treatise on the Scurvy* (London, 1757).

Linebaugh, Peter, and Marcus Rediker. *The Many Headed Hydra: Sailors, Slaves, Commoners, and the Hidden History of the Revolutionary Atlantic* (Boston, MA: Beacon, 2000).

Linklater, Andro. *Measuring America: How an Untamed Wilderness Shaped the United States and Fulfilled the Promise of Democracy* (New York: Walker, 2002).

Linzmayer, Owen. *Apple Confidential 2.0: The Definitive History of the World's Most Colorful Company* (San Francisco, CA: No Starch Press, 2004).

Littlefield, Daniel C. *Rice and Slaves: Ethnicity and the Slave Trade in Colonial South Carolina* (Baton Rouge: Louisiana State University Press, 1981).

Liverani, Mario. "The Bathwater and the Baby," in Lefkowitz and Rogers, eds., *Black Athena Revisited.*

Lohr, Steve. *Go To: The Story of the Math Majors, Bridge Players, Engineers, Chess Wizards, Maverick Scientists and Iconoclasts—the Programmers Who Created the Software Revolution* (New York: Basic Books, 2001).

Long, Pamela O. "Power, Patronage, and the Authorship of *Ars,*" in Shank, ed., *The Scientific Enterprise in Antiquity and the Middle Ages.*

_____, ed. "Science and Technology in Medieval Society." *Annals of the New York Academy of Sciences,* vol. 441 (New York: New York Academy of Sciences, 1985).

Longino, Helen. "Can There Be a Feminist Science?" *Hypatia* 2 (1987).

_____. *Science as Social Knowledge: Values and Objectivity in Scientific Inquiry* (Princeton, NJ: Princeton University Press, 1990).

Lovejoy, Arthur O. *The Great Chain of Being* (New York: Harper & Brothers, 1960; originally published in 1936).

Lyons, Henry G. *The Royal Society, 1660–1940: A History of Its Administration under Its Charters* (Cambridge, UK: Cambridge University Press, 1944).

Lyotard, Jean-François. *The Postmodern Condition: A Report on Knowledge* (Minneapolis, MN: University of Minnesota Press, 1985). Originally published as *La Condition postmoderne: rapport sur le savoir* (Paris, 1979).

MacLachlan, James. "A Test of an 'Imaginary' Experiment of Galileo's," *Isis* 66 (1973).

MacLeod, Roy, ed. "Nature and Empire: Science and the Colonial Enterprise," *Osiris*, 2nd series, 15 (2000).

MacNeish, Richard S. *The Origins of Agriculture and Settled Life* (Norman, OK: University of Oklahoma Press, 1992).

Majno, Guido. *The Healing Hand: Man and Wound in the Ancient World* (Cambridge, MA: Harvard University Press, 1975).

Major, Richard Henry. *The Life of Prince Henry of Portugal, Surnamed the Navigator, and Its Results: Comprising the Discovery, within One Century, of Half the World* (London: A. Asher, 1868).

Markham, Clements R. *The Sea Fathers: A Series of Lives of Great Navigators of Former Times* (London: Cassell, 1884).

Marshack, Alexander. "Lunar Notation on Upper Paleolithic Remains," *Science* 146 (November 6, 1964).

_____. *The Roots of Civilization: The Cognitive Beginnings of Man's First Art, Symbol, and Notation* (New York: McGraw-Hill, 1972).

Martin, Julian. "Natural Philosophy and Its Public Concerns," in Pumphrey, Rossi, and Slawinski, eds., *Science, Culture and Popular Belief in Renaissance Europe*.

Marx, Karl. *Capital* (New York: International Publishers, 1967).

Marx, Karl, and Friedrich Engels. *Correspondence, 1846-1895*, trans. Dona Torr (London: M. Lawrence, 1934).

Mason, Stephen F. *A History of the Sciences* (New York: Collier, 1962).

Mather, Cotton. *The Angel of Bethesda*, ed. Gordon W. Jones (Barre, MA: Barre Publishers, 1972).

McClellan, James. *Science Reorganized: Scientific Societies in the Eighteenth Century* (New York: Columbia University Press, 1985).

McClellan, James, and Harold Dorn. *Science and Technology in World History* (Baltimore, MD: Johns Hopkins University Press, 1999).

McGrew, Roderick E. *Encyclopedia of Medical History* (London: Macmillan, 1985).

Merchant, Carolyn. *The Death of Nature: Women, Ecology and the Scientific Revolution* (New York: Harper & Row, 1983).

Merton, Robert K. "Commentary on the Paper of Rupert Hall," in Clagett, ed., *Critical Problems in the History of Science*.

_____. *On the Shoulders of Giants* (Chicago, IL: University of Chicago Press, 1993).

_____. *Science, Technology and Society in Seventeenth-Century England* (New York: Howard Fertig, 1970; originally published in 1938).

Middleton, W. E. Knowles. *The Experimenters: A Study of the Accademia del Cimento* (Baltimore, MD: Johns Hopkins University Press, 1971).

Midelfort, H. C. E. *Witch Hunting in Southwestern Germany 1562-1684* (Stanford, CA: Stanford University Press, 1972).

Miller, David Philip. "'Puffing Jamie': The Commercial and Ideological Importance of Being a 'Philosopher' in the Case of the Reputation of James Watt (1736–1819)," *History of Science* 38 (2000).

Monardes, Nicholas. *Joyfull Newes out of the Newe Founde Worlde, Written in Spanish by Nicholas Monardes, and Englished by John Frampton, Merchant, anno 1577* (New York: Knopf, 1925).

Montaigne, Michel Eyquem de. *The Essays*, trans. E. J. Trechmann (New York: Modern Library, 1946).

Morgen, Sandra. *Into Our Own Hands: The Women's Health Movement in the United States, 1969–1990* (New Brunswick, NJ: Rutgers University Press, 2002).

Morison, Samuel Eliot. *The European Discovery of America: The Northern Voyages* (New York: Oxford University Press, 1971).

_____. *The European Discovery of America: The Southern Voyages* (New York: Oxford University Press, 1974).

Moritz, Michael. *The Little Kingdom: The Private Story of Apple Computer* (New York: William Morrow, 1984).

Morton, A. L. *A People's History of England* (London: Lawrence & Wishart, 1984).

Motzo, B. R., ed. *Il compasso da navigare* (Cagliari,1947).

Murray, Charles. *Human Accomplishment: The Pursuit of Excellence in the Arts and Sciences, 800 B.C. to 1950* (New York: HarperCollins, 2003).

Murray, M. A. *The God of the Witches* (London: Low, Marston, 1933).

_____. *The Witch-Cult in Western Europe* (Oxford, UK: Oxford University Press, 1962).

Nanda, Meera. "Against Social De(con)struction of Science: Cautionary Tales from the Third World." *Monthly Review* 48, 10 (March 1997).

_____. *Prophets Facing Backward: Postmodern Critiques of Science and Hindu Nationalism in India* (New Brunswick, NJ: Rutgers University Press, 2003).

Neckam, Alexander. *De naturis rerum libri duo* (c. 1187), ed. Thomas Wright (London: Longman, Green, 1863; reprinted by Kraus Reprints, 1967).

Needham, Joseph. *Clerks and Craftsmen in China and the West* (Cambridge, UK: Cambridge University Press, 1970).

_____. *The Grand Titration* (London: Allen & Unwin, 1969).

_____. *Science and Civilisation in China* (Cambridge, UK: Cambridge University Press, 1954–).

_____. *Science in Traditional China* (Cambridge, MA: Harvard University Press, 1981).

_____. *The Shorter Science and Civilisation in China* (Cambridge, UK: Cambridge University Press, 1978–1995).

Nelson, Sarah Milledge. *Gender in Archaeology: Analyzing Power and Prestige* (Walnut Creek, CA: Altamira, 1997).

Nelson, Ted. *Computer Lib: Dream Machines* (Redmond, WA: Microsoft Press, 1987).

Neugebauer, Otto. *The Exact Sciences in Antiquity* (New York: Dover, 1969).

Neumann, A. Lin. "The Great Firewall." Committee to Protect Journalists, http://www.cpj.org/Briefings/2001/China_jan01/China_jan01.html

Newton, Isaac. *The Correspondence of Isaac Newton*, ed. A. R. Hall and Laura Tilling (Cambridge, UK: Cambridge University Press, 1977).

————. *Isaac Newton's Papers and Letters on Natural Philosophy*, ed. I. B. Cohen (Cambridge, MA: Harvard University Press, 1978).

Nordenskiöld, A. E. *Periplus: An Essay on the Early History of Charts and Sailing-Directions*, trans. Francis A. Bather (Stockholm: P. A. Norstedt, 1897).

Norman, Robert. *The Newe Attractive* (London, 1592; first published in 1581).

Novack, George. *The Origins of Materialism* (New York: Merit, 1965).

Nutton, Vivian. "Historical Introduction" to Vesalius, *On the Fabric of the Human Body.*

Osler, Margaret J., ed. *Rethinking the Scientific Revolution* (Cambridge, UK: Cambridge University Press, 2000).

Pagel, Walter. *Paracelsus: An Introduction to Philosophical Medicine in the Era of the Renaissance* (Basel, Switzerland: S. Karger, 1958).

Palissy, Bernard. *Discours Admirables*, trans. Aurèle La Roque (Urbana: University of Illinois Press, 1957).

Palter, Robert. "*Black Athena*, Afrocentrism, and the History of Science," in Lefkowitz and Rogers, eds., *Black Athena Revisited.*

————, ed. *The Annus Mirabilis of Sir Isaac Newton*, 1666-1966 (Cambridge, MA: MIT Press, 1970).

Pannekoek, Anton. *Marxism and Darwin* (Chicago, IL: Charles H. Kerr, 1912).

Paracelsus. *The Diseases that Deprive Man of His Reason*, in Paracelsus, *Four Treatises.*

————. *Four Treatises of Theophrastus von Hohenheim called Paracelsus*, ed. Henry E. Sigerist (Baltimore, MD: Johns Hopkins University Press, 1996).

————. *On the Miners' Sickness and Other Miners' Diseases*, in Paracelsus, *Four Treatises.*

Paré, Ambroise. *Apologie and Treatise* (New York: Dover, 1968; originally published in 1585).

Parry, J. H. *The Age of Reconnaissance* (Berkeley: University of California Press, 1981).

Parsons, William Barclay. *Engineers and Engineering in the Renaissance* (Cambridge, MA: MIT Press, 1976).

Peet, T. Eric. *The Rhind Mathematical Papyrus* (Liverpool, UK: Liverpool University Press, 1923).

Peretti, Aurelio. *Il Periplo di Scilace: Studio sul primo portolano del Mediterraneo* (Pisa: Giardini, 1979).

Peters, Edward. "Science and the Culture of Early Europe," in Dales, *The Scientific Achievement of the Middle Ages.*

Pigafetta, Antonio. *Magellan's Voyage: A Narrative Account of the First Circumnavigation* (New York: Dover, 1969; originally published c. 1525).

Pilger, John. "Australia's Samizdat," September 29, 2004. *Green Left Weekly*, http://www.zmag.org/content/showarticle.cfm?ItemID=6316

Pingree, David. "Hellenophilia versus the History of Science," in Shank, ed., *The Scientific Enterprise in Antiquity and the Middle Ages*.

Plato. *Phaedrus*, trans. Benjamin Jowett (Chicago, IL: Great Books of the Western World, 1952).

_____. *The Republic*, trans. Benjamin Jowett (Chicago, IL: Great Books of the Western World, 1952).

Pliny the Elder. *Natural History*, trans. H. Rackham (Cambridge, MA: Harvard University Press, 1997).

Polo, Marco. *The Travels of Marco Polo*, trans. Ronald Latham (London: Folio Society, 1968).

Polybius. *The Histories*, trans. from the text of F. Hultsch by Evelyn S. Shuckburgh (Bloomington: Indiana University Press, 1962).

Popper, Karl. *The Open Society and Its Enemies* (London: Routledge & K. Paul, 1962).

Porter, Roy. "Gentlemen and Geology: The Emergence of a Scientific Career, 1660–1920," *Historical Journal* 21 (1978).

_____. *The Greatest Benefit to Mankind: A Medical History of Humanity* (New York: W. W. Norton, 1998).

_____. "Introduction," in Pumphrey, Rossi, and Slawinski, eds., *Science, Culture and Popular Belief in Renaissance Europe*.

_____, ed. *Medicine: A History of Healing* (New York: Barnes & Noble Books, 1997).

Porter, Roy, and Mikuláš Teich, eds. *Scientific Revolution in National Context* (Cambridge, UK: Cambridge University Press, 1992).

Porter, Theodore M. "The Promotion of Mining and the Advancement of Science: The Chemical Revolution of Mineralogy," *Annals of Science* 38 (1981).

Poynter, F. N. L. "Nicholas Culpeper and His Books," *Journal of the History of Medicine* 17 (1962).

Price, Derek J. de Solla. *Little Science, Big Science* (New York: Columbia University Press, 1963).

_____. *Little Science, Big Science . . . and Beyond* (New York: Columbia University Press, 1986).

_____. "Of Sealing Wax and String: A Philosophy of the Experimenter's Craft and Its Role in the Genesis of High Technology," in Price, *Little Science, Big Science . . . and Beyond*.

_____. "Proto-Astrolabes, Proto-Clocks and Proto-Calculators: The Point of Origin of High Mechanical Technology," in Schmandt-Besserat, ed., *Early Technologies*.

_____. *Science since Babylon* (New Haven, CT: Yale University Press, 1962).

Price, T. Douglas, and Anne Birgitte Gebauer, eds. *Last Hunters, First Farmers: New Perspectives on the Prehistoric Transition to Agriculture* (Santa Fe, NM: School of American Research Press, 1995).

Ptolemy, Claudius. *The Geography* (New York: Dover, 1991).

Pumphrey, Stephen. "Who Did the Work? Experimental Philosophers and Public Demonstrators in Augustan England," *British Journal for the History of Science* 28 (1995).

Pumphrey, Stephen, Paolo L. Rossi, and Maurice Slawinski, eds. *Science, Culture and Popular Belief in Renaissance Europe* (Manchester, UK: Manchester University Press, 1991).

Pytheas of Massalia. *Pytheas of Massalia, On the Ocean,* ed. C. H. Roseman (Chicago, IL: Ares, 1994).

Rattansi, P. M. "Paracelsus and the Puritan Revolution," *Ambix* 11 (1964).

Raven, Diederick, and Wolfgang Krohn. "Edgar Zilsel: His Life and Work," in Zilsel, *The Social Origins of Modern Science.*

Raymond, Robert. *Out of the Fiery Furnace* (University Park: Pennsylvania State University Press, 1986).

Razzell, Peter. *Edward Jenner's Cowpox Vaccine: The History of a Medical Myth* (Firle, UK: Caliban Books, 1977).

Read, John. *Through Alchemy to Chemistry* (New York: Harper Torchbooks, 1963).

Roberts, H. Edward, and William Yates. "Exclusive! Altair 8800: The Most Powerful Minicomputer Project Ever Presented—Can Be Built for under $400," *Popular Electronics,* January 1975.

Robinson, Eric, and Douglas McKie, eds. *Partners in Science: Letters of James Watt and Joseph Black* (Cambridge, MA: Harvard University Press, 1970).

Rogers, Susan Carol. "Woman's Place: A Critical Review of Archaeological Theory," *Comparative Studies in Society and History* 20 (1978).

Roller, Duane H. D. *The* De magnete *of William Gilbert* (Amsterdam: Hertzberger, 1959).

Rolt, L. T. C., and J. S. Allen. *The Steam Engine of Thomas Newcomen* (New York: Science History Publications, 1977).

Rosen, George. "Introduction" to Paracelsus, *On the Miners' Sickness.*

Rosset, Peter, Joseph Collins, and Frances Moore Lappé. "Lessons from the Green Revolution," *Tikkun* March/April 2000. Available at Third World Network, http://www.foodfirst.org/media/opeds/2000/4-greenrev.html

Rossi, Paolo. *Philosophy, Technology and the Arts in the Early Modern Era* (New York: Harper & Row, 1970).

Rossiter, Margaret. *Women Scientists in America: Struggles and Strategies to 1940* (Baltimore, MD: Johns Hopkins University Press, 1982).

Roule, Louis. *Bernardin de Saint-Pierre et l'harmonie de la nature* (Paris: Flammarion, 1930).

Rousseau, Jean-Jacques. *Discourse on Inequality Among Men,* in *The Essential Rousseau,* trans. Lowell Bair (New York: Meridian, 1983).

Rudgley, Richard. *Lost Civilisations of the Stone Age* (London: Century, 1998).

Russell, Peter. *Prince Henry "the Navigator": A Life* (New Haven, CT: Yale University Press, 2001).

Sabra, A. I. "Situating Arabic Science," in Shank, ed., *The Scientific Enterprise in Antiquity and the Middle Ages*.

Sahlins, Marshall. "The Original Affluent Society," in *Stone Age Economics* (Chicago, IL: Aldine-Atherton, 1972).

Santillana, Giorgio de. "The Role of Art in the Scientific Revolution," in Clagett, ed., *Critical Problems in the History of Science*.

Santillana, Giorgio de, and Edgar Zilsel. *The Development of Rationalism and Empiricism*, vol. 2, no. 8 (Chicago, IL: University of Chicago Press, 1941).

Sarton, George. *A History of Science* (Cambridge, MA: Harvard University Press, 1952–1960).

_____. *Six Wings: Men of Science in the Renaissance* (Bloomington, IN: University of Indiana Press, 1957).

Saunders, J. B. de C. M., and Charles D. O'Malley, eds. "Introduction" to Vesalius, *The Illustrations from the Works of Andreas Vesalius of Brussels*.

Schmandt-Besserat, Denise. *Before Writing*, vol. I: *From Counting to Cuneiform;* vol. II: *A Catalogue of Near Eastern Tokens* (Austin, TX: University of Texas Press, 1992).

_____. *How Writing Came About* (Austin: University of Texas Press, 1996).

_____. "On the Origins of Writing," in Schmandt-Besserat, ed., *Early Technologies*.

_____, ed. *Early Technologies* (Malibu, CA: Undena, 1979).

Schneer, Cecil J. "William Smith's Geological Map of England and Wales and Part of Scotland, 1815–1817." University of New Hampshire, http://www.unh.edu/esci/mapexplan.html

Schofield, Robert E., ed. *A Scientific Autobiography of Joseph Priestley* (Cambridge, MA: MIT Press, 1966).

Scriba, Christoph J., ed. "The Autobiography of John Wallis, F.R.S.," *Notes and Records of the Royal Society* 25, 1 (June 1970).

Shank, Michael H., ed. *The Scientific Enterprise in Antiquity and the Middle Ages* (Chicago, IL: University of Chicago Press, 2000).

Shapin, Steven. *A Social History of Truth: Civility and Science in Seventeenth-Century England.* (Chicago, IL: University of Chicago Press, 1994).

_____. *The Scientific Revolution* (Chicago, IL: University of Chicago Press, 1996).

Shapin, Steven, and Barry Barnes. "Science, Nature and Control: Interpreting Mechanics' Institutes," *Social Studies of Science* 7 (1977).

Sharp, Andrew. *Ancient Voyagers in the Pacific* (Harmondsworth, UK: Penguin, 1957).

_____. *Ancient Voyagers in Polynesia* (Berkeley: University of California Press, 1964).

Skelton, R. A., Thomas E. Marston, and George D. Painter. *The Vinland Map and the Tartar Relation* (New Haven, CT: Yale University Press, 1965).

Smiles, Samuel. *Self-Help* (1858). eBookMall, http://www.ebookmall.com

Smith, Bruce D. *The Emergence of Agriculture* (New York: Scientific American Library, 1995).

Smith, Cyril Stanley. "The Development of Ideas on the Structure of Metals," in Clagett, ed., *Critical Problems in the History of Science*.

Smith, Pamela H. *The Body of the Artisan* (Chicago, IL: University of Chicago Press, 2004).

————. "Vital Spirits: Redemption, Artisanship, and the New Philosophy in Early Modern Europe," in Osler, ed., *Rethinking the Scientific Revolution*.

Snowden, Frank M., Jr. "Bernal's 'Blacks,' and Other Classical Evidence," *Arethusa* 22 (1989).

Sobel, Dava. *Longitude* (New York: Walker, 1995).

Souriau, Maurice. *Bernardin de Saint-Pierre d'après ses manuscrits* (Geneva: Slatkine Reprints, 1970; first published in 1905).

Spence, Jonathan D. *The Question of Hu* (New York: Vintage Books, 1989).

Sprat, Thomas. *The History of the Royal-Society of London* (London: 1667).

Steadman, Philip. *Vermeer's Camera* (Oxford, UK: Oxford University Press, 2001).

Stewart, Larry. "The Edge of Utility: Slaves and Smallpox in the Early Eighteenth Century," *Medical History* 29 (1985).

Strabo. *The Geography of Strabo*, trans. Horace Leonard Jones (London: Loeb Classical Library, 1917–1933).

Straus, William L., Jr., and A. J. E. Cave. "Pathology and the Posture of Neanderthal Man," *Quarterly Review of Biology* (December 1957).

Swetz, Frank J. *Capitalism and Arithmetic: The New Math of the 15th Century* (La Salle, IL: Open Court, 1987).

Tang, Anthony M., Fred M. Westfield, and James S. Worley, eds. *Evolution, Welfare, and Time in Economics* (Lexington, MA: Lexington Books, 1976).

Tarn, W. W. *Alexander the Great and the Unity of Mankind* (London: H. Milford, 1933).

Taylor, E. G. R., *The Haven-Finding Art: A History of Navigation from Odysseus to Captain Cook* (London: Hollis & Carter, 1971).

————. *The Mathematical Practitioners of Hanoverian England, 1714–1840* (Cambridge, UK: Cambridge University Press, 1966).

————. *The Mathematical Practitioners of Tudor & Stuart England* (Cambridge, UK: Cambridge University Press, 1954).

Taylor, Frederick W. *Scientific Management* (New York: Harper & Row, 1964).

Temple, Robert. *The Genius of China: 3,000 Years of Science, Discovery, and Invention* (New York: Simon & Schuster, 1986).

Teresi, Dick, *Lost Discoveries* (New York: Simon & Schuster, 2002).

Thiel, J. H. "Eudoxus of Cyzicus," in *Historische Studies* 23 (Utrecht, Netherlands: University of Utrecht, 1966).

Thomas, Keith. *Religion and the Decline of Magic* (New York: Scribner's, 1971).

Thomas, Steve. *The Last Navigator* (Camden, ME: International Marine, 1997).

Thoren, Victor E. *The Lord of Uraniborg: A Biography of Tycho Brahe* (Cambridge, UK: Cambridge University Press, 1990).

Thorndike, Lynn. *The Place of Magic in the Intellectual History of Europe* (New York: AMS Press, 1967).

Tourneux, Maurice. *Bibliographie de l'Histoire de Paris pendant la Révolution Française* (Paris, 1890-1913; reprinted by AMS Press, 1976).

Trevor-Roper, H. R. "The European Witch-Craze of the Sixteenth and Seventeenth Centuries," in *The European Witch-Craze of the Sixteenth and Seventeenth Centuries and Other Essays* (New York: Harper and Row, 1969).

Union of Concerned Scientists. *The Fallacy of Star Wars* (New York: Vintage Books, 1984).

Ure, Andrew. *The Philosophy of Manufactures, or, an Exposition of the Scientific, Moral, and Commercial Economy of the Factory System of Great Britain* (London: C. Knight, 1835).

Van Doren, Charles. *A History of Knowledge* (New York: Ballantine, 1991).

Vasari, Giorgio. *The Lives of the Artists* (Harmondsworth, UK: Penguin, 1965).

Vesalius, Andreas. *On the Fabric of the Human Body.* An annotated translation by Daniel Garrison and Malcolm Hast of the 1543 and 1555 editions of *De humani corporis fabrica.* Northwestern University, http://vesalius.north-western.edu/index.html

―――――. *The Illustrations from the Works of Andreas Vesalius of Brussels*, ed. J. B. de C. M. Saunders and Charles D. O'Malley (New York: Dover, 1950).

Vess, Donald M. *Medical Revolution in France, 1789–1795* (Gainesville: University Presses of Florida, 1975).

Vogel, Virgil J. *American Indian Medicine* (Norman: University of Oklahoma Press, 1970).

Vogt, David. "Medicine Wheel Astronomy," in Clive N. Ruggles and Nicholas J. Saunders, eds., *Astronomies and Cultures* (Boulder: University Press of Colorado, 1993).

von Bodenstein, Adam. "Foreword" to Paracelsus, *The Diseases that Deprive Man of His Reason.*

Wallace, Anthony F. C. *The Social Context of Innovation* (Princeton, NJ: Princeton University Press, 1982).

Ware, John Nottingham. "The Vocabulary of Bernardin de Saint-Pierre and Its Relation to the French Romantic School," in *The Johns Hopkins Studies in Romance Literatures and Languages*, vol. IX (Baltimore, MD: Johns Hopkins University Press, 1927).

Waters, David W. "Nautical Astronomy and the Problem of Longitude," in Burke, ed., *The Uses of Science in the Age of Newton.*

Watson, David. *Beyond Bookchin: Preface for a Future Social Ecology* (Brooklyn, NY: Autonomedia, 1996).

Watson, Patty Jo. "Explaining the Transition to Agriculture," in Price and Gebauer, eds., *Last Hunters, First Farmers*.

Weatherford, Jack. *Indian Givers: How the Indians of the Americas Transformed the World* (New York: Crown, 1988).

_____. *Native Roots: How the Indians Enriched America* (New York: Crown, 1991).

Webster, Charles. *From Paracelsus to Newton: Magic and the Making of Modern Science* (Cambridge, UK: Cambridge University Press, 1982).

Weinberg, Alvin. *Reflections on Big Science* (Cambridge, MA: MIT Press, 1966).

Werskey, Gary. *The Visible College* (New York: Holt, Rinehart and Winston, 1979).

Wertime, Theodore A. "Pyrotechnology: Man's Fire-Using Crafts," in Schmandt-Besserat, ed., *Early Technologies*.

Westfall, Richard S. *Never at Rest: A Biography of Isaac Newton* (Cambridge, UK: Cambridge University Press, 1980).

_____. "Science and Technology during the Scientific Revolution: An Empirical Approach," in Field and James, eds., *Renaissance and Revolution: Humanists, Scholars, Craftsmen and Natural Philosophers in Early Modern Europe*.

_____. "The Scientific Revolution Reasserted," in Osler, ed., *Rethinking the Scientific Revolution*.

White, Lynn, Jr. *Medieval Technology and Social Change* (Oxford, UK: Oxford University Press, 1962).

_____. "Pumps and Pendula: Galileo and Technology," in Golino, ed., *Galileo Reappraised*.

Wightman, W. P. D. *The Growth of Scientific Ideas* (New Haven, CT: Yale University Press, 1953).

Wilford, John Noble. "Debate Is Fueled on When Humans Became Human," *New York Times*, February 26, 2002.

Wilson, E. O. *Sociobiology: The New Synthesis* (Cambridge, MA: Harvard University Press, 1975).

Wilson, L. G. *Sir Charles Lyell's Scientific Journals on the Species Question* (New Haven, CT: Yale University Press, 1970).

Winchester, Simon. *The Map That Changed the World: William Smith and the Birth of Modern Geology* (New York: HarperCollins, 2001).

Winsor, Justin. *Christopher Columbus: And How He Received and Imparted the Spirit of Discovery* (Boston, MA: Houghton Mifflin, 1892).

Withering, William. *An Account of the Foxglove, and Some of Its Medical Uses* (1785), reproduced in J. K. Aronson, *An Account of the Foxglove and Its Medical Uses, 1785–1985* (Oxford, UK: Oxford University Press, 1985).

Wooldridge, Adrian. *Measuring the Mind: Education and Psychology in England, c. 1860–c. 1990* (Cambridge, UK: Cambridge University Press, 1994).

Worsley, Peter. *Knowledges* (New York: New Press, 1997).

Worster, Donald. *Nature's Economy* (Cambridge, UK: Cambridge University Press, 1977).

Wråkberg, Urban. "The Northern Space: Reflections on Arctic Landscapes." Centrum för Vetenskapshistoria (Center for History of Science, Royal Swedish Academy of Sciences), http://www.cfvh.kva.se

Xenophon. *The Economist*, trans. Alexander D. O. Wedderburn and W. Gershom Collingwood (New York: Burt Franklin, 1971).

Yong, Lam Lay, and Ang Tian Se. *Fleeting Footsteps: Tracing the Conception of Arithmetic and Algebra in Ancient China* (River Edge, NJ: World Scientific, 1992).

Young, Sidney. *The Annals of the Barber-Surgeons of London* (London: Blades, East & Blades, 1890).

Zilsel, Edgar. "The Genesis of the Concept of Physical Law," in Zilsel, *The Social Origins of Modern Science*. Originally published in *Philosophical Review* 60 (1942).

————. "The Genesis of the Concept of Scientific Progress and Scientific Cooperation," in Zilsel, *The Social Origins of Modern Science*. Originally published in *Journal of the History of Ideas* 6 (1945).

————. "The Origins of Gilbert's Scientific Method," in Zilsel, *The Social Origins of Modern Science*. Originally published in *Journal of the History of Ideas* 2 (1941).

————. "Problems of Empiricism," in Zilsel, *The Social Origins of Modern Science*. Originally published in Giorgio de Santillana and Edgar Zilsel, *The Development of Rationalism and Empiricism* (1941).

————. *The Social Origins of Modern Science*, ed. Diederick Raven, Wolfgang Krohn, and Robert S. Cohen (Dordrecht, Boston, and London: Kluwer Academic, 2000).

————. "The Sociological Roots of Science," in Zilsel, *The Social Origins of Modern Science*. Originally published in *American Journal of Sociology* 47 (1942).

Zinn, Howard. *A People's History of the United States, 1492–Present* (New York: Harper & Row, 1995).

INDEX

Mercier, Louis Sébastien, 417n*141*
Mercury, 164, 379
Mercury-sulfur hypothesis, 164
Mersenne, Marin, 318
Merton, Robert, 461; quoted, 287
Mesmer, Franz Anton, 401, 407
Mesmerists, Mesmerism, 400, 402, 403, 404, 406, 407; therapy, 404; social dangers feared, 405. *See also* Animal magnetism
Mesopotamia, 68, 80, 83, 118, 119, 216
Metallurgy, metallurgists, 3, 14, 75, 77, 83, 119, 133, 159, 165, 252, 257, 279, 288, 298, 300, 306, 307, 315, 316, 318, 353, 378, 426, 431, 432, 434; history, 79; Chinese, 177
Metals, Age of: beginning, 78
Metals: use, 41
Metalwork, metalworkers, 8, 82, 83, 158, 261, 262, 315, 423, 429
Metaphysics, metaphysicians, 146, 276, 285, 290, 361
Meteorology, 15, 298
Meyer, Albert, 273
Michaelangelo, 9, 262, 264, 266, 271, 192
Microcomputers, 486, 491
Microscope, 325–329 *passim*; invention, 325, 329
Microsoft, 491, 492, 494, 496
Middle ages, 161, 262, 279, 308
Middleton, W. E. Knowles: quoted, 412n*16*
Midelfort, H. C. E.: quoted, 367
Midwives, 4, 5, 291, 472; male, 370
Miletus, 133, 135, 136, 149, 210
Military-industrial complex, 472, 485, 491
Milkmaids, 105, 106, 323
Miller, David Philip: quoted, 431
Millwrights, 280, 425; Chinese, 175
Mimicry, biological, 399
Mineralogy, mineralogists, 133, 298, 316, 318; and hunter-gatherers, 107; Chinese, 174

Miners, mining, 2, 3, 79, 80, 251, 253, 254, 256, 276, 281, 283, 287, 288, 305–307, 315–318, 329, 356, 377, 423, 424, 430, 432, 433, 434
Mines, 305, 306, 424, 427, 316; antimony, 379; coal, 434; copper, 379; gold, 379; iron, 288; silver, 434; uranium, 483; vitriol, 379; Egyptian, 75; Melbourne, 430; and beer-drinking, 431
Ming dynasty, 170, 217
Mining schools and academies, 306, 434
Modern science, 17, 19, 88, 107, 137, 159, 165, 240, 241, 248, 287, 288, 365, 381, 393, 396, 411, 422, 423, 449, 450, 497, 499; origins, 4, 6, 274, 282, 321; ideology, 13; development, 15; foundation myths, 275; rise, 293, 376; programmatic foundation, 319; and teamwork, 321, 325; and capitalism, 350, 456; distrust of, 466–467
Moldboards: Chinese innovation, 179
Monarchy, 356, 358, 359, 360, 403; constitutional, 141
Monardes, Nicholas: quoted, 94
Monsoon winds, 201, 202, 203, 243n*40*
Montagu, Lady Mary Wortley: quoted, 104, 105
Montaigne, Michel de, 240; quoted, 240; Montaigne's servant, 240–241
Montesquieu, Baron de, 366
Moon, Earth's, 327; erratic motions, 230
Moons, Jupiter's, 229, 230, 327
More, Henry, 368, 369
Morison, Samuel Eliot: quoted, 243n*25*, 243n*26*
Mosaic (Internet browser), 492
Mu'allim (Sanscrit document): quoted, 213–214
Muller, Hermann J., 450
Müller, Karl Otfried, 127
Murdoch, William, 425
Museum of Alexandria, 143, 150, 156, 157, 158
Mutually Assured Destruction, 482

Nagasaki: obliteration of, 449, 467
Nanda, Meera: quoted, 462
Napier, John, 16
Napoleonic empire, 382, 390, 396, 408, 409
Nasser, Gamal Abdel, 135
National Assembly, French, 403
National Convention, French, 388, 389
National Science Foundation, 478
National Union of the Working Classes, 439
Natural philosophers, 276, 279, 319, 350, 391; ancient, 241; medieval, 284
Natural philosophy, 117, 253, 254, 270, 300, 327, 330, 363, 394, 396, 402; Aristotelian, 201; Baconian, 303; Mesmerist, 400, 401, 407
Natural Resources Defense Council, 477
Natural selection, 86, 93, 376, 437, 455
Nature (British journal), 10
Nature: as female, 364; vs. nurture, 454
Navigation, 15, 41–58 *passim*, 120, 133, 190–247 *passim*, 254, 258; direction-finding, 48, 53–56 *passim*, 226, 235; position-finding, 53, 54; land-finding, 56, 57; charts, 54; Western compared to Pacific islanders', 46–47, 49–50, 53, 55, 56; Chinese, 179–180; manuals, 258
Navigators, 16; indigenous, 3; Arab, 202; Indian, 202; Genoese, 217
Navvies, 435
Nazis, 121, 143, 278, 451
Neanderthal man, 108n*3*
Neckam, Alexander 235
Needham Institute, 187n*142*
"Needham problem," the, 189n*195*
Needham, Joseph, 460, 461; bibliographic note, 187n*142*; quoted, 165–171 *passim*, 174, 175, 187n*142*, 187n*143*, 187n*144*, 188n*163*
Nelson, Sarah Milledge: quoted, 30, 113n*141*, 113n*144*
Nelson, Ted, 493
Neolithic era, 33, 34, 75; end, 77